Exploring Tomorrow

The Comprehensive Guide to Robotics and Automation

Prof. A. B. Bhattacharya

M. Sc., Ph. D. (Cal Univ.), Pot Doc. (MIT, USA), FIETE
Former: Professor & Head, Department of Physics, Kalyani University;
Dean of Research Techno India University;
Pro Vice Chancellor, JIS University;
Director of Brainware University

&

Dr. Debasish Roy, IPS

BE, MCA, M Tech, Ph. D., MB, IPM, PPM
Director General & Inspector General of Police
Railways, West Bengal

KHANNA BOOK PUBLISHING CO. (P) LTD.

Publisher of Engineering and Computer Books

4C/4344, Ansari Road, Darya Ganj, New Delhi-110002
Phone : 011-23244447-48 **Mobile:** +91-9910909320
E-mail : contact@khannabooks.com
Website : www.khannabooks.com

Price: ₹ 575.00

ISBN: 978-93-5538-725-7

Exploring Tomorrow:
The Comprehensive Guide to Robotics and Automation
by Prof. A. B. Bhattacharya, Dr. Debasish Roy

First Edition: 2025

To view complete list of books, please scan the QR Code:

Published by:
Khanna Book Publishing Co. (P) Ltd.
CIN: U22110DL1998PTC095547

Visit us at: www.khannabooks.com
Write us at: contact@khannabooks.com

Printed in Bharat.

Preface

In the ever-evolving landscape of technology, few fields have captivated our imagination and transformed our reality quite like robotics and automation. From the intricacies of human-robot collaboration to the ethical dilemmas posed by their integration into society, this book serves as your compass through the vast and dynamic world of robotics and automation. As we embark on this journey together, we delve into the myriad facets of this multifaceted domain. We begin with an exploration of the various types of robots, from industrial machines to humanoid companions, illuminating their capabilities and potential applications. Through this exploration, we uncover the intricate dance between humans and robots, shedding light on the ethical considerations and social implications that accompany their collaboration.

Continuing our voyage, we navigate the realm of autonomous vehicles and drones, recognizing their integral role within the broader field of robotics and artificial intelligence. From revolutionizing transportation to enhancing agricultural practices, these technologies epitomize the innovative spirit driving our collective quest for progress. Our odyssey further unfolds as we delve into the transformative impact of robotics in healthcare, agriculture, education, space exploration, disaster response, and beyond. We witness the remarkable advancements that have revolutionized these industries, empowering us to confront challenges with ingenuity and resilience.

Moreover, we explore the fascinating world of biologically-inspired robotics, drawing inspiration from nature's ingenuity to engineer innovative solutions. We also examine the role of robotics in entertainment and service industries, where creativity converges with functionality to enrich our lives in unexpected ways. As we peer into the future, we confront the pressing realities of automation and its profound implications for the workforce. Emerging trends underscore the need for adaptability and foresight, as we navigate the interconnected domains of industrial automation and robotics.

Additionally, we confront the sobering intersection of robotics and modern warfare, grappling with the ethical complexities and strategic implications of autonomous technologies. Challenges abound, yet so too do opportunities for innovation and progress. In "Exploring Tomorrow," we invite you to embark on a voyage of discovery, where curiosity knows no bounds and the possibilities are as limitless as our imagination. Whether you're a seasoned technologist, a curious enthusiast, or simply an explorer of the unknown, this book offers insights and perspectives that transcend boundaries and inspire action. So, join us as we embark on this odyssey into the future—a future shaped by the remarkable synergy of human ingenuity and technological innovation. Together, let us explore tomorrow, one page at a time.

Prof. A. B. Bhattacharya
Dr. Debasish Roy

Acknowledgments

The creation of "Exploring Tomorrow: The Comprehensive Guide to Robotics and Automation" has been a journey fueled by the collective effort and dedication of numerous individuals and organizations. As we embark on this acknowledgment, we extend our deepest gratitude to those who have contributed to the realization of this endeavor.

First and foremost, we express our heartfelt appreciation to the researchers, engineers, and experts whose pioneering work serves as the foundation of this book. Your unwavering commitment to advancing the field of robotics and automation has inspired us and enriched the pages of this guide with invaluable insights and knowledge. We extend our sincere thanks to the educators and practitioners who generously shared their expertise and experiences, enriching our understanding of the diverse applications and implications of robotics and automation. Your contributions have helped bridge the gap between theory and practice, empowering readers to navigate this complex landscape with confidence and clarity.

We are deeply grateful to the organizations and institutions that have supported our research and provided access to resources essential for the development of this book. Your collaboration has been instrumental in ensuring the comprehensiveness and accuracy of the content presented herein. Furthermore, we extend our appreciation to the reviewers and editors whose meticulous feedback and guidance have enhanced the quality and coherence of this manuscript. Your dedication to excellence has been instrumental in shaping the final product and elevating its impact.

We also acknowledge the invaluable support of friends, family, and colleagues who have offered encouragement and understanding throughout the writing process. Your unwavering belief in our vision has fueled our determination and inspired us to persevere in the face of challenges. Last but certainly not least, we express our deepest gratitude to the readers whose curiosity and thirst for knowledge motivate us to continue pushing the boundaries of exploration and discovery. It is our sincere hope that this book serves as a beacon of inspiration and enlightenment, igniting passion and curiosity in the hearts and minds of all who dare to envision the possibilities of tomorrow.

Together, we embark on this journey into the future, guided by the spirit of collaboration, innovation, and relentless pursuit of progress. Thank you for joining us on this adventure as we explore the boundless horizons of robotics and automation, together shaping the destiny of tomorrow.

Prof. A. B. Bhattacharya

Dr. Debasish Roy

A Special Note from the Authors

As we stand at the precipice of an era defined by unprecedented technological advancement, we are humbled and honored to present to you "Exploring Tomorrow: The Comprehensive Guide to Robotics and Automation." In the pages that follow, we invite you to embark on a voyage of discovery—a journey that transcends the boundaries of time and space, illuminating the boundless potential of robotics and automation to shape the course of our collective future.

For us, the journey of writing this book has been nothing short of exhilarating—an odyssey fueled by curiosity, passion, and an unwavering commitment to unraveling the mysteries of tomorrow. With each chapter, we have endeavored to capture the essence of this rapidly evolving field, offering insights and perspectives that span the breadth and depth of robotics and automation.

At the heart of our exploration lies a profound appreciation for the diversity of applications and implications that define this domain. From the intricate dance between humans and robots to the transformative impact of autonomous vehicles and drones, we have sought to weave together a tapestry of knowledge that reflects the multifaceted nature of robotics and its profound influence on virtually every aspect of our lives. Yet, our journey extends far beyond the realm of technological innovation. It is also a journey of introspection—a quest to confront the ethical dilemmas, social implications, and existential questions that accompany the rise of intelligent machines. As we navigate the complex interplay between man and machine, we are compelled to confront our own humanity—to ponder the very essence of what it means to be human in an increasingly automated world.

In each chapter, we delve into a distinct facet of robotics and automation, exploring its nuances, challenges, and potential for transformative impact. From the life-saving applications of robotics in healthcare to the awe-inspiring endeavors of space exploration, we invite you to witness the remarkable convergence of human ingenuity and technological innovation. Needless to say, our exploration would be incomplete without acknowledging the profound uncertainties and challenges that lie ahead. As we peer into the future, we are confronted with the sobering realities of automation and its profound implications for the workforce, society, and the very fabric of our existence. It is a future fraught with complexity, yet brimming with opportunity—for those bold enough to seize it.

Prof. A. B. Bhattacharya

Prof. A. B. Bhattacharya did his M Sc and Ph D degrees from the University of Calcutta in Physics and Post-doc from the Massachusetts Institute of Technology (MIT), Cambridge, USA. He worked in close collaboration with the leading laboratories like Lincoln Laboratory; Milestone Hill Observatory; Earth, Atmospheric and Planetary Sciences division of MIT and many other well recognized laboratories in India and abroad. He served as Professor & Head of the Department of Physics, University of Kalyani; Dean of Research and Head of the ECE Department, Techno India University; Pro Vice Chancellor of JIS University and Director of Brainware University. He has published 32 text books for Science and Engineering students, two popular books for general readers and over 400 research articles in reputed International Journals. Under his supervision 24 scholars received their doctoral degree. He is a fellow of the Institute of Electronics and Telecommunication Engineers and member of many scientific organizations. India International Friendship Society awarded him the prestigious "Shiksha Ratna Award" and Certificate of Excellence for meritorious services, outstanding performance and significant contribution to education. The International Institute of Success Awareness honored him with their globally recognized "Glory of India Gold Medal" for his outstanding contribution to the national stature of India. A few words from the certificate issued by Prof. Williams of MIT may be quoted "Our collaborative effort was directed at the wavelength dependence of radar echoes observations with the Millstone Hill UHF radar in Westford. He made use of the Charles River Data Systems computer in the Weather Radar Laboratory for these calculations. His results help to make a convincing case that the volume reflectivity associated with lightning is a diminishing function of radar wavelength, contrary to contemporary thinking".

Dr. Debasish Roy, IPS

Dr. Debasish Roy is a highly accomplished and motivated professional with a robust academic background in Information Technology, including an MCA, MBA, MTech, and a PhD. He was currently the Director General (Railways) at a top-level position in the Indian Police Service (IPS). He received the Indian Police Medal and President's Police Medal in 2000 and 2010. He is well known for pioneering contributions in land mobile radio systems. He is highly recognized for spearheading strategic planning and execution of security measures for the railway network, implementing innovative technologies to enhance railway security and counteract emerging threats, overseeing a team of professionals, and managing day-to-day operations efficiently. He has provided expert guidance on implementing land mobile radio systems for public safety agencies and conducted groundbreaking research in cryptanalysis, leading to the development of advanced techniques for enhancing cyber security. He successfully led a team in designing and implementing secure communication protocols for land mobile radio systems, ensuring robust and encrypted communication. He has published numerous research papers in reputed journals and presented many works at international conferences. He mentored and guided students at the postgraduate and doctoral levels. He has a proven track record of effective leadership and team management, besides all his strategic planning and execution in high-stakes environments. He received many honors and awards for his outstanding contributions to the field and was recognized for exceptional leadership and innovation in technology-driven security initiatives. He completed an Executive Program in Quantum Computing and Machine Learning from IIT Delhi in 2023. He participated in the Workshop on Machine Learning for Cryptology (ML4Crypto 2023) at ISI, Kolkata, from November 8-10, 2023, organized by SAG, DRDO. His recent paper on "Iterative Quantum Algorithm for Linear System Solving" has been accepted at the International Conference SAI, Computing, 2024 London, organized by Microsoft Corporation.

Contents

Chapter 11: Automation and the Workforce: Emerging Trends 307

Chapter 12: Interconnected Domains of Industrial Automation and Robotics 336

Various Types of Robots

1.1 INTRODUCTION

Robots are physical or virtual machines that are programmed to carry out independently or partially autonomous tasks. They are available in a variety of varieties, each suitable for a certain application.

1.2 MAIN CATEGORIES OF ROBOTS

Let us explore some of the main categories of robots in detail:

1. **Industrial Robots:** These are made for use in manufacturing and other industrial applications. In workplaces, they are frequently employed to carry out repetitive and accurate activities including welding, painting, assembling, and material handling. These robots often work in fixed positions and adhere to predetermined workflows. In industrial contexts, they can significantly raise productivity, consistency, and safety.
2. **Service Robots:** Service robots are designed to help and communicate with people in a variety of settings. They can also be broken down into other categories:
   (*a*) **Domestic Robots:** These robots are created to carry out domestic duties like cooking, cleaning (such as robotic vacuum cleaners), and mowing lawns. They seek to improve the efficiency and comfort of our daily life.
   (*b*) **Healthcare Robots:** When utilized in the healthcare industry, robots can help with duties including patient lifting and moving, drug delivery, and precise surgical procedures. Some of them are also employed in therapy and rehabilitation.
   (*c*) **Education Robots:** The purpose of these robots is to support educational environments by assisting with tasks like teaching programming or language skills, providing individualised tutoring, and promoting interactive learning.
   (*d*) **Entertainment Robots:** People are intended to be amused and entertained by entertainment robots. Robotic animals, humanoid robots that can dance and play music, and interactive companions all fall under this category.

3. **Autonomous Vehicles:** Autonomous vehicles, also referred to as self-driving or driverless cars, use sensors, artificial intelligence, and sophisticated algorithms to manoeuvre without the assistance of a driver. They are made for logistics, transportation, and even space exploration. Autonomous vehicles, drones, and unmanned underwater vehicles are a few examples.
4. **Military and Defense Robots:** Military robots are utilized for jobs like bomb disposal, reconnaissance, and surveillance that could be hazardous for humans. Drones, also referred to as unmanned aerial vehicles (UAVs), are frequently used in military operations for both defensive and offensive objectives.
5. **Social Robots:** Social robots are created to interact socially and emotionally with people. They frequently behave and seem like humans in an effort to promote social interaction, emotional support, and companionship. Social robots are occasionally employed in the medical field to provide emotional support and companionship to the elderly or those with particular disabilities.
6. **Humanoid Robots:** Robots created to mimic humans in look and behavior are called humanoid robots. They frequently have a head, torso, arms, and legs, and some of them are able to mimic human expressions and movements. Robots that resemble people are being created for a variety of uses, such as research, entertainment, and customer service.
7. **Agricultural Robots:** Also referred to as agribots or agri-drones, agricultural robots are created to maximize farming activities. They can help with jobs like planting, harvesting, and keeping an eye on the health of the crops. Agricultural robots are designed to boost productivity, lower labor costs, and raise agricultural yields.
8. **Space Robots:** Space robots are used in space exploration missions to carry out activities that would be challenging or risky for humans to complete. They can be remotely controlled or only partially autonomous, and are employed for tasks such

These are only a few of the many different kinds of robots that exist today. As technology develops, we can anticipate seeing even more advanced and varied robots in the future, which will further revolutionize a variety of businesses and facets of our life.

1.3 INDUSTRIAL ROBOTS

Machines called industrial robots are made to automate and carry out particular duties in industrial environments. To increase effectiveness, accuracy, and safety throughout production and manufacturing operations, they are frequently utilized. The following are some crucial facts about industrial robots:

Setup: Industrial robots often feature a robotic arm made up of numerous articulated joints, which enables them to move and carry out tasks in a variety of directions. While industrial robots can have a variety of joints, most have at least three to six axes of motion, giving them a wide range of motion and versatility.

End Effectors: The end effector, also known as the robot's "hand," is the part of the robotic arm that interacts with the work piece or object. It can be equipped with various tools or grippers, depending on the specific task the robot is designed to perform. Common end effectors include grippers, welding torches, paint sprayers, and suction cups.

Programming: Industrial robots can be programmed in different ways:

(*a*) Robots can be programmed using specialized software on a computer, independent of their actual location.

(*b*) **Offline Programming:** This method makes programming simpler and safer without interfering with ongoing activities.

(*c*) **Teach Pendant:** A teach pendant is a handheld device that many industrial robots have that enables operators to physically move the robot through the appropriate actions. The robot "learns" these motions and stores them for replay.

(*d*) **CAD/CAM Integration:** In some sophisticated systems, the programming of the robot can be directly integrated with CAD/CAM software, speeding the automation process.

Applications: Industrial robots are used in various manufacturing processes, such as:

(*a*) **Welding:** In the automotive, aerospace, and other industries, robots are widely utilized for arc welding, spot welding, and laser welding.

(*b*) **Material Handling:** On assembly lines and in warehouses, robots are used to pick, place, and move products and commodities.

(*c*) **Painting:** For consistent and accurate painting in the automobile and other industries, robots with paint spraying end effectors are used.

(*d*) **Assembly:** Complex assembly jobs like inserting components, tightening screws, and inspecting the quality of the finished product can be handled by robots.

(*e*) **Machine Tending:** For automated manufacturing, robots can load and unload workpieces from equipment like CNC (Computer Numerical Control) machines.

Robots may check products for flaws and make sure that quality requirements are met by using sensors and cameras.

1. **Safety:** When deploying industrial robots, safety must be taken into account. The protection of adjacent personnel is ensured by safety features including barriers and sensors that can detect human presence and halt or slow down the robot's motion.
2. **Benefits:** Industrial robots provide a number of advantages, including faster production, better product quality, fewer human errors, increased workplace safety, and the capacity to operate continuously without getting tired.
3. **Cobots, or Collaborative Robots:** Collaborative robots, often known as cobots, are a recent advancement in industrial robotics. These robots are made to function safely next to human employees, frequently without the need for safety barriers. Cobots are more flexible for smaller production runs and shifting workflows and are designed for jobs that call for human-robot collaboration.

Overall, industrial robots are essential to modern manufacturing and help to improve product quality by streamlining production processes, increasing productivity, and boosting efficiency.

1.3.1 Configuration of Industrial Robots

Although industrial robots can be built in many different ways, the articulated robot is the most typical design. Let's examine the articulated configuration in greater depth and present a straightforward graphic to show its composition:

1. **Articulated Arrangement:** The arm of the robot can move in several axes using rotary joints in this arrangement. Each joint offers a certain amount of flexibility, allowing the robot to take on various positions and orientations. This arrangement allows for flexibility in carrying out numerous duties and resembles the motion of a human arm.

2. **Parts of a Mobile Robot with Joints:** The following elements are commonly seen in an articulated robot:
   (*a*) **Robot's Base:** It is in the bottom portion and a source of stability and support. It is secured to a stationary platform or the ground.
   (*b*) **Waist (Axis 1):** The robot's arm can rotate horizontally around the vertical axis thanks to the waist joint. This motion is comparable to how a human waist twists.
   (*c*) **Shoulder (Axis 2):** Just like a human shoulder, the robot's arm may move up and down thanks to the shoulder joint.
   (*d*) **Elbow (Axis 3):** The robot's forearm may flex and extend thanks to the elbow joint, which works similarly to a human elbow.
   (*e*) **Wrist (Axis 4 and 5):** The wrist joints offer a range of rotational motions that enable the robot's end effector to tilt and rotate. The robot can position the end effector at different angles thanks to its flexibility.
   (*f*) **End Effector:** The robot's "hand" or end effector can be outfitted with various tools or grippers depending on the particular task it is intended to carry out.
3. **Diagram of an Articulated Robot:** A simple diagram illustrating the basic structure of an articulated robot is shown in Figure 1.1

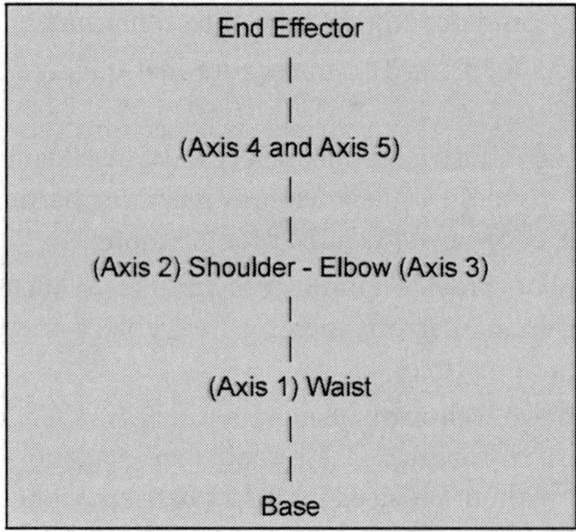

Fig. 1.1. Simplified representations of the Articulated Robots

The robot's arm may rotate horizontally thanks to the waist joint (Axis 1), which is where the robot's base is seen connected to the ground. The arm can move up and down through the shoulder joint (Axis 2) and can bend and stretch through the elbow joint (Axis 3). For precise positioning of the end effector, additional rotational movements are provided via the wrist joint (Axis 4 and Axis 5).

You should be aware that industrial robots may have more axes than five, depending on their sophistication and the particular tasks they are intended to carry out. Six or even more axes can be found in advanced industrial robots, which provide better dexterity and manoeuvrability.

1.4 SERVICE ROBOTS

A broad group of robots known as "service robots" are created to work with people in a variety of settings and situations. They were developed to improve the comfort, safety, and ease of our lives. The following information relates to service robots:

1. **Service Robot Types 1:** Service robots can be broadly divided into the following categories:
   (*a*) **Domestic Robots:** These robots are made to perform duties around the house. Robotic lawn mowers, window washers, and floor scrubbers are a few examples. They are designed to automate tedious processes and free up homeowners' time.

   (*b*) **Healthcare Robots:** To help patients and healthcare personnel, healthcare robots are employed in medical and healthcare environments. They can assist with duties such as patient lifting and transporting, medicine delivery, and companionship for the elderly or those in need of emotional support.

   Education robots are designed to support and improve educational activities. They can help with the teaching of a variety of subjects, including social skills, foreign languages, and programming and mathematics. Some educational robots are made to offer individualised instruction and adaptive learning opportunities.

   (*c*) **Entertainment Robots:** People are amused and entertained by entertainment robots. They can be humanoid robots that can dance and play music, interactive toys, robotic animals, or even robots used in amusement parks or interactive displays.

   (*d*) **Customer Service Robots:** These machines help and support customers in a variety of service sectors, including hotels, airports, and retail establishments. They are able to enlighten clients, respond to inquiries, and direct clients to various destinations.

   (*e*) **Social Robots:** Social robots are created to interact socially and emotionally with people. They frequently have artificial intelligence that allows them to understand and react to human emotions, fostering a sense of engagement and camaraderie.
2. **Effective Human-Robot Interaction:** Service robôts are equipped with advanced sensors and algorithms. To comprehend spoken commands and comply, they can use speech recognition and natural language processing. Some highly developed service robots can even recognize gestures and facial expressions to interact with people more effectively.
3. **Autonomous Navigation:** A lot of service robots have the ability to navigate on their own, allowing them to manoeuvre through changing situations without continual human assistance. To sense their surroundings and avoid impediments, they make use of a variety of sensors, including cameras, LiDAR, and ultrasonic sensors.
4. **Safety Features:** When service robots engage closely with humans, safety is a key design consideration. To protect users and maintain secure interactions, service robots frequently contain safety features like collision avoidance sensors and soft materials.
5. **Applications:** The range and scope of service robot applications is constantly growing. Homes, hospitals, schools, airports, hotels, retail establishments, and public areas are just a few places where you might find them.
6. **Collaborative Robots (Cobots):** A subset of service robots known as cobots are created to work securely alongside people. These robots are designed to work in tandem with humans to complete jobs like helping with manual labour or serving customers in busy areas.

Service robots are still evolving and taking on more and more significance in many facets of our life, from home tasks to medical assistance and beyond. Service robots will likely grow much more capable, versatile, and incorporated into our daily lives as technology develops.

1.4.1 Configuration of Service Robot

Depending on the environments and duties they are intended to perform, service robots can be configured in different ways. Let's examine several typical arrangements for service robots and show illustrations of their internal components.

1. **Household robot (robotic hoover cleaner)**

   Configuration: Robotic hoover cleaners are intended to clean floors and carpets autonomously in residential settings. They typically have a small, circular or D-shaped body with wheels for mobility. On top, they might have sensors for navigation and obstacle detection.

   Diagram:

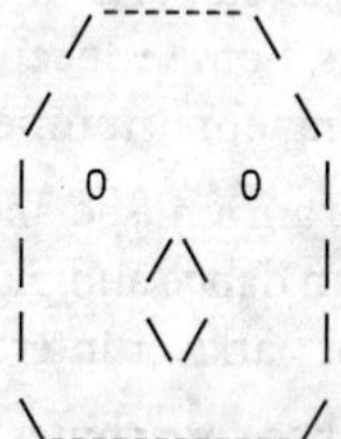

Fig. 1.2. Simplified representations of the Domestic Robot

   In this simple diagram, the "O" represents the wheels, and the "V" and "^" symbols represent the sensors for detecting obstacles and walls.

2. **A healthcare robot that serves as a patient assistant**
   - Humanoid or semi-humanoid-looking healthcare robots may be used to aid patients and healthcare workers.
   - They might have a torso, a head with a mouth and expressive features, arms, and perhaps a mobile base.
   - Advanced healthcare robots might have touchscreens or patient-interaction interfaces.

   Diagram:

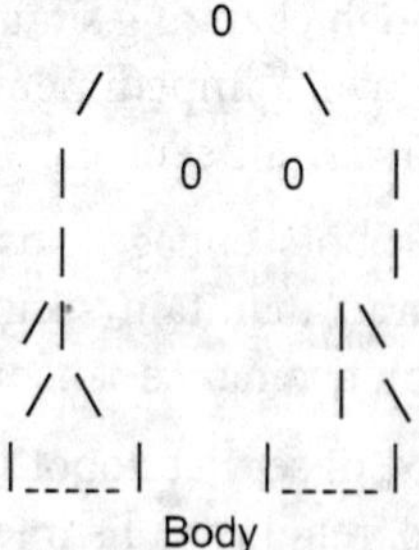

Fig. 1.3. Simplified representations of the Healthcare Robot

   In this diagram, the "O" represents the robot's arms and hands, while the lines and "Body" indicate the torso and base.

3. **Education Robot (Programmable Companion):**

   Configuration:
   - To promote a favorable learning environment, education robots frequently feature an approachable and captivating appearance.

- They might resemble a streamlined humanoid with a head, arms, and perhaps a movable base.
- Some educational robots have interactive screens that can be used for learning activities and programming.

Diagram:

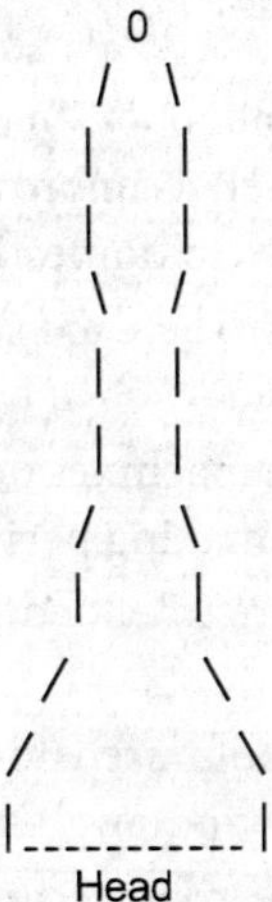

Fig. 1.4. Simplified representations of the Education Robot

In this diagram, the "O" represents the robot's arms, and the lines and "Head" indicate the robot's head and interactive display.

4. **Customer Service Robot (Information Guide):**

Configuration: Robots that provide customer service are frequently made to look professional and accessible, and they may have a tall, slender body with a touchscreen or display for information. They are able to travel swiftly in crowded areas thanks to wheels or a mobile base.

Diagram:

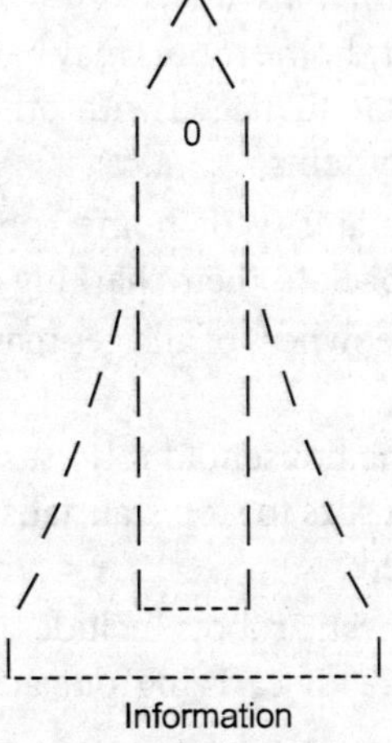

Fig. 1.5. Simplified representations of the Customer Service Robot

In this diagram, the "O" represents the robot's touch screen display, and the lines indicate the robot's body.

The various service robot setups are depicted in these diagrams, which are simplified versions of them. Based on the particular jobs, capabilities, and user interactions needed for each sort of service robot, the actual designs can differ greatly. Advanced service robots frequently combine artificial intelligence, actuators, and sensors to carry out their intended tasks correctly and securely.

1.5 DOMESTIC ROBOTS

Domestic robots, sometimes referred to as home robots or domestic robots, are made to help with various household duties and responsibilities. They seek to improve the comfort, efficiency, and convenience of our daily life. The following information relates to domestic robots:

1. **Different Kinds of Domestic Robots:** There are several different types of domestic robots, each created for a particular household task:
   (*a*) **Robotic Hoover Cleaners:** These machines operate on their own to clean carpets and floors while avoiding obstructions and making periodic trips back to charging stations.
   (*b*) **Robotic Lawn Mowers:** These are intended to cut lawns automatically while keeping the grass at a particular length.
   (*c*) **Window Cleaning Robots:** Robots used for window cleaning have suction or cleaning systems for cleaning mirror and window glass.
   (*d*) **Floor Scrubbing Robots:** These robots can mop and scour hard floors to keep them spotless.
   (*e*) **Robotic Pet Feeders:** Robotic pet feeders may automatically feed animals at predetermined intervals, ensuring that they are nourished even while their owners are not home.
   (*f*) **Smart Home Assistants:** Smart home assistants like Amazon Echo (Alexa) and Google Home use voice commands to control a variety of smart devices, give information, and handle domestic duties even if they are not physically mobile.
2. **Navigation and Sensors:** To move around the house and interact with their surroundings, domestic robots employ a combination of sensors, cameras, and algorithms. They can avoid obstructions, see drop-offs, and map their environment thanks to their sensors.
3. **Connectivity and Control:** Many home robots may be operated remotely using voice commands or Smartphone apps. They can be managed with other smart devices by users when they are incorporated into smart home systems.
4. **Autonomous Charging:** The majority of home robots have rechargeable batteries and are configured to automatically go back to their charging stations when their battery levels are low.
5. **User-Friendly Interfaces:** Homeowners of all ages may easily operate domestic robots thanks to their user-friendly interfaces.
6. **Safety Features:** When designing household robots, safety comes first. These robots frequently have safety features like drop sensors to stop stair falls and collision sensors to stop furniture or other things from being damaged.
7. **Benefits:** The advantages of domestic robots include time savings, less physical effort, enhanced cleanliness, and the convenience of carrying out activities even when homeowners are not present.
8. **Restrictions:** Despite the fact that domestic robots can be quite practical, they do have some restrictions. Their powers can be constrained by their design and technology, and they might not be appropriate for particular house types or floor plans.

Domestic robots are constantly improving, and when artificial intelligence and machine learning are integrated, they are becoming more capable and versatile to meet different household needs. We may anticipate seeing increasingly more advanced domestic robots as technology develops, which will be able to perform a larger range of jobs and be more productive in our homes.

1.5.1 Configuration of Domestic Robot

Domestic robots are available in a variety of designs, each one intended for a particular household task. Let's examine the configuration of a few prevalent household robot types, together with figures that show how they are built:

1. **Robotic Vacuum Cleaner:**

   Configuration: Robotic hoover cleaners typically feature a circular or D-shaped body with wheels for mobility, and they are made to independently clean floors and carpets in homes. They might have sensors for obstacle detection and navigation on top.

   Figure:

Fig. 1.6. (a) Line drawing and (b) Photograph of Robotic Vacuum Cleaner

2. **Robotic Lawn Mower:**

   Configuration: Robotic lawn mowers typically have a small, low-profile design with wheels or tracks for movement. Some models incorporate sensors to detect obstructions and boundary wires to define the mowing area. Robotic lawn mowers are designed to autonomously mow lawns, maintaining the grass at a set length.

   Figure:

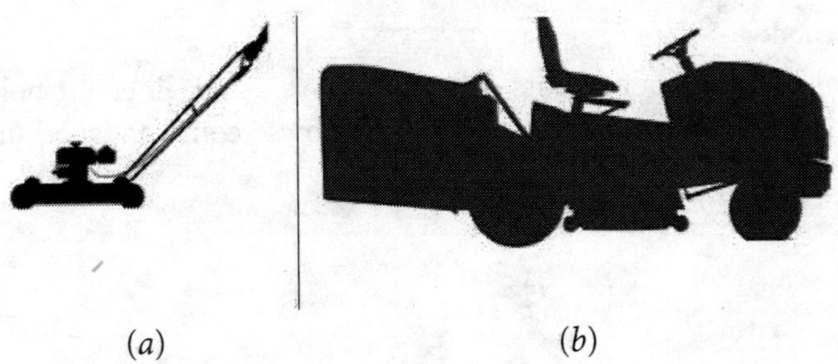

Fig. 1.7. Two different types of Robotic Lawn Mower

3. **Window Cleaning Robot:**

   Configuration: Window cleaning robots frequently have a square or rectangular shape with cleaning pads on the bottom for efficient cleaning. They are outfitted with suction or cleaning motors to clean glass surfaces on windows and mirrors.

Figure:

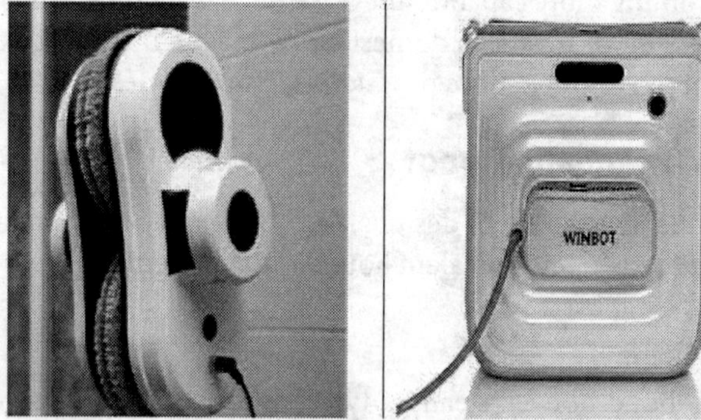

Fig. 1.8. Two different types of Window Cleaning Robot

4. **Floor Scrubbing Robot:**

   Configuration: Robotic floor cleaners often feature a flat, disc-like form with revolving brushes or scrubbing pads and are made to scrub and wash hard floors to keep them clean and stain-free.

   Figure:

Fig. 1.9. Photograph of a Floor Scrubbing Robot

5. **Robotic Pet Feeder:**

   Configuration: Robotic pet feeders frequently have a container for holding pet food and a dispensing mechanism controlled by timers or remote commands, ensuring that pets are fed even when the owners are away.

   Figure:

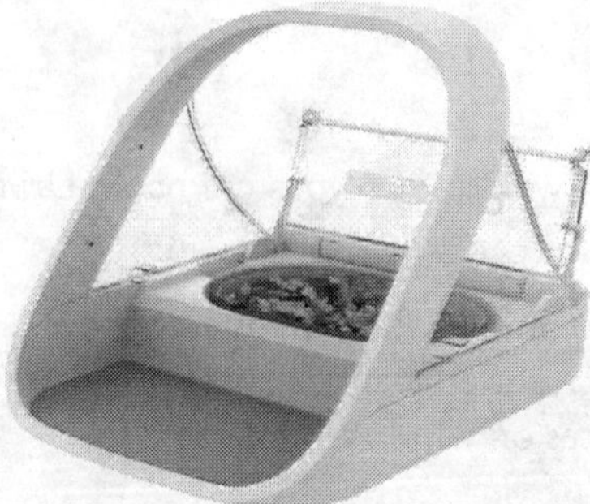

Fig. 1.10. Robotic Pet Feeders

It should be emphasized that the images presented here are illustrations of typical home robot configurations. Domestic robot designs might differ according on the maker, model, and particular functions they offer. As technology develops, domestic robots' look and capabilities can also change, bringing with them fresh and improved configurations for various household chores.

1.6 HEALTHCARE ROBOTS

One category of robots called "healthcare robots" is intended to support and help medical staff and patients. These robots are designed to boost patient outcomes, enhance the quality of care, and address the problems brought on by ageing populations and rising healthcare costs. The following information relates to healthcare robots:

Types of Healthcare Robot

(*a*) **Surgery Robots:** During minimally invasive procedures, surgical robots are utilized to aid doctors. They enable more accurate and minimally invasive operations by providing precision, dexterity, and enhanced visualization. Some surgical robots can be remotely managed by the surgeon, who does so from a console.

(*b*) **Rehabilitation Robots:** Patients who are recovering from accidents or physical disabilities can use rehabilitation robots to help them. To support mobility and motor function, they can help with exercises, gait training, and other therapeutic activities.

(*c*) **Robotic Telemedicine:** It is used to provide remote patient care and consultations. These robots, which include cameras, screens, and audio capabilities, enable medical staff to communicate with patients remotely, especially in regions with poor access to healthcare facilities.

(*d*) **Medication Dispensing Robots:** Robots that accurately and quickly deliver medications are known as "medication dispensing robots," and they are employed in hospitals and pharmacies. They lessen pharmaceutical errors and free up mundane work for the medical team.

(*e*) **Old Care Robots:** Elderly care robots help and comfort old people, especially those who live alone or in nursing homes. These robots may assist with daily tasks, keep an eye on vital signs, and send notifications when it's time for medication or an appointment.

Robotic prostheses: These use cutting-edge technology to restore amputated limbs, giving amputees more mobility and functionality.

1. **Healthcare Robot Advantages**

   Accuracy: Robots can accomplish jobs with an accuracy that may be higher than human capacity, lowering the possibility of mistakes during procedures and treatments.

   Consistency: By reliably providing care and according to instructions, healthcare robots can guarantee treatment uniformity.

   Lessened Demand on Healthcare Staff: Since robots can do monotonous activities, healthcare workers are free to concentrate on more difficult and crucial facets of patient care.

   Remote Care and Monitoring: Telemedicine robots allow for remote patient care and monitoring, enhancing access to healthcare services in off-the-grid or underserved locations.

Better Patient Results: Using medical robots can result in better patient results, quicker recovery periods, and higher levels of patient satisfaction.

2. **Safety Features:** Safety safeguards are built into healthcare robots to protect both patients and healthcare staff. These characteristics could include fail-safe mechanisms, strict adherence to safety regulations, and collision detection and obstacle avoidance.
3. **Ethical Considerations:** Ethics issues related to patient privacy, data security, permission, and the possible impact on human-human interactions in medical contexts are brought up by the use of robots in healthcare. Implementing ethics involves making sure that robots enhance human care rather than taking its place.

With continued research and development in the fields of robotics and medical technology, healthcare robots continue to advance quickly. Healthcare robots have the potential to revolutionize patient care and the healthcare sector as a whole as they become more advanced and extensively used. To achieve appropriate and successful integration in healthcare settings, a balance must be struck between the advantages they offer and the ethical issues they create.

1.6.1 Configuration of Healthcare Robot

Healthcare robots are available in a variety of designs, each tailored for certain medical functions and purposes. Let's examine how two popular types of healthcare robots are configured, along with simplified schematics that show their internal components:

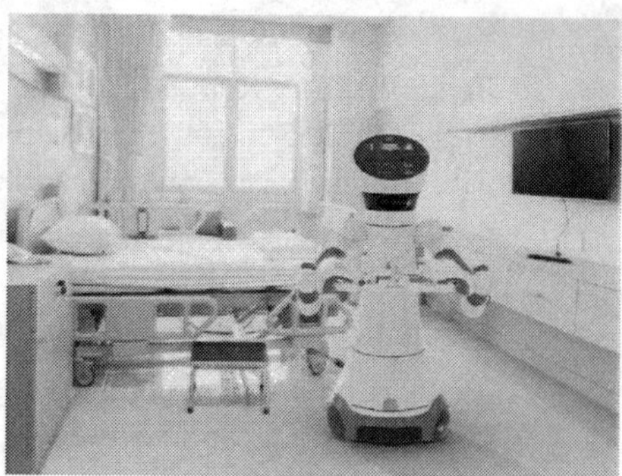

Fig. 1.11. Healthcare robots

1. **Surgical Robot (da Vinci Surgical System):**

   Configuration: Surgical robots, like the da Vinci Surgical System, are created to help surgeons do minimally invasive procedures. They are made up of robotic arms with end effectors (surgical tools) and a control panel where the surgeon runs the robot.

   Diagram:

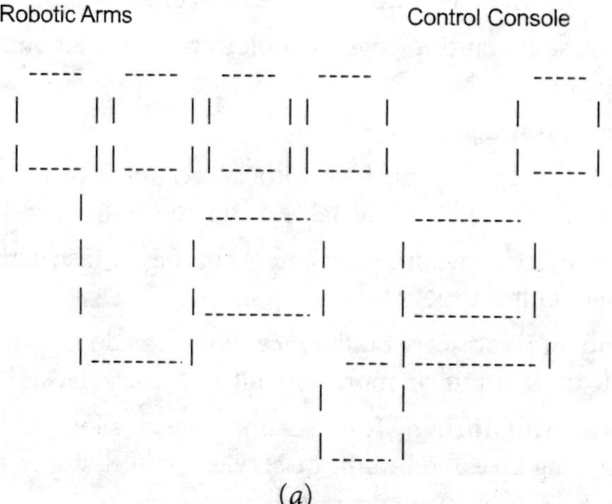

(*a*)

(*b*)

Fig. 1.12. (a) Line diagram and (b) Photograph of Surgical Robot

In this diagram, the "Control Console" is the console where the surgeon sits to operate the surgical robot, and the "Robotic Arms" are the arms of the surgical robot with end effectors (not shown in full).

2. **Rehabilitation Robot (Exoskeleton):**

   Configuration: Exoskeletons are wearable robotic systems that support and assist patients' limbs during exercises and rehabilitation. Rehabilitation robots like these are designed to help patients recover from injuries or physical disabilities.

   Diagram:

(*a*) (*b*)

Fig. 1.13. (a) Line diagram and (b) Photograph of Rehabilitation Robot

The robotic apparatus that the patient wears and uses to support their limbs while performing rehabilitation activities is represented by the term "Wearable Exoskeleton" in this diagram.

Please note that the schematics presented here are condensed depictions of how healthcare robot configurations might look. The actual designs of healthcare robots could be more intricate and could have more parts, sensors, and security measures. Healthcare robot configurations may change as technology develops, adding new and better structures to improve patient care and medical processes.

1.7 EDUCATION ROBOTS

Robotic tools created to support and improve educational activities include "educational robots," "teaching robots," and other similar terms. To engage pupils and encourage interactive and experience learning, these robots are utilized in schools, educational facilities, and private residences. The following information relates to educational robots:

1. **Types of Education Robots:**
    (*a*) **Programmable Robots:** Robots that can be programmed by students using graphical user interfaces or coding languages are known as programmable robots. While ordering the robot to carry out various actions and manoeuvres, students can learn programming fundamentals.
    (*b*) **STEM Robots:** Robots that teach STEM (Science, Technology, Engineering, and Mathematics) principles are known as STEM robots. They frequently include different motors, sensors, and building materials for in-depth exploration and problem-solving.
    (*c*) **Language Learning Robots:** By engaging in interactive discussions, pronunciation drills, and language games, language learning robots assist students in honing their language skills.
    (*d*) **Social-Emotional Learning (SEL) Robots:** Students' social and emotional skills are encouraged through SEL robots. They can engage in role-playing exercises, show empathy, and offer emotional support.
    (*e*) **Robotic Kits and Platforms:** Students can assemble and programme a variety of parts and modules from educational robotic kits and platforms to build the unique robots they want.
2. **Learning Objectives:** Specific learning goals are considered when designing education robots, including:
    - **Improving Computational Thinking:** By decomposing problems into step-by-step instructions, robots assist pupils in developing computational thinking abilities.
    - **Promoting Creativity:** Students can explore their creativity in design and problem-solving by building and programming robots.
    - **Fostering Critical Thinking:** Students are frequently required to analyse problems and come up with creative solutions when using educational robots.
    - **Promoting Collaboration:** Robotics projects frequently need teamwork, which encourages the development of collaboration and communication skills.
3. **Sensors and Actuators:** Education robots have a variety of sensors, including proximity sensors, color sensors, and touch sensors, as well as actuators, such motors and servos, that enable them to interact with their surroundings and react to stimuli.
4. **Programming Interfaces:** Many educational robots provide simple programming interfaces that are appropriate for users of all ages and skill levels. These interfaces are user-friendly for beginners since they frequently employ visual programming languages or block-based coding.
5. **Classroom Integration:** Education robots can be included into both formal and casual learning environments. They can be included by teachers into lesson plans to promote certain learning goals and involve students in hands-on learning activities.

6. **Real-World Uses:** Some educational robots are made to resemble actual uses, including autonomous automobiles or robotic arms used in manufacturing. Students are better able to comprehend how their information is put to use in real-world situations because of this relationship.
7. **Age Appropriateness:** From preschool to university-level instruction, education robots are available for pupils of all age groups. Depending on the intended age range, the complexity and features of the robots frequently change.

Through their promotion of active learning, problem-solving techniques, and digital literacy, education robots are essential to the modern educational system. They encourage students to major in STEM disciplines and help them get ready for the difficulties of the modern workforce. Education robots are anticipated to develop in sophistication and contribute even more to changing how students around the world learn as technology progresses.

1.7.1 Information on the Configuration of Education Robot

Education robots come in a variety of configurations as of my knowledge cutoff date in September 2021, each one created to meet a different learning aim and include students in interactive and experiential learning. We describe how two popular varieties of teaching robots are configured below, along with simplified illustrations that show how they are built.

1. **Programmable Robot:**

   Configuration: Programmable robots typically include a movable base with wheels or tracks, a programmable control unit, and several sensors and actuators. They are made to be programmed by students using coding languages or graphical user interfaces

   Diagram:

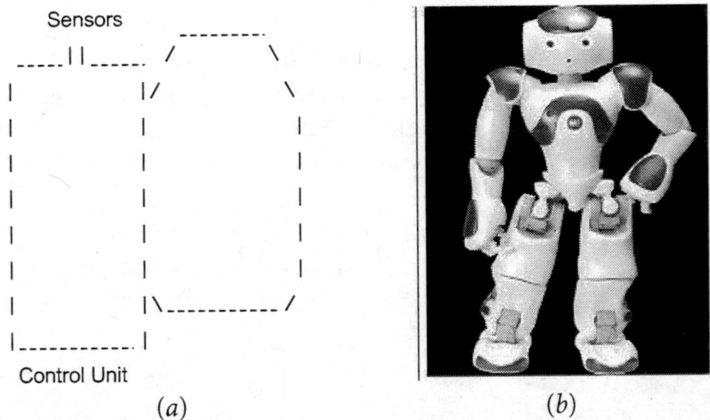

Fig. 1.14. (a) Line diagram and (b) Photograph of Programmable Robot

The term "Sensors" in this diagram refers to a variety of sensors, including line-following, color, and proximity sensors. The "Control Unit" is the robot's brain, and it is here that students may program the machine's actions. Due to its mobility, the robot can move about and engage with its surroundings in accordance with the instructions it has been given.

2. **STEM Robot (Building Kit):**

Configuration:

- STEM robots, which are frequently distributed as building kits, concentrate on imparting knowledge of STEM topics.
- Typically, they contain building elements like blocks, wheels, gears, and motors that kids may combine to create different robot configurations.

Diagram:

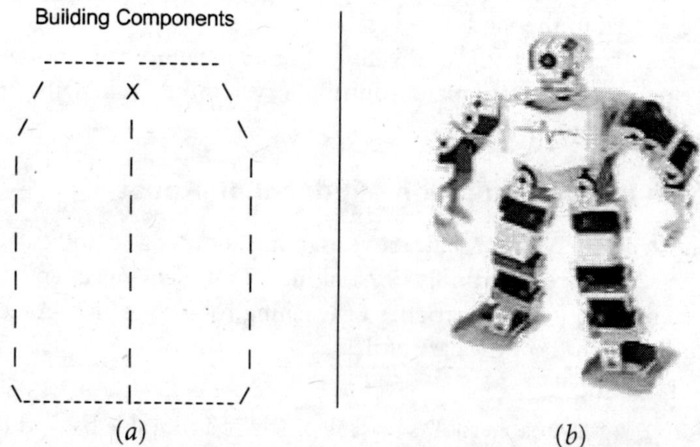

Fig. 1.15. (a) Line diagram (b) Artificial Intelligent Visual Humanoid Robot

The many elements and components in the STEM robot building kit are represented by the "Building Components" in this graphic. These parts can be used by students to build various robot designs, fostering exploration and creativity via hands-on learning.

Please note that the schematics presented here are condensed depictions of how educational robots are set up. Depending on the particular type of robot and its intended educational purpose, the actual designs of education robots might differ greatly. Simple programmable toys to more sophisticated robotic kits created for specific educational goals are all examples of education robots.

Education robots are anticipated to advance in sophistication and variety as technology and educational strategies do as well, offering students priceless learning opportunities across a range of STEM fields and beyond.

1.8 ENTERTAINMENT ROBOTS

Robotic devices primarily intended for amusement, entertainment, and recreational use are referred to as entertainment robots. These robots are designed to connect with people in enjoyable and fun ways, offering friendship, fun exchanges, and interactive experiences. The following information relates to entertainment robots:

1. **Types of Entertainment Robots:**

   (*a*) **Robotic Pets:** These machines mimic the actions and appearance of pets while offering companionship without the upkeep that comes with actual animals. They can move and sound like living things and react to touch.

(b) **Robot Toys:** These are interactive are made for play and interaction. They can carry out tricks, reply to vocal orders, and take part in games with people.

(c) **Humanoid Dancing Robots:** Designed to dance, these robots frequently exhibit human-like choreography and motions. They can amuse viewers with their dance skills and occasionally engage the audience.

(d) **Theme Park Robots:** To keep guests entertained, amusement parks and theme parks frequently deploy entertainment robots. They might be included in interactive displays, performances, or rides.

(e) **Robotic Performers:** A few robots are created expressly to do performances on stages or at public gatherings. They might dance, play musical instruments, or take part in theatrical productions.

2. **Interaction and Engagement:** Entertainment robots can engage with people and interact with them thanks to a variety of sensors and actuators. They are able to recognize voice instructions, gestures, and touch, and they may react in ways that make an experience interesting and enjoyable.
3. **Artificial Intelligence and Machine Learning:** To enhance their interactions, advanced entertainment robots may employ artificial intelligence (AI) and machine learning techniques. This gives them the ability to learn from user feedback and modify their behavior to produce experiences that are more individualized.
4. **Mobility and Manipulation:** Depending on how they are made, entertainment robots may be able to move around and engage with their environment. Some persons may also have manipulators or arms that they can use for duties or for entertainment.
5. **Emotional Expression:** Many entertainment robots are built with the ability to convey emotions through body language, sounds, and facial expressions. This improves their capacity to connect emotionally with users.
6. **Use Cases:** Homes, schools, entertainment venues, and theme parks all frequently employ entertainment robots. They can act as interactive teaching instruments, as well as companionship and amusement for both adults and children.
7. **Limitations:** Even while they might be delightful and entertaining, entertainment robots are mostly made to be amusing and social. They are not designed to carry out difficult jobs or meaningfully take the place of human connection.

The usage of entertainment robots is essential for demonstrating how robotics and AI can enthral and please users. They continue to develop as technology does, giving users of all ages increasingly complex and engaging experiences.

1.8.1 Information on the Configuration of Entertainment Robot

As of my knowledge cutoff in September 2021, entertainment robots are available in a variety of forms, each of which is intended to entertain and engage humans. I'll outline the setup of two popular categories of entertainment robots below, along with simplified schematics that show how they are built:

1. **Robotic Pet:**

Configuration: Robotic pets often have a charming and small design, with sensors and actuators to react to user interactions. They are designed to mimic the behaviour and appearance of genuine animals, bringing companionship and delight to users without the responsibility of caring for a living pet.

Diagram:

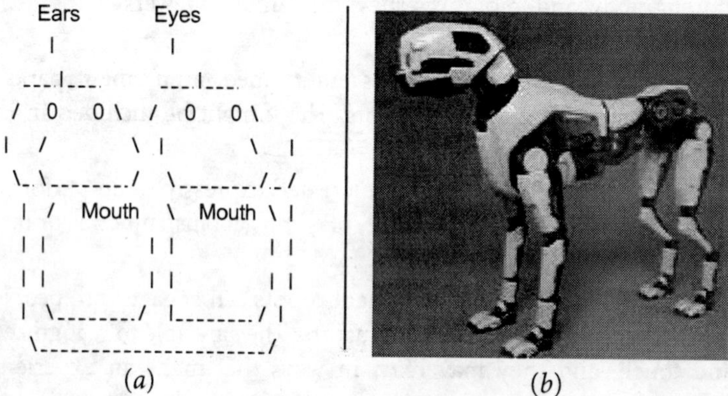

Fig. 1.16. (a) Line diagram and (b) Photograph of Robotic dog

The sensors and actuators that the robotic pet uses to react to user inputs are represented by the "Ears," "Eyes," and "Mouth" in this schematic. To recognize touching and interaction, these robots may incorporate touch sensors on their bodies. To communicate emotions, the mouth and eyes may make a variety of expressions and movements.

2. **Humanoid Dancing Robot:**

Configuration: Humanoid dancing robots typically have a humanoid form with movable limbs and joints to conduct dance routines, and they're made to amuse viewers with their dancing skills and human-like gestures.

Diagram:

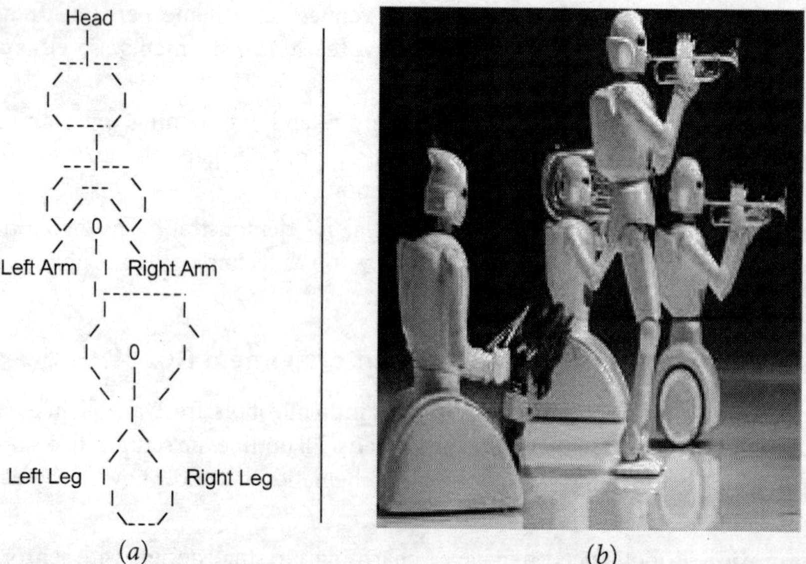

Fig. 1.17. (a) Line diagram and (b) Photograph of Humanoid Dancing Robot

The various body parts in this diagram stand in for the dancing robot's humanoid structure. The robot's head may contain expressive characteristics for a more engaging performance, and its arms and legs have flexible joints for dancing moves.

Please take note that the diagrams presented here are condensed depictions of how entertainment robot configurations might look. The actual designs of entertainment robots can be more complex, and depending on the technology employed and the intended usage, their capabilities can differ greatly. As technology develops, entertainment robots become more lifelike and interactive, providing consumers with entertaining and enjoyable experiences.

1.9 AUTONOMOUS VEHICLES

Autonomous vehicles, sometimes referred to as self-driving automobiles or driverless cars, are autos that can navigate and run without a driver by using sophisticated sensors, cameras, radar systems, and artificial intelligence algorithms. These cars can sense their environment, make decisions, and manage their movements, allowing them to drive safely and effectively without a driver's assistance. Here are some significant facts regarding autonomous vehicles:

1. **Levels of Automation:** From Level 0 (no automation) to Level 5 (complete automation), the Society of Automotive Engineers (SAE) has identified six levels of automation for vehicles:
   - **Level 0:** No Automation – All aspects of driving are handled by the human driver.
   - **Level 1:** Driver assistance allows the car to help with steering or acceleration/deceleration, but not both at once.
   - **Level 2:** Partial Automation – Under certain circumstances, the vehicle can simultaneously control steering and acceleration/deceleration. The motorist needs to pay attention to the surroundings and stay focused.
   - **Level 3:** Conditional Automation – Under certain circumstances, the vehicle can do all driving-related functions, but the driver may still need to take control when the system signals it's time.
   - **Level 4:** High Automation – Under some circumstances, the vehicle can execute all driving responsibilities without human participation, even if the driver doesn't acknowledge a takeover request.
   - **Level 5:** Full Automation – The vehicle is totally autonomous and able to operate in any situation without any input from or supervision from a human.
2. **Sensors and Perception:** To sense their surroundings, autonomous cars use a variety of sensors, such as cameras, LiDAR (Light Detection and Ranging) sensors, radar, and ultrasonic sensors. These sensors offer a 360-degree view of the environment, picking up on other cars, people walking or cycling, road signs, and obstructions.
3. **Decision-Making and Control:** Machine learning and artificial intelligence algorithms process the sensor data to provide real-time decision-making and movement control for the vehicle. To ensure safe and effective driving, the AI system takes into account a variety of aspects, including traffic conditions, traffic laws, and safety considerations.
4. **Safety Features:** In autonomous vehicles, safety comes first. To prevent crashes and react effectively in an emergency, they have redundant systems, fail-safe mechanisms, and enhanced driver aid capabilities.

5. **Testing and Development:** The technology for autonomous vehicles is put through a lot of testing and development, including simulation, testing on closed courses, and testing in controlled conditions. The goal of this iterative procedure is to increase the dependability and security of autonomous driving systems.
6. **Potential Benefits:** Autonomous cars have the potential to provide a number of advantages, including a decrease in traffic accidents, an improvement in mobility for the elderly and disabled, a decrease in congestion, and a reduction in fuel consumption.
7. **Regulatory and Legal Obstacles:** The widespread use of autonomous vehicles poses a number of regulatory and legal obstacles, such as worries about cybersecurity, liability and insurance, and the need to develop uniform laws and regulations for their safe operation.
8. **Current Deployment:** At the time of September 2021, firms like Waymo, Tesla, Uber (previously), and others were paving the ground for the testing and gradual deployment of autonomous vehicles on public roads in a number of areas across the world.

Autonomous vehicle development and deployment continue to advance quickly thanks to continued research and innovation in the automotive and technological sectors. Autonomous vehicles are anticipated to proliferate in the transportation sector and revolutionize mobility and how we commute in the future as legislation and technology advance.

1.9.1 Information on the Configuration of Autonomous Vehicles

Different parts that work together to allow self-driving capabilities are fitted in autonomous vehicles. I'll go into more depth on how autonomous vehicles are set up below, along with simplified diagrams that show how they are built:

1. **The perceptual system**

   Sensors: To perceive their surroundings and learn about their environment, autonomous vehicles make use of a variety of sensors. Radar, ultrasonic, LiDAR, and cameras are examples of common sensors.

   Diagram:

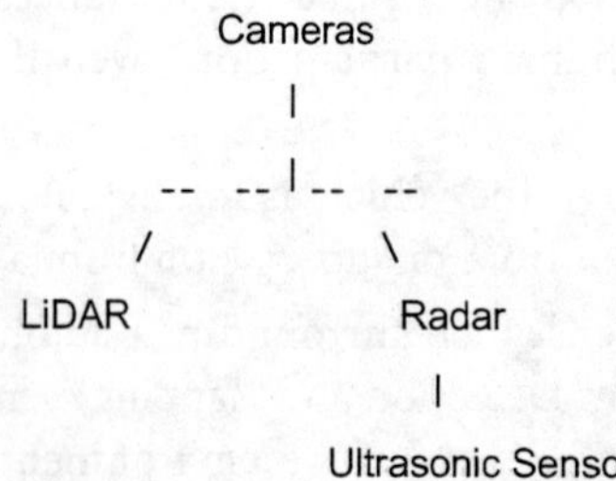

Fig. 1.18. Configuration of Autonomous vehicles

   The cameras, LiDAR, radar, and ultrasonic sensors are each shown as a separate component in this diagram. Together, these sensors offer a complete picture of the area around the vehicle.

2. **Localization System:**

   GPS and Localization Technologies: The localization system aids in pinpointing the exact location of the car on the road. It makes use of GPS as well as other localization tools like RTK or INS.

Diagram:

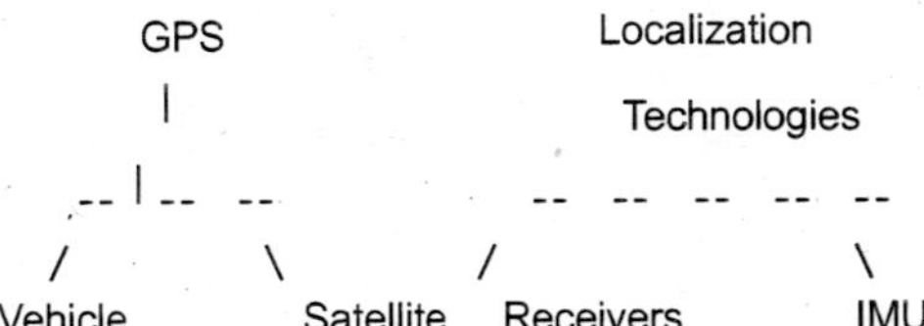

Fig. 1.19. Localization systems for helping the vehicle to determine its precise location

The GPS receiver in this diagram uses signals from satellites to determine the general location of the vehicle. High-precision localization is offered by localization technologies like RTK (Real-Time Kinematic) and INS (Inertial Navigation System).

3. **Decision-Making Algorithms:**

   Artificial intelligence (AI) and machine learning: To make decisions for driving in the present, decision-making algorithms analyse inputs from the perceptual system.

   Diagram:

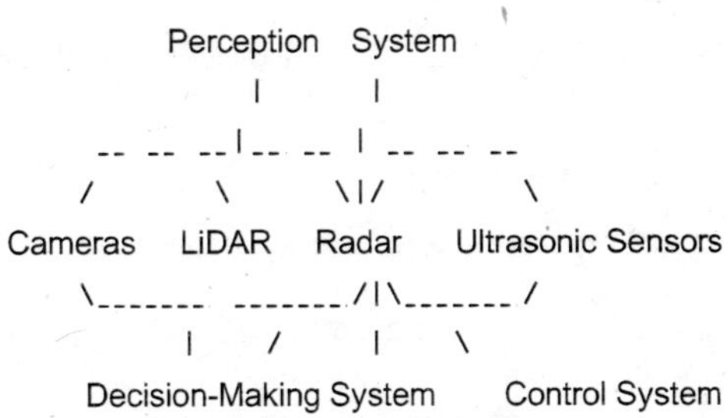

Fig. 1.20. Decision-making algorithms to take input from the perception system, including data from cameras, LiDAR, radar, and ultrasonic sensors

The decision-making algorithms in this figure incorporate information from the perception system, such as images from cameras, LiDAR, radar, and ultrasonic sensors. To decide on the best course of action for driving, the decision-making system analyses this data using AI and machine learning methods.

4. **Control System:** The throttle, brakes, and steering are all controlled by the control system, which translates the decisions generated by the decision-making algorithms into actual vehicle motions.

   Diagram:

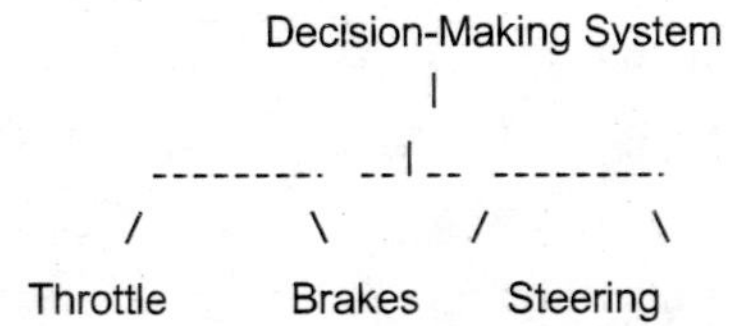

Fig. 1.21. Decision-making system to send control signals to the vehicle's throttle, brakes, and steering systems

In this picture, the decision-making system controls the vehicle's steering, braking, and throttle systems to help it travel safely and effectively.

Please be aware that the simplified representations of autonomous vehicle configurations in the given illustrations are just that—simplified. Autonomous vehicle designs in practice are more intricate and may include extra parts, redundant systems, and safety features. Since autonomous vehicle technology is developing quickly, more recent configurations might include sophisticated features and technologies that are outside the purview of these designs.

Mathematical Preliminaries with Answers

Q.1. What is the significance of mathematics in robotics?

Ans. Mathematics is essential in robotics for modeling robot kinematics, dynamics, control systems, and trajectory planning.

Q.2. What are the key mathematical concepts used in robot kinematics?

Ans. Robot kinematics involves concepts like coordinate systems, transformations, rotations, and translations.

Q.3. How is trigonometry used in robotics?

Ans. Trigonometry is used to calculate angles and distances, which are crucial for robot navigation and path planning.

Q.4. What is the role of linear algebra in robotics?

Ans. Linear algebra is used to represent and manipulate robot transformations, positions, and orientations.

Q.5. How are matrices and vectors used in robot control?

Ans. Matrices and vectors are used to represent robot dynamics and control inputs, facilitating motion planning and stability analysis.

Q.6. How is calculus applied in robot motion control?

Ans. Calculus is used to derive and optimize robot trajectories, velocities, and accelerations.

Q.7. What is the importance of probability and statistics in robotics?

Ans. Probability and statistics are crucial for sensor fusion, localization, and handling uncertainties in robot perception and control.

Q.8. How are differential equations relevant to robot control systems?

Ans. Differential equations are used to model robot dynamics and design control algorithms.

Q.9. What is forward kinematics in robotics?

Ans. Forward kinematics deals with determining the position and orientation of the robot's end-effector given the joint angles.

Q.10. What is inverse kinematics in robotics?

Ans. Inverse kinematics involves computing the joint angles needed to achieve a desired end-effector position and orientation.

Q.11. How are quaternions used in representing robot orientations?

Ans. Quaternions are used to represent rotations in 3D space and are popular for efficient orientation calculations in robotics.

Q.12. What is the role of transformation matrices in robotics?

Ans. Transformation matrices are used to represent spatial relationships between robot links and coordinate frames.

Q.13. How is matrix multiplication used in robot transformations?

Ans. Matrix multiplication combines multiple transformations to find the overall robot pose in a specific coordinate system.

Q.14. What is the concept of the Jacobian matrix in robotics?

Ans. The Jacobian matrix relates joint velocities to end-effector velocities and is crucial for robot control and motion planning.

Q.15. How is the distance formula used in robot perception?

Ans. The distance formula is used to calculate distances between the robot and objects in its environment, enabling obstacle detection.

Q.16. What is the significance of probability distributions in robot localization?

Ans. Probability distributions help represent uncertainties in robot localization, such as in the Kalman filter or particle filter.

Q.17. How is the Monte Carlo method used in robot mapping?

Ans. The Monte Carlo method, particularly the Monte Carlo localization (MCL) algorithm, is used for probabilistic robot mapping.

Q.18. What is the use of vector calculus in robot motion planning?

Ans. Vector calculus is used to optimize robot trajectories, velocities, and accelerations to achieve smooth and efficient motion.

Q.19. How are eigenvalues and eigenvectors applied in robotics?

Ans. Eigenvalues and eigenvectors are used to analyze the stability and dynamic behavior of robot systems.

Q.20. How is optimization theory used in robot control?

Ans. Optimization theory is used to find optimal robot trajectories and control inputs that minimize costs or achieve specific objectives.

Q.21. How is Laplace transform used in robot control systems?

Ans. Laplace transform is used to analyze the dynamics of linear time-invariant systems, which are common in robot control.

Q.22. What is the role of calculus of variations in robotics?

Ans. Calculus of variations is used to find the optimal path or motion of a robot by minimizing a cost function.

Q.23. How is linear programming used in robot motion planning?

Ans. Linear programming is used for problems with linear constraints, such as collision avoidance or optimal path planning.

Q.24. What is the concept of Bayesian estimation in robot perception?

Ans. Bayesian estimation is used to update the robot's beliefs about its environment based on sensor measurements and prior knowledge.

Q.25. How are differential equations used in modeling robot dynamics?

Ans. Differential equations help represent the relationship between joint torques, accelerations, and velocities in robot motion.

Q.26. How is statistical regression used in robot learning?

Ans. Statistical regression is used to learn mappings between inputs and outputs, enabling robots to adapt and improve their performance.

Q.27. What is the role of Fourier analysis in signal processing for robot sensors?

Ans. Fourier analysis is used to analyze and process signals from robot sensors, such as filtering noise or extracting features.

Q.28. How are probability distributions used in robot motion planning?

Ans. Probability distributions help model uncertainties in robot motion planning and collision avoidance.

Q.29. What is the significance of convex optimization in robot control?

Ans. Convex optimization algorithms are used to find optimal control inputs that satisfy linear constraints and cost functions.

Q.30. How are numerical methods used in solving robot control equations?

Ans. Numerical methods like Runge-Kutta or numerical integration are used to solve robot dynamics equations and control algorithms.

Q.31. How is statistics used in evaluating robot performance?

Ans. Statistics are used to analyze experimental data and evaluate the performance of robot systems and algorithms.

Q.32. What is the use of Bayesian networks in robot decision-making?

Ans. Bayesian networks help represent and update probabilities in robot decision-making processes.

Q.33. How is geometry applied in robot path planning?

Ans. Geometry is used to calculate distances, angles, and geometric properties necessary for planning robot trajectories.

Q.34. What is the significance of linear programming in robot optimization?

Ans. Linear programming is used in optimization problems where the objective function and constraints are linear.

Q.35. How is coordinate geometry used in robot localization?

Ans. Coordinate geometry is used to represent and transform the robot's position and orientation in various coordinate frames.

Q.36. What is the role of Markov chains in robot state estimation?

Ans. Markov chains are used to model the transition probabilities between different robot states in state estimation processes.

Q.37. How is statistical hypothesis testing used in robot perception?

Ans. Statistical hypothesis testing helps make decisions about sensor data reliability and object recognition.

Q.38. What is the significance of matrix algebra in robot control systems?

Ans. Matrix algebra is used to represent and manipulate robot dynamics, kinematics, and control equations.

Q.39. How are statistics used in designing robot experiments?

Ans. Statistics are used to determine sample sizes, confidence intervals, and hypothesis testing in experimental design.

Q.40. What is the role of calculus in analyzing robot motion trajectories?

Ans. Calculus is used to analyze velocity, acceleration, and curvature along robot motion paths.

Q.41. How are neural networks used in robot learning and control?

Ans. Neural networks are used for pattern recognition, control approximation, and learning in robot systems.

Q.42. What is the significance of numerical optimization in robot control?

Ans. Numerical optimization algorithms are used to find optimal control inputs and trajectory planning.

Q.43. How is linear algebra used in robot manipulator analysis?

Ans. Linear algebra is used to analyze robot manipulator kinematics, Jacobians, and end-effector positions.

Q.44. What is the use of integral calculus in robot perception?

Ans. Integral calculus is used to calculate areas and volumes from sensor data, such as in environment mapping.

Q.45. How are eigenvalues used in analyzing robot stability?

Ans. Eigenvalues are used to determine the stability of robot control systems.

Q.46. What is the significance of optimization algorithms in robot motion planning?

Ans. Optimization algorithms help find the best paths and control inputs for robot motion planning.

Q.47. How is coordinate geometry used in robot calibration?

Ans. Coordinate geometry is used to calibrate robot manipulators and align coordinate systems.

Q.48. What is the role of statistical filtering techniques in robot localization?

Ans. Statistical filtering techniques, such as the Kalman filter or particle filter, are used for sensor fusion and robot localization.

Q.49. How is probability theory applied in robot decision-making under uncertainty?

Ans. Probability theory helps assess risks and make decisions in uncertain environments, such as in robot path planning.

Q.50. What are the mathematical concepts used in robot control optimization algorithms?

Ans. Robot control optimization algorithms use concepts like gradient descent, Newton's method, and Lagrange multipliers to find optimal control inputs and trajectories.

Numerical Problems with Solutions

Problem 1. A line-following robot needs to travel 2 meters along a curved path. If the robot's wheels have a diameter of 5 cm, how many wheel rotations are required?

Solution: Wheel circumference = π * diameter = π * 5 cm ≈ 15.71 cm Number of wheel rotations = 200 cm/15.71 cm ≈ 12.73 rotations (approximately 13 rotations)

Problem 2. An obstacle avoidance robot detects an obstacle 1.5 meters away. If the robot's maximum speed is 0.5 m/s, how long will it take to stop before colliding with the obstacle?

Solution: Time to stop = Distance/Maximum speed = 1.5 m/0.5 m/s = 3 seconds

Problem 3. A robotic arm has three joints, and their joint angles are given as 30°, 60°, and 90°. Calculate the forward kinematics to find the position of the end-effector.

Solution: The forward kinematics formula depends on the robot arm's geometry and joint types.

Problem 4. A drone has a flight time of 15 minutes and a maximum speed of 30 km/h. How far can it travel in a single flight?

Solution: Distance = Maximum speed * Flight time = 30 km/h * (15 minutes/60 minutes) = 7.5 km.

Problem 5. A robotic hand has five fingers, each with three joints. How many total degrees of freedom does the hand have?

Solution: Total degrees of freedom = Number of fingers * Number of joints per finger = 5 fingers * 3 joints/finger = 15 degrees of freedom

Problem 6. A robot arm has a reach of 80 cm. If it needs to pick an object located 60 cm away, can it reach the object?

Solution: Yes, the robot arm can reach the object since 60 cm is within the arm's reach of 80 cm.

Problem 7. A robotic vehicle covers 500 meters in 2 minutes. Calculate its average speed in meters per second.

Solution: Average speed = Distance/Time = 500 meters/(2 minutes * 60 seconds/minute) = 4.17 m/s.

Problem 8. A humanoid robot walks a distance of 10 meters with a step length of 50 cm. How many steps does it take to cover this distance?

Solution: Number of steps = Distance/Step length = 10 meters/0.5 meters = 20 steps

Problem 9. A cleaning robot's battery capacity is 4000 mAh, and it consumes power at a rate of 1000 mA. How long can the robot operate on a single charge?

Solution: Operating time = Battery capacity/Power consumption = 4000 mAh/1000 mA = 4 hours

Problem 10. A robotic arm moves from a starting position $(x_1, y_1) = (2, 3)$ to an end position $(x_2, y_2) = (6, 8)$. Calculate the distance traveled by the arm.

Solution: Distance = $\sqrt{((x_2 - x_1) + (y_2 - y_1)^2)} = \sqrt{((6-2)^2 + (8-3)^2)} = \sqrt{(16+25)} = \sqrt{41} \approx 6.40$ units

Problem 11. A drone is flying at an altitude of 100 meters. It descends 20 meters and then ascends 30 meters. What is the drone's final altitude?

Solution: Final altitude = Initial altitude – Descent + Ascent = 100 m – 20 m + 30 m = 110 meters

Problem 12. A robotic arm takes 5 seconds to complete one full rotation. Calculate its rotational speed in revolutions per minute (RPM).

Solution: Rotational speed (RPM) = 1 revolution/Time taken = 1/5 seconds * 60 seconds/minute ≈ 12 RPM

Problem 13. A drone flies at a speed of 20 m/s for 10 minutes, then it flies at 30 m/s for the next 5 minutes. Calculate the average speed of the drone during this time.

Solution: Total distance = $Speed_1$ * $Time_1$ + $Speed_2$ * $Time_2$ = 20 m/s * (10 minutes * 60 seconds/minute) + 30 m/s * (5 minutes * 60 seconds/minute) = 12,000 meters Total time = $Time_1$ + $Time_2$ = 10 minutes + 5 minutes = 15 minutes Average speed = Total distance/Total time = 12,000 meters/(15 minutes * 60 seconds/minute) = 13.33 m/s

Problem 14. A robot arm needs to lift an object weighing 50 kg. If the robot applies a force of 200 N, will it be able to lift the object?

Solution: To lift the object, the robot needs to apply a force greater than or equal to the object's weight. Since 200 N < 50 kg * 9.8 m/s^2 ≈ 490 N, the robot cannot lift the object.

Problem 15. A robot moves 5 meters forward, 3 meters to the right, and then 2 meters backward. Calculate the robot's net displacement from its starting point.

Solution: Net displacement = $\sqrt{(5^2 + 3^2)} \approx \sqrt{34} \approx 5.83$ meters

Problem 16. A cleaning robot moves at a speed of 0.2 m/s and cleans a 3-meter wide path. How long will it take to clean a 50-meter long corridor?

Solution: Time taken = Distance/Speed = 50 meters/0.2 m/s = 250 seconds = 4 minutes and 10 seconds

Problem 17. A drone flies from point A to point B, which is 1 km away. If it maintains a constant speed of 30 km/h, how long will it take to reach point B?

Solution: Time taken = Distance/Speed = 1 km/30 km/h = 1/30 hours = 2 minutes

Problem 18. A robot has an angular velocity of 2 rad/s. How long will it take to complete one full rotation (2π radians)?

Solution: Time taken = Angle/Angular velocity = 2π radians/2 rad/s = π seconds ≈ 3.14 seconds

Problem 19. A robotic arm has a link of length 1 meter and rotates at a speed of 60 degrees per second. Calculate the linear velocity of the end-effector at a given point.

Solution: Linear velocity = Angular velocity * Radius = (60 degrees/second) * (1 meter) * (π/180) ≈ 1.05 m/s

Problem 20. A drone is flying at an altitude of 100 meters. It descends at a rate of 2 m/s. How long will it take to reach the ground?

Solution: Time taken = Altitude/Descent rate = 100 meters/2 m/s = 50 seconds

Problem 21. A robotic arm has a payload of 10 kg and a link of length 0.5 meters. If the arm is fully extended horizontally, what torque is required to hold the payload?

Solution: Torque = Force * Distance = 10 kg * 9.8 m/s^2 * 0.5 meters = 49 Nm.

Problem 22. A robot arm has a maximum payload capacity of 20 kg. If the arm is carrying an object weighing 15 kg, how much additional weight can it lift?

Solution: Additional weight capacity = Maximum payload capacity – Current weight = 20 kg – 15 kg = 5 kg

Problem 23. A drone travels at a speed of 25 m/s for 40 seconds. How far does the drone travel during this time?

Solution: Distance = Speed * Time = 25 m/s * 40 seconds = 1000 meters = 1 km

Problem 24. A robotic arm moves in a circular path with a radius of 2 meters. If it completes one full revolution in 10 seconds, calculate its angular velocity.

Solution: Angular velocity = 2π radians/Time taken = 2π radians/10 seconds = 0.628 rad/s

Problem 25. A cleaning robot's battery voltage is 12 V, and it consumes power at a rate of 50 W. Calculate the current drawn by the robot from the battery.

Solution: Current = Power/Voltage = 50 W/12 V ≈ 4.17 A

Problem 26. A drone battery has a capacity of 3000 mAh and operates at a voltage of 11.1 V. How much energy does it store in watt-hours (Wh)?

Solution: Energy (Wh) = Battery capacity (mAh) * Voltage (V)/1000 = 3000 mAh * 11.1 V/1000 = 33.3 Wh

Problem 27. A robot's arm is 1.5 meters long. If it rotates at 30 degrees per second, calculate the linear velocity of its end-effector at a given point.

Solution: Linear velocity = Angular velocity * Radius = (30 degrees/second) * (1.5 meters) * (π/180) ≈ 0.262 m/s

Problem 28. A drone has a maximum flight range of 5 km. If it travels at a speed of 20 m/s, how long can it fly continuously?

Solution: Time taken = Flight range/Speed = 5000 meters/20 m/s = 250 seconds = 4 minutes and 10 seconds

Problem 29. A robot arm has a payload of 5 kg and a link of length 0.8 meters. If the arm is fully extended horizontally, what torque is required to hold the payload?

Solution: Torque = Force * Distance = 5 kg * 9.8 m/s^2 * 0.8 meters = 39.2 Nm

Problem 30. A robotic arm moves in a circular path with a radius of 3 meters. If it completes one full revolution in 15 seconds, calculate its angular velocity.

Solution: Angular velocity = 2π radians/Time taken = 2π radians/15 seconds = 0.419 rad/s

Problem 31. A drone battery has a capacity of 2500 mAh and operates at a voltage of 14.8 V. How much energy does it store in watt-hours (Wh)?

Solution: Energy (Wh) = Battery capacity (mAh) * Voltage (V)/1000 = 2500 mAh * 14.8 V/1000 = 37 Wh

Problem 32. A robot's arm is 2 meters long. If it rotates at 45 degrees per second, calculate the linear velocity of its end-effector at a given point.

Solution: Linear velocity = Angular velocity * Radius = (45 degrees/second) * (2 meters) * (π/180) ≈ 1.57 m/s

Problem 33. A drone has a maximum flight range of 8 km. If it travels at a speed of 25 m/s, how long can it fly continuously?

Solution: Time taken = Flight range/Speed = 8000 meters/25 m/s = 320 seconds = 5 minutes and 20 seconds

Problem 34. A robot arm has a payload of 7 kg and a link of length 1 meter. If the arm is fully extended horizontally, what torque is required to hold the payload?

Solution: Torque = Force * Distance = 7 kg * 9.8 m/s^2 * 1 meter = 68.6 Nm

Problem 35. A robotic arm moves in a circular path with a radius of 4 meters. If it completes one full revolution in 20 seconds, calculate its angular velocity.

Solution: Angular velocity = 2π radians/Time taken = 2π radians/20 seconds = 0.314 rad/s

Problem 36. A drone battery has a capacity of 1500 mAh and operates at a voltage of 12 V. How much energy does it store in watt-hours (Wh)?

Solution: Energy (Wh) = Battery capacity (mAh) * Voltage (V)/1000 = 1500 mAh * 12 V/1000 = 18 Wh

Problem 37. A robot's arm is 3 meters long. If it rotates at 60 degrees per second, calculate the linear velocity of its end-effector at a given point.

Solution: Linear velocity = Angular velocity * Radius = (60 degrees/second) * (3 meters) * (π/180) ≈ 3.14 m/s

Problem 38. A drone has a maximum flight range of 10 km. If it travels at a speed of 30 m/s, how long can it fly continuously?

Solution: Time taken = Flight range/Speed = 10,000 meters/30 m/s = 333.33 seconds ≈ 5 minutes and 33 seconds

Problem 39. A robot arm has a payload of 9 kg and a link of length 1.2 meters. If the arm is fully extended horizontally, what torque is required to hold the payload?

Solution: Torque = Force * Distance = 9 kg * 9.8 m/s^2 * 1.2 meters = 105.84 Nm

Problem 40. A robotic arm moves in a circular path with a radius of 5 meters. If it completes one full revolution in 25 seconds, calculate its angular velocity.

Solution: Angular velocity = 2π radians/Time taken = 2π radians/25 seconds = 0.251 rad/s

Problem 41. A drone battery has a capacity of 2000 mAh and operates at a voltage of 10 V. How much energy does it store in watt-hours (Wh)?

Solution: Energy (Wh) = Battery capacity (mAh) * Voltage (V)/1000 = 2000 mAh * 10 V/1000 = 20 Wh

Problem 42. A robot's arm is 4 meters long. If it rotates at 90 degrees per second, calculate the linear velocity of its end-effector at a given point.

Solution: Linear velocity = Angular velocity * Radius = (90 degrees/second) * (4 meters) * (π/180) ≈ 8.38 m/s

Problem 43. A drone has a maximum flight range of 12 km. If it travels at a speed of 35 m/s, how long can it fly continuously?

Solution: Time taken = Flight range/Speed = 12,000 meters/35 m/s = 342.86 seconds ≈ 5 minutes and 43 seconds.

Problem 44. A robot arm has a payload of 6 kg and a link of length 1.5 meters. If the arm is fully extended horizontally, what torque is required to hold the payload?

Solution: Torque = Force * Distance = 6 kg * 9.8 m/s^2 * 1.5 meters = 88.2 Nm

Problem 45. A robotic arm moves in a circular path with a radius of 6 meters. If it completes one full revolution in 30 seconds, calculate its angular velocity.

Solution: Angular velocity = 2π radians/Time taken = 2π radians/30 seconds = 0.209 rad/s

Problem 46. A drone battery has a capacity of 1800 mAh and operates at a voltage of 9 V. How much energy does it store in watt-hours (Wh)?

Solution: Energy (Wh) = Battery capacity (mAh) * Voltage (V)/1000 = 1800 mAh * 9 V/1000 = 16.2 Wh

Problem 47. A robot's arm is 5 meters long. If it rotates at 120 degrees per second, calculate the linear velocity of its end-effector at a given point.

Solution: Linear velocity = Angular velocity * Radius = (120 degrees/second) * (5 meters) * (π/180) ≈ 13.09 m/s

Problem 48. A drone has a maximum flight range of 15 km. If it travels at a speed of 40 m/s, how long can it fly continuously?

Solution: Time taken = Flight range/Speed = 15,000 meters/40 m/s = 375 seconds ≈ 6 minutes and 15 seconds

Problem 49. A robot arm has a payload of 8 kg and a link of length 1.8 meters. If the arm is fully extended horizontally, what torque is required to hold the payload?

Solution: Torque = Force * Distance = 8 kg * 9.8 m/s^2 * 1.8 meters = 141.12 Nm

Problem 50. A robotic arm moves in a circular path with a radius of 7 meters. If it completes one full revolution in 35 seconds, calculate its angular velocity.

Solution: Angular velocity = 2π radians/Time taken = 2π radians/35 seconds = 0.180 rad/s

Short Questions and Answers

Q.1. What is a robot?

Ans. A robot is a programmable machine that can carry out tasks autonomously or semi-autonomously.

Q.2. What is an industrial robot?

Ans. An industrial robot is designed to perform repetitive tasks in manufacturing and assembly lines.

Q.3. What is a humanoid robot?

Ans. A humanoid robot is designed to resemble and mimic human movements and actions.

Q.4. What is a medical robot?

Ans. A medical robot is used in healthcare settings to assist with surgeries, rehabilitation, and patient care.

Q.5. What is a drone?

Ans. A drone is an unmanned aerial vehicle that can be controlled remotely or autonomously.

Q.6. What is an underwater robot?

Ans. An underwater robot, also known as an ROV (Remotely Operated Vehicle), is designed to operate underwater.

Q.7. What is a military robot?

Ans. A military robot is used for defense and combat-related purposes, such as bomb disposal or reconnaissance.

Q.8. What is a service robot?

Ans. A service robot is designed to assist humans in various tasks, such as cleaning, cooking, or companionship.

Q.9. What is a self-driving car?

Ans. A self-driving car is an autonomous vehicle capable of navigating without human intervention.

Q.10. What is a surgical robot?

Ans. A surgical robot is used by surgeons to perform minimally invasive surgeries with enhanced precision.

Q.11. What is an exoskeleton robot?

Ans. An exoskeleton robot is a wearable robotic device that enhances the physical abilities of the wearer.

Q.12. What is a telepresence robot?

Ans. A telepresence robot allows users to remotely interact with a location using a mobile robotic platform.

Q.13. What is a toy robot?

Ans. A toy robot is a robotic device designed for entertainment and play purposes.

Q.14. What is a delivery robot?

Ans. A delivery robot is used to transport goods from one location to another, often autonomously.

Q.15. What is a firefighting robot?

Ans. A firefighting robot is used to assist firefighters in hazardous situations.

Q.16. What is a companion robot?

Ans. A companion robot is designed to provide emotional support and companionship to users.

Q.17. What is a vacuum cleaning robot?

Ans. A vacuum cleaning robot autonomously cleans floors and carpets.

Q.18. What is an agricultural robot?

Ans. An agricultural robot assists in farming tasks, such as planting, harvesting, or monitoring crops.

Q.19. What is a rescue robot?

Ans. A rescue robot is used in search and rescue operations to locate and assist victims in disaster scenarios.

Q.20. What is a swarm robot?

Ans. A swarm robot refers to a group of robots that work together to achieve a common goal.

Q.21. What is an AI robot?

Ans. An AI robot integrates artificial intelligence to enhance its decision-making and problem-solving capabilities.

Q.22. What is a climbing robot?

Ans. A climbing robot is capable of navigating vertical surfaces, like walls or ceilings.

Q.23. What is a cleaning robot?

Ans. A cleaning robot is designed to clean surfaces, including floors, windows, and gutters.

Q.24. What is a social robot?

Ans. A social robot is designed to interact and communicate with humans in social settings.

Q.25. What is a space exploration robot?

Ans. A space exploration robot is used for scientific research and data collection in outer space.

Q.26. What is a warehouse robot?

Ans. A warehouse robot helps with tasks like picking, packing, and moving goods in a warehouse setting.

Q.27. What is a remote-controlled robot?

Ans. A remote-controlled robot is operated by a human using a remote control device.

Q.28. What is a walking robot?

Ans. A walking robot moves by mimicking the locomotion of animals or humans.

Q.29. What is a flying robot?

Ans. A flying robot is capable of aerial movements and can include drones and aerial vehicles.

Q.30. What is a marine robot?

Ans. A marine robot operates in underwater environments and can include submersibles and autonomous underwater vehicles (AUVs).

Q.31. What is a bomb disposal robot?

Ans. A bomb disposal robot is used to defuse and dispose of explosive devices safely.

Q.32. What is a toy robot?

Ans. A toy robot is a robotic device designed for entertainment and play purposes.

Q.33. What is an assistive robot?

Ans. An assistive robot aids people with disabilities or limited mobility in their daily activities.

Q.34. What is a surveillance robot?

Ans. A surveillance robot is used to monitor and gather information in security-related scenarios.

Q.35. What is a microbot?

Ans. A microbot is a tiny robot designed for applications at a miniature scale, often in medical or industrial contexts.

Q.36. What is a teleoperated robot?

Ans. A teleoperated robot is controlled by a human operator from a distance.

Q.37. What is a rescue snake robot?

Ans. A rescue snake robot is a flexible, elongated robot that can access tight spaces during rescue missions.

Q.38. What is a nano-robot?

Ans. A nano-robot is a robot that operates at the nanoscale, typically used in medical and scientific research.

Q.39. What is a soft robot?

Ans. A soft robot is made from flexible materials, allowing it to deform and adapt to its environment.

Q.40. What is an autonomous underwater glider?

Ans. An autonomous underwater glider is a robot that can glide through the water without using traditional propellers.

Q.41. What is a lawn-mowing robot?

Ans. A lawn-mowing robot automatically mows grass and maintains lawns without human intervention.

Q.42. What is a construction robot?

Ans. A construction robot assists in building tasks, such as bricklaying or 3D printing structures.

Q.43. What is a robot swarm?

Ans. A robot swarm refers to a large group of simple robots that work together to accomplish complex tasks.

Q.44. What is a snake-like robot?

Ans. A snake-like robot mimics the movements of a snake and is useful in navigating confined spaces.

Q.45. What is a firefighting drone?

Ans. A firefighting drone assists in extinguishing fires from the air.

Q.46. What is a warehouse sorting robot?

Ans. A warehouse sorting robot automatically organizes and sorts packages in a distribution center.

Q.47. What is a bomb-defusing drone?

Ans. A bomb-defusing drone is used to disarm explosives safely and remotely.

Q.48. What is a painting robot?

Ans. A painting robot is used for automated painting applications in industries like automotive manufacturing.

Q.49. What is a robot kit?

Ans. A robot kit is a package containing components and instructions to build and program a robot.

Q.50. What is a bipedal robot?

Ans. A bipedal robot is a two-legged robot capable of walking on its own.

Multiple Choice Questions and Answers

Q.1. What is an industrial robot?

(*a*) A robot used for space exploration
(*b*) A robot designed to perform repetitive tasks in manufacturing and assembly lines
(*c*) A robot that assists in medical surgeries
(*d*) A robot used for underwater exploration

Q.2. What is a humanoid robot?

(*a*) A robot designed to resemble and mimic human movements and actions
(*b*) A robot used for agricultural tasks
(*c*) A robot used for surveillance
(*d*) A robot designed for space exploration

Q.3. What is a medical robot?

(*a*) A robot used for space exploration
(*b*) A robot used for military purposes
(*c*) A robot used in healthcare settings to assist with surgeries and patient care
(*d*) A robot designed for household cleaning

Q.4. What is a drone?

(*a*) A robot designed for household cleaning
(*b*) A robot used for space exploration
(*c*) An unmanned aerial vehicle that can be controlled remotely or autonomously
(*d*) A robot designed for underwater exploration

Q.5. What is an underwater robot?

(*a*) A robot used for military purposes
(*b*) A robot used for space exploration
(*c*) A robot designed for agricultural tasks
(*d*) A robot designed to operate underwater

Q.6. What is a military robot?

(*a*) A robot used for space exploration
(*b*) A robot designed for agricultural tasks
(*c*) A robot designed for household cleaning
(*d*) A robot used for defense and combat-related purposes

Q.7. What is a service robot?

(*a*) A robot designed for agricultural tasks
(*b*) A robot used for military purposes
(*c*) A robot used for space exploration
(*d*) A robot designed to assist humans in various tasks

Q.8. What is a self-driving car?

(*a*) A robot designed to assist with surgeries
(*b*) A robot designed for agricultural tasks
(*c*) An autonomous vehicle capable of navigating without human intervention
(*d*) A robot used for military purposes

Q.9. What is a surgical robot?

(*a*) A robot used for space exploration
(*b*) A robot used for agricultural tasks
(*c*) A robot designed for household cleaning
(*d*) A robot used by surgeons to perform minimally invasive surgeries

Q.10. What is an exoskeleton robot?

(*a*) A robot used for military purposes (*b*) A robot used for agricultural tasks

(*c*) A wearable robotic device that enhances the physical abilities of the wearer

(*d*) A robot designed for space exploration

Q.11. What is a telepresence robot?

(*a*) A robot designed to assist with surgeries (*b*) A robot used for agricultural tasks

(*c*) A robot used for space exploration

(*d*) A robot that allows users to remotely interact with a location using a mobile robotic platform

Q.12. What is a toy robot?

(*a*) A robot used for agricultural tasks (*b*) A robot designed for household cleaning

(*c*) A robot used for military purposes

(*d*) A robotic device designed for entertainment and play purposes

Q.13. What is a delivery robot?

(*a*) A robot used for space exploration (*b*) A robot designed for agricultural tasks

(*c*) A robot used for military purposes

(*d*) A robot used to transport goods from one location to another

Q.14. What is a firefighting robot?

(*a*) A robot used for space exploration (*b*) A robot used for military purposes

(*c*) A robot designed for agricultural tasks

(*d*) A robot used to assist firefighters in hazardous situations

Q.15. What is a companion robot?

(*a*) A robot used for military purposes (*b*) A robot used for space exploration

(*c*) A robot designed to provide emotional support and companionship to users

(*d*) A robot used for agricultural tasks

Q.16. What is a vacuum cleaning robot?

(*a*) A robot used for military purposes (*b*) A robot designed for agricultural tasks

(*c*) A robot used for space exploration

(*d*) A robot that autonomously cleans floors and carpets

Q.17. What is an agricultural robot?

(*a*) A robot used for military purposes (*b*) A robot used for space exploration

(*c*) A robot used to assist with surgeries

(*d*) A robot that assists in farming tasks, such as planting and harvesting

Q.18. What is a rescue robot?

(*a*) A robot used for space exploration (*b*) A robot designed for agricultural tasks

(*c*) A robot used for military purposes

(*d*) A robot used in search and rescue operations to locate and assist victims in disaster scenarios

Q.19. What is a swarm robot?

(*a*) A robot used for space exploration
(*b*) A robot designed for agricultural tasks
(*c*) A robot used for military purposes
(*d*) A group of robots that work together to achieve a common goal

Q.20. What is an AI robot?

(*a*) A robot designed for agricultural tasks
(*b*) A robot used for space exploration
(*c*) A robot used for military purposes
(*d*) A robot that integrates artificial intelligence to enhance decision-making and problem-solving capabilities

Q.21. What is a climbing robot?

(*a*) A robot used for space exploration
(*b*) A robot designed for agricultural tasks
(*c*) A robot used for military purposes
(*d*) A robot capable of navigating vertical surfaces, like walls or ceilings

Q.22. What is a cleaning robot?

(*a*) A robot used for space exploration
(*b*) A robot designed for agricultural tasks
(*c*) A robot used for military purposes
(*d*) A robot designed to clean surfaces, including floors, windows, and gutters

Q.23. What is a social robot?

(*a*) A robot used for space exploration
(*b*) A robot used for military purposes
(*c*) A robot designed for agricultural tasks
(*d*) A robot designed to interact and communicate with humans in social settings

Q.24. What is a space exploration robot?

(*a*) A robot used for military purposes
(*b*) A robot designed for agricultural tasks
(*c*) A robot designed for household cleaning
(*d*) A robot used for scientific research and data collection in outer space

Q.25. What is a warehouse robot?

(*a*) A robot used for military purposes
(*b*) A robot used for space exploration
(*c*) A robot designed for agricultural tasks
(*d*) A robot that helps with tasks like picking, packing, and moving goods in a warehouse setting

Q.26. What is a remote-controlled robot?

(*a*) A robot used for space exploration
(*b*) A robot designed for agricultural tasks
(*c*) A robot used for military purposes
(*d*) A robot operated by a human using a remote control device

Q.27. What is a walking robot?

(*a*) A robot used for space exploration
(*b*) A robot designed for agricultural tasks
(*c*) A robot used for military purposes
(*d*) A robot that moves by mimicking the locomotion of animals or humans

Q.28. What is a flying robot?

(*a*) A robot designed for agricultural tasks (*b*) A robot used for military purposes

(*c*) A robot used for space exploration

(*d*) A robot capable of aerial movements, including drones and aerial vehicles

Q.29. What is a marine robot?

(*a*) A robot used for space exploration (*b*) A robot designed for agricultural tasks

(*c*) A robot used for military purposes

(*d*) A robot that operates in underwater environments, including submersibles and autonomous underwater vehicles (AUVs)

Q.30. What is a bomb disposal robot?

(*a*) A robot used for space exploration (*b*) A robot designed for agricultural tasks

(*c*) A robot used for military purposes

(*d*) A robot used to defuse and dispose of explosive devices safely

Q.31. What is a rescue snake robot?

(*a*) A robot used for space exploration (*b*) A robot used for military purposes

(*c*) A robot designed for agricultural tasks

(*d*) A flexible, elongated robot that can access tight spaces during rescue missions

Q.32. What is a nano-robot?

(*a*) A robot used for space exploration (*b*) A robot designed for agricultural tasks

(*c*) A robot used for military purposes

(*d*) A robot that operates at the nanoscale, often used in medical and scientific research

Q.33. What is a soft robot?

(*a*) A robot used for space exploration (*b*) A robot designed for agricultural tasks

(*c*) A robot used for military purposes

(*d*) A robot made from flexible materials, allowing it to deform and adapt to its environment

Q.34. What is an autonomous underwater glider?

(*a*) A robot used for space exploration (*b*) A robot used for military purposes

(*c*) A robot designed for agricultural tasks

(*d*) A robot that can glide through the water without using traditional propellers

Q.35. What is a lawn-mowing robot?

(*a*) A robot used for space exploration (*b*) A robot used for military purposes

(*c*) A robot designed for agricultural tasks

(*d*) A robot that automatically mows grass and maintains lawns without human intervention

Q.36. What is a construction robot?

(*a*) A robot used for space exploration (*b*) A robot used for military purposes

(*c*) A robot designed for agricultural tasks

(*d*) A robot that assists in building tasks, such as bricklaying or 3D printing structures

Q.37. What is a robot swarm?

(*a*) A robot used for space exploration (*b*) A robot used for military purposes

(*c*) A robot designed for agricultural tasks

(*d*) A large group of simple robots that work together to accomplish complex tasks

Q.38. What is a snake-like robot?

(*a*) A robot used for space exploration (*b*) A robot designed for agricultural tasks

(*c*) A robot used for military purposes

(*d*) A robot that mimics the movements of a snake and is useful in navigating confined spaces

Q.39. What is a firefighting drone?

(*a*) A robot used for space exploration (*b*) A robot used for military purposes

(*c*) A robot designed for agricultural tasks

(*d*) A drone that assists in extinguishing fires from the air

Q.40. What is a warehouse sorting robot?

(*a*) A robot used for space exploration (*b*) A robot used for military purposes

(*c*) A robot designed for agricultural tasks

(*d*) A robot that automatically organizes and sorts packages in a distribution center

Q.41. What is a bomb-defusing drone?

(*a*) A robot used for space exploration (*b*) A robot used for military purposes

(*c*) A robot designed for agricultural tasks

(*d*) A drone used to disarm explosives safely and remotely

Q.42. What is a painting robot?

(*a*) A robot used for space exploration (*b*) A robot used for military purposes

(*c*) A robot designed for agricultural tasks

(*d*) A robot used for automated painting applications in industries like automotive manufacturing

Q.43. What is a robot kit?

(*a*) A robot used for space exploration

(*b*) A robot used for military purposes

(*c*) A robot designed for agricultural tasks

(*d*) A package containing components and instructions to build and program a robot

Q.44. What is a bipedal robot?

(*a*) A robot used for space exploration (*b*) A robot used for military purposes

(*c*) A robot designed for agricultural tasks

(*d*) A two-legged robot capable of walking on its own

Q.45. What is a quadruped robot?

(*a*) A robot used for space exploration (*b*) A robot used for military purposes

(*c*) A robot designed for agricultural tasks

(*d*) A four-legged robot capable of walking on its own

Q.46. What is a swarm of quadcopters called?

(*a*) Quadroid (*b*) QuadSwarm
(*c*) QuadPod (*d*) Quadrotor Swarm

Q.47. Which robot is designed to assist people with disabilities or limited mobility?

(*a*) Companion Robot (*b*) Medical Robot
(*c*) Assistive Robot (*d*) Surgical Robot

Q.48. What is a robot capable of navigating underwater currents without the use of propulsion systems called?

(*a*) Underwater Glider (*b*) Aquatic Rover
(*c*) Submersible Robot (*d*) Diving Drone

Q.49. Which robot is controlled by a human operator from a remote location?

(*a*) Autonomous Robot (*b*) Teleoperated Robot
(*c*) AI Robot (*d*) Swarm Robot

Q.50. What type of robot can be used to explore and collect data in environments with high radiation or toxic substances?

(*a*) Space Robot (*b*) Hazardous Environment Robot
(*c*) Nuclear Robot (*d*) Medical Robot

Key to Multiple Choice Questions

1. (*b*)	**2.** (*a*)	**3.** (*c*)	**4.** (*c*)	**5.** (*d*)	**6.** (*d*)	**7.** (*d*)	**8.** (*c*)	**9.** (*d*)
10. (*c*)	**11.** (*d*)	**12.** (*d*)	**13.** (*d*)	**14.** (*d*)	**15.** (*c*)	**16.** (*d*)	**17.** (*d*)	**18.** (*d*)
19. (*d*)	**20.** (*d*)	**21.** (*d*)	**22.** (*d*)	**23.** (*d*)	**24.** (*d*)	**25.** (*d*)	**26.** (*d*)	**27.** (*d*)
28. (*d*)	**29.** (*d*)	**30.** (*d*)	**31.** (*d*)	**32.** (*d*)	**33.** (*d*)	**34.** (*d*)	**35.** (*d*)	**36.** (*d*)
37. (*d*)	**38.** (*d*)	**39.** (*d*)	**40.** (*d*)	**41.** (*d*)	**42.** (*d*)	**43.** (*d*)	**44.** (*d*)	**45.** (*d*)
46. (*d*)	**47.** (*c*)	**48.** (*a*)	**49.** (*b*)	**50.** (*b*)				

Project Works

1. **Line-following Robot:** Build a robot capable of following a line on the ground using sensors and feedback control mechanisms.
2. **Obstacle Avoidance Robot:** Create a robot that can navigate its way through an environment while avoiding obstacles using ultrasonic or infrared sensors.
3. **Voice-controlled Home Assistant Robot:** Develop a robot that responds to voice commands to perform tasks like turning on lights, playing music, or providing weather updates.
4. **Robotic Arm with Object Recognition:** Build a robotic arm that can pick up and manipulate objects based on their shape and color using a camera and image recognition.
5. **Gesture-controlled Robot:** Create a robot that responds to hand gestures or body movements for navigation and performing specific actions.
6. **Maze-solving Robot:** Design a robot capable of navigating through a maze and finding the most efficient path to the exit.

7. **Remote-controlled Surveillance Robot:** Build a robot equipped with a camera that can be remotely controlled to monitor areas like home or office spaces.
8. **Autonomous Lawn Mowing Robot:** Develop a robot that autonomously mows the lawn without human intervention, using sensors to detect boundaries and obstacles.
9. **Robotic Fish:** Create a robotic fish capable of swimming and maneuvering underwater mimicking the movements of real fish.
10. **Robotic Pet Feeder:** Design a robot that automatically dispenses pet food at scheduled times to feed pets.
11. **Rubik's Cube Solving Robot:** Build a robot that can solve a Rubik's Cube using algorithms and mechanical manipulators.
12. **Home Security Robot:** Develop a robot equipped with sensors and a camera to patrol and monitor the home for security purposes.
13. **Robotic Hand with Haptic Feedback:** Create a robotic hand that can mimic human hand movements and provide haptic feedback to the user.
14. **Dance-bot:** Build a robot that can dance to music, either pre-programmed or based on real-time beats and rhythms.
15. **Soccer-playing Robot:** Design a robot capable of playing soccer and competing against other robots or human players.
16. **Humanoid Robot with Emotional Expressions:** Create a humanoid robot that can display emotional expressions through facial movements and body language.
17. **Robotic Chess Player:** Build a robot that can play chess against a human opponent or another robot.
18. **3D Printing Robot:** Develop a robot capable of 3D printing objects using additive manufacturing techniques.
19. **Self-balancing Segway-style Robot:** Design a two-wheeled robot that can balance itself while carrying a load or a rider.
20. **Robotic Drummer:** Build a robot that can play drums and follow a musical beat.
21. **Robotic Bartender:** Create a robot capable of mixing and serving drinks to guests.
22. **Robotic Companion for Elderly:** Design a robot that can provide companionship and assistance to the elderly, such as reminders for medication and daily tasks.
23. **Trash-sorting Robot:** Develop a robot that can sort different types of trash (e.g., recyclables and non-recyclables) using sensors and mechanical arms.
24. **Mars Rover Replica:** Build a robot inspired by Mars rovers to navigate and explore a simulated Martian terrain.
25. **Robotic Guitar Player:** Create a robot capable of playing guitar and performing songs.
26. **Robotic Fruit Picker:** Develop a robot that can identify ripe fruits and pick them from trees without damaging the produce.
27. **Solar-powered Robot:** Design a robot that operates using solar energy and can recharge itself in sunlight.
28. **Swarm Robotics:** Create a group of robots that can communicate and collaborate to achieve a common task, like cooperative navigation or pattern formation.

29. **Robotic Drawing Artist:** Build a robot capable of creating artworks by drawing on paper or canvas.
30. **Robotic Waiter:** Develop a robot capable of serving food and drinks in a restaurant setting.
31. **Mars Rover Rover-Operator Interface:** Design a user interface to remotely control a simulated Mars rover on a computer.
32. **Robotic Snake for Search and Rescue:** Create a flexible snake-like robot capable of accessing tight spaces during search and rescue operations.
33. **Robotic Cocktail Maker:** Build a robot that can mix and serve cocktails based on user preferences.
34. **Firefighting Drone Swarm:** Design a swarm of drones that can coordinate efforts to extinguish fires in a controlled environment.
35. **Robotic Wheelchair:** Develop a robot-assisted wheelchair with autonomous navigation and obstacle avoidance capabilities.
36. **Robotic Guitar Tuner:** Build a robot that can tune a guitar automatically.
37. **Robotic Book Sorter**: Create a robot that can organize and categorize books in a library.
38. **Robotic Car Parking System**: Design a robot-assisted car parking system that autonomously parks and retrieves vehicles in a parking lot.
39. **Robotic Voice Translator**: Develop a robot that can listen to and translate speech in real-time across different languages.
40. **Robotic Suitcase**: Build a robot suitcase that can follow its owner and navigate through airports or other crowded places.
41. **Robotic Cocktail Stirrer:** Create a robot that can stir cocktails with different recipes and flavors.
42. **Robotic Ball Collector for Golf Courses**: Design a robot that collects golf balls from the course and returns them to a designated location.
43. **Robotic Book Reader for the Visually Impaired**: Build a robot that can scan and read aloud books or documents for visually impaired individuals.
44. **Robotic Bodyguard**: Develop a robot with security features that can follow and protect a person during a walk or commute.
45. **Robotic Weather Station**: Create a robot with sensors to measure and report weather parameters like temperature, humidity, and wind speed.
46. **Robotic Food Delivery System**: Design a system of delivery robots that can transport food from the kitchen to customers' tables in a restaurant.
47. **Robotic Bike Rider**: Build a robot capable of riding a bicycle and maintaining balance.
48. **Robotic Bike Parking Valet**: Develop a robot that can autonomously park and retrieve bicycles in crowded bike racks.
49. **Robotic Language Tutor:** Create a robot that helps individuals learn new languages through interactive lessons and conversations.
50. **Robotic Sports Player**: Build a robot capable of playing various sports like basketball, table tennis, or soccer.

2

Human-Robot Collaboration: Ethics and Social Implications of Robotics

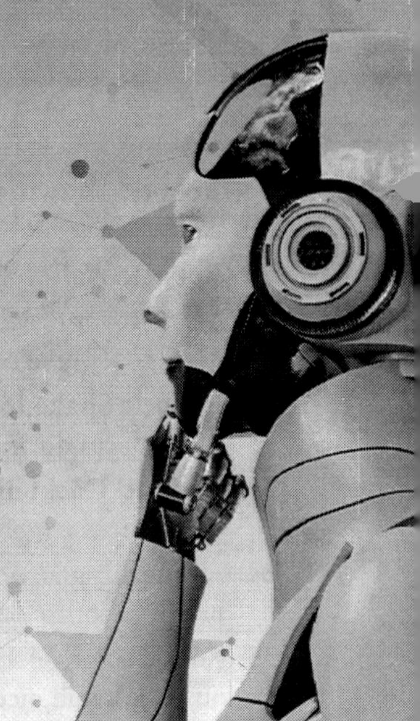

2.1 INTRODUCTION

At the nexus of technology, robotics, and human interaction is the revolutionary idea of human-robot collaboration. It entails the seamless fusion of human abilities and robot capabilities to boost innovation, productivity, and efficiency across a range of industries. By combining the strengths of humans and robots, this partnership ushers in a new era of synergy and invention that has the potential to transform several sectors of the economy, workplaces, and daily life. A notable breakthrough in recent years has been the integration of robots into many human occupations. Artificial intelligence, sensor, and automation advancements have made robots more capable, flexible, and versatile. They can work in circumstances that are dangerous, repetitious, or need precise activities to be performed with constant accuracy. It's crucial to investigate the most recent technological advancements, scientific discoveries, practical applications, and societal ramifications that result from humans and robots collaborating to build a brighter future as we delve into the field of human-robot collaboration.

2.2 ROBOTS WORKING IN A SYNERGISTIC MANNER

As robots advance in sophistication, it becomes obvious that human collaboration is the key to maximizing their potential. Robots aren't just working beside humans in this collaboration; rather, technologies are being developed that enable human and robot interaction, communication, and skill-complementation. This can happen in a variety of contexts:

1. **Manufacturing:** While people may concentrate on complicated decision-making, quality control, and creativity, robots can do repetitive and physically demanding work on assembly lines.
2. **Healthcare:** While medical experts concentrate on diagnosis, treatment planning, and patient care, robots can help surgeons with precision, do mundane jobs in hospitals, and give rehabilitation support.
3. **Logistics and Warehousing:** Humans handle logistics, customer service, and complicated problem-solving, while robots can simplify order fulfillment and inventory management.

4. **Research and Exploration:** While human researchers analyze data, reach important conclusions, and create new experiments, robots can explore dangerous or remote environments, collect data, and carry out experiments.
5. **Agriculture:** While farmers concentrate on crop planning, disease management, and sustainable practices, robots can automate the planting, harvesting, and monitoring of crops.
6. **Education and Entertainment:** Interactive teachers, companions, or performers can be provided by robots to improve the learning or entertainment process as a whole.

Human-robot collaboration offers both exciting prospects and difficult problems. Among the complicated issues that need to be handled include ensuring effective communication, creating user-friendly interfaces, following safety measures, thinking ethically, and tackling job displacement issues. The potential rewards, however, are enormous as we overcome these difficulties: more production, increased safety, expanded creativity, and the capacity to address issues that were previously intractable by humans.

2.3 CONCEPT OF HUMAN-ROBOT COLLABORATION

The term "human-robot collaboration" (HRC) refers to the integration of human and robotic capabilities for the purpose of achieving common objectives. In order to develop a more effective working relationship, it goes beyond simple coexistence and requires task collaboration, communication, and the leveraging of each entity's capabilities.

Here are some key aspects of the concept:

1. **Complementary Skills:** Robots and people each have their own strengths. Robots excel in precision, repetitive jobs, data processing, and physical strength; humans excel at sophisticated decision-making, flexibility, creativity, and emotional intelligence. In order to establish a powerful synergy, HRC seeks to utilize these complimentary skills.
2. **Task Distribution:** In HRC, tasks are distributed according to each participant's skills. Humans do jobs that call for decision-making, inventiveness, and situational awareness, whereas robots undertake duties that are hazardous, repetitive, or require precision. Performance is improved overall as a result of this allocation.
3. **Joint Work:** In some instances, people and robots physically collaborate. Collaborative robots (cobots), for instance, can support human operators in manufacturing by helping with heavy lifting or repetitive activities, improving the production process.
4. **Cognitive Collaboration:** In different circumstances, the collaboration is more sophisticated. Robots can analyze enormous volumes of data fast and execute computations, giving human decision-makers insightful data in areas like data analysis, medical diagnosis, and financial forecasts.
5. **Shared Control:** Shared control, where people and robots cooperatively make decisions, is a common component of HRC. Interfaces and communication techniques that enable both parties to comprehend each other's intents and actions are necessary for this.
6. **Advanced Automation:** Automation technologies, such as machine learning and artificial intelligence, are essential for enabling HRC. Robots are capable of imitating human behavior, adapting to shifting circumstances, and continuously improving.
7. **Versatility:** HRC systems are made to be versatile and flexible. This is crucial in dynamic contexts where activities must be completed quickly. Robots can be reprogrammed or reconfigured to handle new tasks, and humans can provide guidance when faced with unexpected situations.

8. **Safety:** The protection of people using robots in the workplace is of the utmost importance. Safety features of HRC systems include sensors, collision avoidance, and the capacity to recognize and react to human presence.
9. **Impact on Society:** HRC has substantial effects on the labour force, the economy, and society in general. It can enhance productivity and open up new options, but it also raises concerns about job loss, retraining, and the morality of using technology.
10. **Research and Innovation:** Human-robot interaction, machine learning, human aspects, and ethical issues are just a few of the fields under HRC that are currently undergoing active research. The creation and wide-scale deployment of HRC systems are driven by innovations in these fields.

The idea of human-robot collaboration will spread over a variety of industries and an area as technology develops, revolutionizing the way people produce and interact with machines.

2.4 HOW ROBOTS CAN WORK ALONGSIDE HUMANS IN VARIOUS DOMAINS

Robots may collaborate with people in a variety of fields by building on their strengths and enhancing their skills. Here are some instances of how robots and people can work together in many fields:

1. **Manufacturing:** In manufacturing facilities, collaborative robots, or cobots, can work alongside human operators. They are capable of handling monotonous and physically taxing activities including lifting bulky objects, putting parts together, and carrying out exact processes. Human workers may now concentrate on jobs that call for creativity, quality assurance, and problem-solving.
2. **Healthcare:** Robots can help medical staff with a variety of activities. More accurate and minimally invasive operations are made possible by surgical robots. In hospitals, autonomous robots can transfer supplies and medications. Robotic exoskeletons help people with physical disabilities with rehabilitation and mobility.
3. **Warehousing and Logistics:** Robots can increase the effectiveness of warehousing and logistics operations. In order to improve order fulfillment, they can autonomously pick, pack, and sort things in warehouses. In large facilities, autonomous drones and robots can be employed for delivery and inventory management.
4. **Agriculture:** By automating crop planting, harvesting, and monitoring, robots can revolutionize agriculture. Drones used in agriculture can survey fields, evaluate the health of crops, and precisely administer pesticides or fertilizers. Harvesting crops with robotic harvesters requires less manual labor because they can work nonstop.
5. **Research and Exploration:** Robotic exploration of difficult and dangerous settings is crucial. Drones inspect the landscape and collect data in off-the-grid or hazardous regions while autonomous underwater vehicles (AUVs) explore the ocean's depths. Mars rovers and other robots
6. **Education:** By acting as interactive tutors or teaching aids, robots can contribute to education. Social robots can help children learn math, language, and other disciplines. They can offer personalized instruction and help teachers run their classrooms.
7. **Entertainment:** Robots can improve the quality of entertainment. They can be used to captivate and entertain people at theme parks, museums, and interactive displays. Robots may potentially work in the gaming sector, offering fresh game play.

8. **Construction:** Robots can help with heavy lifting, bricklaying, welding, and other construction-related chores. Construction robots increase productivity and safety on construction sites by doing repetitive jobs with accuracy and without becoming tired.
9. **Retail:** Tasks like inventory management, shelf replenishment, and customer support can all be handled by robots in retail settings. Robots that operate on their own can assist customers find products by navigating retail aisles.
10. **Finance:** Robots can help with customer care, fraud detection, and financial analysis. Human financial professionals can concentrate on sophisticated financial strategies since chatbots powered by AI can connect with customers, respond to frequently asked questions, and complete transactions.

Designing user-friendly user interfaces, guaranteeing safety, offering suitable training, and addressing ethical issues all go a long way towards fostering a positive and effective work environment in each of these fields.

2.5 HUMAN-ROBOT COLLABORATION: SAFETY

In order for humans and robots to collaborate safely and effectively, safety must come first in human-robot collaboration (HRC). To reduce risks, avoid accidents, and foster confidence in these cooperative networks, appropriate safety measures must be put in place. The following are important safety factors for HRC:

1. **Risk Assessment:** Before deploying collaborative robots, conduct in-depth risk analyses. Determine any dangers that may exist, such as moving parts, large weights, and places where there may be interactions between people and robots. Determine how serious the hazards are and put the right protections in place.
2. **Safety Standards:** Follow industry-specific safety regulations and standards like ISO 10218 and ISO 15066. For robot design, system integration, and safety needs, these standards offer guidelines.
3. **Collision Detection and Avoidance:** Collaborative robots need sensors, such as force/torque sensors, vision systems, and proximity sensors, to find people or obstacles in their workplace. When the robots notice a barrier, they should be trained to halt or slow down.
4. **Safe Speed and Force Limiting:** In order to protect people in the event of contact, collaborative robots should be built with safe speed and force limits. This minimizes the danger of injury by ensuring that the robot's motions are slow and gentle when it is close to people.
5. **Isolation and Guarding:** When direct physical collaboration is not required, use physical barriers, cages, or light curtains to keep robots and human workers apart. These safeguards can avoid unintentional interaction.
6. **Emergency Stop:** Include readily available emergency stop buttons that, in the event of any perceived risk, can rapidly cease robot operations. Every employee that interacts with robots should be aware of where these buttons are.
7. **Education and Awareness:** Inform those who work with robots—human operators, technicians, and others—about safety measures, emergency protocols, and possible threats. It is crucial to raise safety awareness.
8. **User-Friendly Interfaces:** Create intuitive, simple-to-use user interfaces. The state, motions, and goals of the robot should be clear to the operators. Uncertainties can be decreased with the use of clear and visible signs.

9. **Fail-Safe Design:** Use fail-safe methods that let the robot return to a secure state in the event of a serious system breakdown. This stops the robot from acting erratically and hurting people.
10. **Routine Maintenance:** Check that all safety systems are operating properly by doing routine maintenance and inspections on robots. In order to preserve, swiftly address any wear and tear.

Building trust between people and robots and putting safety first in the design, implementation, and use of collaborative robots can help us make the workplace safer, minimize accidents, and enable productive collaboration between people and robots in a variety of disciplines.

2.6 ETHICAL CHALLENGES AND IMPLICATIONS OF WIDESPREAD ROBOT ADOPTION

To ensure the ethical development and application of these technologies, it is important to carefully address the substantial ethical problems and implications that come with widespread robot adoption. The following ethical issues emerge as robots are increasingly incorporated into various facets of our lives:

1. **Job Replacement:** One of the main worries is the possibility of robots and automation replacing human workers. Unemployment, income disparity, and social instability may result from this. To lessen these effects, it is essential to get the workforce ready for retraining and upgrading.
2. **Privacy:** Robots with sensors and cameras may gather a tone of information. It is crucial to make sure that this information is utilized appropriately and that people's privacy rights are upheld. Surveillance, data breaches, and the possibility for misuse of personal information are all causes for concern.
3. **Fairness and Bias:** There is a chance that biases contained in the data used to train these systems will be perpetuated if robots are utilized in decision-making processes. To prevent bias, openness, and accountability must be upheld in algorithmic judgments.
4. **Autonomy and Control:** As robots grow increasingly autonomous, issues of who is responsible for their acts, who has control over their decisions, and whether or not we can overrule them in dire circumstances become very important ethical conundrums.
5. **Ethical Dilemmas in Decision Making:** When robots make decisions in the present (like self-driving cars), they may face moral choices between competing values, such as pedestrian safety against passenger safety. It is difficult to create frameworks for resolving such problems.
6. **Societal Impact:** It is important to consider how widely used robots will affect culture, society, and interpersonal interactions. It might alter societal conventions, the nature of labour, and how we communicate with one another.
7. **Transparency and Accountability:** It can be difficult to comprehend how robots make decisions, particularly in complicated AI systems. For accountability and trust-building, it is crucial to ensure transparency in how robots arrive at decisions.
8. **Human Identity:** As robots become more prevalent in our daily lives, concerns have been raised about how this may affect our connections with others and our understanding of what it means to be human.
9. **Dependency:** As we rely more on robots to complete a variety of jobs, there is a risk of becoming too reliant on this technology, which could have negative repercussions if it malfunctions or is improperly handled.

10. **Ethical Employment in Military and Security:** The employment of robots in military applications, such as autonomous weapons, raises severe ethical concerns concerning the possibility of indiscriminate injury, lack of human supervision, and the possibility of cyber attacks.

Collaboration between technologists, policymakers, ethicists, and society at large is necessary to address these ethical issues. To successfully traverse the ethical complications of broad robot adoption, it is crucial to develop and abide by clear ethical standards, participate in open conversations, establish suitable rules, and make sure that the rewards of robot adoption are distributed fairly.

2.7 ROBOTICS INCLUDING ISSUES RELATED TO PRIVACY AND JOB DISPLACEMENT

The appropriate development and application of robotic technology depends heavily on the ethics and societal consequences of robotics, including worries about privacy and job displacement. Here is a look at these concerns:

Privacy Issues: Robots with sensors, cameras, and data processing skills have the potential to gather a lot of private data. This presents the following privacy issues:

- **Surveillance:** Robots can invade people's privacy when utilized for monitoring, security, or surveillance reasons, raising worries about ongoing monitoring and the potential misuse of gathered data.
- **Data Protection:** It is crucial to make sure that the information collected by robots is safely sent, stored, and utilized. Unauthorized access to this data and data breaches can have serious repercussions.
- **Informed Consent:** Individuals should be made aware of how robots acquire and use their data. For confidence to remain intact, data usage, sharing, and storage must be transparent.
- **Automated Recognition:** Facial recognition and other biometric technologies can be made possible by advanced robots.

Job Displacement: Robotics and automation technology can cause job displacement in a variety of sectors and occupations, including:

- **Automation in Manufacturing:** Highly developed robotic systems can take the place of human workers in labor-intensive or repetitive operations, which will result in the loss of jobs in the manufacturing industry.
- **Service Industry:** In the service industry (such as customer support and data entry), robots and AI systems may replace some human workers, which may result in job losses in these sectors.
- **Re-skilling and Training:** As some tasks are automated, the workforce must be re-trained in order to adapt to new roles. To lessen the detrimental consequences of employment relocation, this requirement must be addressed.
- **Income Inequality:** Automation's improved productivity and lower costs could worsen income inequality if they are not allocated equitably. This could also lead to societal inequities.

Ethical Considerations: Several broader ethical considerations arise from the integration of robots into society:

- **Fairness and Bias:** A major difficulty is ensuring that robotic algorithms and systems are impartial and produce fair results, particularly in decision-making processes.

- **Human Control and Autonomy:** As robots grow more independent, deciding the proper amount of human supervision and control is essential, especially in applications that need safety.
- **Unexpected Effects:** The sophistication of robotics and AI systems may have unintended effects. It's crucial to prepare for and reduce these possible negative effects.
- **Social Acceptance:** For these technologies to be widely accepted, it is essential that society is ready for the integration of robots and that ethical problems are effectively handled.

Collaboration between developers, politicians, ethicists, and the general public is necessary to address these ethical and social ramifications. It is possible to ensure that robotics technology serves society while minimizing its negative effects by establishing explicit regulations, guidelines, and open conversations.

2.8 HUMAN-ROBOT INTERACTIONS

The term "human-robot interactions" (HRI) refers to the different ways that humans and robots interact, cooperate, and communicate. These contacts span a variety of industries, from manufacturing and entertainment to personal support and healthcare. The successful integration of robots into society depends on our understanding of and improvements to HRI. Here are some features of interactions between humans and robots:

1. **Communication:** HRI requires effective communication. Robots must be able to comprehend spoken language, written language, gestures, and facial expressions as well as the context in which they are used. Systems for human-robot dialogue and natural language processing (NLP) are essential for enabling seamless communication.
2. **Sensing:** Robots with sensors can detect human presence and their surroundings. Robots can avoid collisions and interact securely by using this sensory capability and respond to human gestures or commands.
3. **Collaboration:** In collaborative scenarios, people and machines cooperate to accomplish common objectives. Coordination, task distribution, and mutual understanding of each other's abilities are necessary for this. This sort of contact is exemplified by collaborative robots (cobots), which are robots made to operate alongside people.
4. **User Interfaces:** For HRI, user interface design is crucial. User-friendly, straightforward interfaces that deliver feedback that people can comprehend are ideal. Augmented reality, voice instructions, visual clues, and touch screens can all improve the engagement process.
5. **Acceptance and Trust:** For humans and robots to collaborate effectively, trust is a prerequisite. This trust is reliant on the safety, predictability, and overall user experience of robots. It is crucial to make sure that people accept and feel at ease with robots performing a variety of tasks.
6. **Social Interaction:** Social robots are made to interact socially and emotionally with people. In contexts like healthcare, education, or friendship, they can recognize emotions, react sympathetically, and offer a more engaging and comfortable experience.
7. **Shared Control:** Achieving the ideal level of shared control is essential in situations where humans and robots both have decision-making authority (such as autonomous vehicles). This entails allowing robots to perform routine activities while allowing humans to step in when needed.
8. **Ethical Considerations:** Interactions between humans and robots create ethical issues include accountability in the event of accidents, decision-making openness, and potential biases in robot-use algorithms.

9. **Flexibility:** Robots should be able to change their behavior to accommodate different user contexts and preferences. Personalized interactions can improve user experience and increase the utility of robots in a variety of applications.
10. **Learning and Improvement:** Over time, robots that can learn from their interactions with people will be able to perform better and better suit the needs of their users.

Interdisciplinary research in robotics, psychology, human factors, languages, and ethics is needed to improve human-robot interactions. Realising the full potential of HRI requires striking a balance between technology potential, user needs, safety, and ethical considerations.

Numerical Problems with Solutions

Job Displacement

Problem 1. In a manufacturing plant, 100 human workers are replaced by collaborative robots (cobots) to increase efficiency. Each human worker was earning an average monthly wage of $3,000. Calculate the potential monthly savings due to the implementation of cobots.

Solution: Potential Monthly Savings = (Number of Workers Replaced) * (Average Monthly Wage per Worker) Potential Monthly Savings = 100 * $3,000 Potential Monthly Savings = $300,000

Privacy Concerns

Problem 2. A company deploys security robots equipped with cameras to monitor a public area. Each camera records video at a resolution of 1080p and consumes 1 GB of data per hour. If the cameras operate 24/7, how much data will be generated in a week?

Solution: Data Generated in a Week = (Data per Hour) * (Hours in a Day) * (Days in a Week) Data Generated in a Week = 1 GB/hour * 24 hours/day * 7 days/week Data Generated in a Week = 168 GB

Ethical Dilemmas

Problem 3. A self-driving car is programmed to prioritize passenger safety. In a critical situation, the car detects an unavoidable collision with either a pedestrian or another vehicle carrying two passengers. If the car must make a decision, what should it prioritize, and why?

Solution: The car should prioritize minimizing harm to all parties. It's essential to consider the greater good and try to avoid causing harm to any party. The ethical dilemma involves finding the best compromise to minimize overall harm.

Safety Measures

Problem 4. A collaborative robot operates in a factory, and its maximum speed is 2 meters per second. If the robot's sensors detect a human entering its work area, it must stop within 0.5 seconds to avoid collision. Calculate the minimum stopping distance required for the robot to ensure safety.

Solution: Minimum Stopping Distance = (Maximum Speed) * (Stopping Time) Minimum Stopping Distance = 2 m/s * 0.5 s Minimum Stopping Distance = 1 meter

Income Inequality

Problem 5. A company introduces robotic automation in its warehouse, leading to the elimination of 30% of the manual labor positions. If the affected workers had an average annual salary of $40,000, calculate the potential decrease in the company's annual labor expenses due to automation.

Solution: Decrease in Annual Labor Expenses = (Percentage of Workers Replaced) * (Total Labor Expenses) Decrease in Annual Labor Expenses = 0.30 * (Number of Workers Replaced) * (Average Salary per Worker) Decrease in Annual Labor Expenses = 0.30 * (30% * Total Number of Workers) * $40,000

Transparency in Decision-making

Problem 6. A medical diagnosis AI system uses deep learning to diagnose diseases. It provides a diagnosis with a confidence score ranging from 0 to 1. If the system recommends a treatment with a confidence score of 0.8, what does this score indicate to the healthcare provider?

Solution: A confidence score of 0.8 indicates that the AI system is 80% confident in its recommendation. It suggests a high level of certainty in the proposed treatment based on the available data and analysis.

Human Oversight

Problem 7. An autonomous delivery robot operates in a busy city center. It's programmed to stop and request human intervention if it encounters a situation it can't handle safely. On average, the robot requests human intervention once every 100 kilometers. If the robot has traveled 500 kilometers, how many times has it requested human intervention?

Solution: Number of Requests for Human Intervention = (Total Distance Traveled)/(Distance per Request) Number of Requests for Human Intervention = 500 km/100 km/request Number of Requests for Human Intervention = 5 requests

Ethical Guidelines

Problem 8. A company is developing a healthcare robot that interacts with patients in hospitals. They want to ensure the robot's behavior aligns with ethical standards. They have identified 15 key ethical guidelines the robot should follow. If the robot complies with 13 out of the 15 guidelines, what is the compliance rate as a percentage?

Solution: Compliance Rate = (Number of Compliant Guidelines)/(Total Number of Guidelines) * 100% Compliance Rate = (13/15) * 100% Compliance Rate = 86.67%

Personal Data Protection

Problem 9. A service robot in a hotel collects guest data for personalized services. It records 50 MB of data per guest, including preferences and stay details. If the hotel hosts 300 guests per month, calculate the total data generated in a year.

Solution: Total Data Generated in a Year = (Data per Guest) * (Guests per Month) * (Months in a Year) Total Data Generated in a Year = 50 MB/guest * 300 guests/month * 12 months/year

Ethical Decision-making

Problem 10. An AI-powered chatbot handles customer complaints for an online retailer. If the chatbot encounters a complex issue it can't resolve, it escalates the complaint to a human customer service representative. On average, the chatbot escalates 5% of all complaints. If the chatbot receives 1,000 complaints in a month, how many of these are escalated to humans?

Solution: Number of Escalated Complaints = (Percentage Escalated) * (Total Number of Complaints) Number of Escalated Complaints = 5% * 1,000 complaints

Economic Impact

Problem 11. A small town introduces a robotic waste collection system. The system reduces the need for human garbage collectors by 50%. If the annual salary of a garbage collector was $35,000, calculate the potential annual savings for the town.

Solution: Potential Annual Savings = (Percentage of Workers Replaced) * (Total Annual Salary Savings) Potential Annual Savings = 50% * (Total Number of Garbage Collectors) * $35,000

Transparency and Trust

Problem 12. A social robot interacts with children in a daycare setting. It's programmed to provide explanations for its actions to help build trust with the children. If the robot performs 80 actions during a day, and it provides explanations for 60% of these actions, how many actions have explanations?

Solution: Number of Actions with Explanations = (Percentage with Explanations) * (Total Number of Actions) Number of Actions with Explanations = 60% * 80 actions

Environmental Impact

Problem 13. A warehouse deploys a fleet of autonomous forklift robots to handle inventory. The forklifts are electric and consume 15 kWh of energy per day. If the warehouse operates 365 days a year, calculate the annual energy consumption of the forklift fleet.

Solution: Annual Energy Consumption = (Energy per Day) * (Days in a Year) Annual Energy Consumption = 15 kWh/day * 365 days/year

Acceptance and Integration

Problem 14. A company introduces a customer service chatbot on its website. It aims to increase customer engagement and reduce response time. If, after one month of implementation, the chatbot receives 1,500 customer interactions, and it successfully resolves 80% of them, how many interactions are effectively handled by the chatbot?

Solution: Number of Interactions Handled by the Chatbot = (Percentage of Interactions Resolved) * (Total Number of Interactions) Number of Interactions Handled by the Chatbot = 80% * 1,500 interactions

Ethical Considerations in Design

Problem 15. A research team is developing a social robot for elderly care. They want to ensure the robot respects the privacy and dignity of the elderly. They identify 10 ethical principles the robot should follow. If the robot successfully adheres to 8 of these principles, what is the compliance rate?

Solution: Compliance Rate = (Number of Compliant Principles)/(Total Number of Principles) * 100% Compliance Rate = (8/10) * 100%

Human-Robot Interaction Efficiency

Problem 16. A manufacturing company introduces cobots to work alongside human operators. The cobots can complete a task in 4 minutes, while a human operator takes 12 minutes for the same task. If the company needs to complete 200 tasks, how many cobots and human operators should they use to finish the work in the least amount of time?

Solution: Total Time with Cobots = (Number of Tasks) * (Time per Cobot Task) Total Time with Humans = (Number of Tasks) * (Time per Human Task)

Robot Autonomy in Healthcare

Problem 19. A hospital deploys an AI-powered diagnostic system that can analyze medical images. On average, the system correctly identifies 90% of the abnormalities in the images it analyzes. If the system analyzes 1,000 images, how many abnormalities will it likely miss?

Solution: Number of Missed Abnormalities = (1 – Percentage of Correct Identifications) * (Total Number of Images Analyzed) Number of Missed Abnormalities = (1 – 90%) * 1,000 images

Social Acceptance and Adaptation

Problem 20. A community center introduces a programmable robot for educational workshops. The robot is designed to engage with children and teach them basic programming concepts. In the first week, 40 children attend the workshop, and 75% of them show a high level of enthusiasm and interest in working with the robot. How many children showed enthusiasm?

Solution: Number of Children Showing Enthusiasm = (Percentage Showing Enthusiasm) * (Total Number of Children) Number of Children Showing Enthusiasm = 75% * 40 children

Safety Impact

Problem 21. An industrial robot is used in a factory to handle heavy materials. The robot can lift up to 500 kilograms safely. If the factory has a load of 2,500 kilograms that needs to be moved, how many times does the robot need to lift the load to complete the task?

Solution: Number of Lifts Required = (Total Load)/(Robot's Maximum Lifting Capacity) Number of Lifts Required = 2,500 kg/500 kg/lift

Ethical Training for Robot Operators

Problem 22. A company operates a fleet of autonomous delivery drones. Each drone is operated by a human operator who monitors the drone's performance and intervenes if necessary. If the company has 25 drone operators, and it provides ethics training to 80% of them, how many operators have received the training?

Solution: Number of Operators with Ethics Training = (Percentage with Training) * (Total Number of Operators) Number of Operators with Ethics Training = 80% * 25 operators

Short Questions and Answers

Q.1. What is human-robot collaboration (HRC)?

Ans. HRC is the integration of human and robot capabilities to achieve shared goals.

Q.2. What are the benefits of HRC?

Ans. Improved productivity, safety, and creativity in various domains.

Q.3. What's an example of HRC in manufacturing?

Ans. Collaborative robots (cobots) working alongside human operators on assembly lines.

Q.4. What's the ethical concern with job displacement due to robots?

Ans. It can lead to unemployment and income inequality.

Q.5. How can society mitigate job displacement?

Ans. By investing in reskilling and upskilling programs for affected workers.

Q.6. What are privacy concerns in HRC?

Ans. Robots with sensors can collect personal data, raising surveillance and data protection issues.

Q.7. How can HRC impact personal privacy?

Ans. Robots with cameras and sensors may intrude on individuals' private spaces.

Q.8. What's an essential principle for handling personal data in HRC?

Ans. Informed consent: individuals should know how their data will be used.

Q.9. What's a potential ethical dilemma in autonomous robots?

Ans. Robots may face life-and-death decisions (e.g., self-driving cars). How should these be resolved?

Q.10. How can we ensure fair outcomes in robotic decisions?

Ans. Address biases in algorithms and ensure transparency in decision-making.

Q.11. What's a key safety measure in HRC?

Ans. Implementing collision detection and avoidance mechanisms.

Q.12. How can robots avoid harming humans in shared spaces?

Ans. Robots can be designed with safe speed and force limits.

Q.13. How should robots handle unexpected situations?

Ans. Flexibility is crucial; robots must adapt to changing conditions.

Q.14. What's an essential aspect of human-robot interaction (HRI)?

Ans. Effective communication between humans and robots.

Q.15. How do collaborative robots (cobots) differ from traditional robots?

Ans. Cobots are designed to work safely alongside humans, often without the need for physical guarding.

Q.16. What's the role of user interfaces in HRI?

Ans. They should be intuitive and user-friendly, facilitating interaction.

Q.17. How can society ensure the ethical use of robots?

Ans. By establishing clear guidelines and regulations for robotic technology.

Q.18. What's the potential impact of robots on job roles in the service industry?

Ans. Robots and AI systems can reduce the need for human employees in customer support, data entry, etc.

Q.19. What's the importance of transparency in HRI?

Ans. It builds trust between humans and robots by clarifying robot behavior.

Q.20. What's a significant ethical challenge in HRC?

Ans. Ensuring robots and AI systems make unbiased decisions.

Q.21. How can we address the potential misuse of personal data collected by robots?

Ans. Implement strong data protection measures and cybersecurity protocols.

Q.22. What's the ethical dilemma of robots making life-or-death decisions?

Ans. Deciding how robots should prioritize between different lives in a critical situation.

Q.23. Why is it crucial to involve ethicists in the development of HRC systems?

Ans. Ethicists can provide insights into potential moral challenges and help guide responsible development.

Q.24. How can we ensure that the benefits of HRC are equitably distributed?

Ans. By designing policies that prevent income inequality and by providing access to technology.

Q.25. What's the role of human oversight in autonomous robots?

Ans. Ensuring that humans have control and can intervene when necessary to avoid harmful outcomes.

Q.26. How can we ensure that robots understand human intentions?

Ans. By advancing natural language processing (NLP) and human-robot dialogue systems.

Q.27. How does HRC impact the concept of work?

Ans. It may change the nature of jobs, requiring humans to focus on tasks that demand creativity and decision-making.

Q.28. What's the societal impact of robots in education?

Ans. Robots can serve as interactive tutors, enhancing the learning experience.

Q.29. Why is it important to involve the public in discussions about HRC?

Ans. Public input ensures that diverse perspectives are considered in shaping robot technology.

Q.30. How can we ensure robots are accountable for their actions?

Ans. By establishing clear guidelines for responsibility and consequences in case of errors.

Q.31. How can robots be designed to adapt to different user preferences?

Ans. By creating customizable interfaces and behavior patterns.

Q.32. What's the role of safety standards in HRC?

Ans. Standards ensure that robotic systems are designed with safety in mind.

Q.33. What's an example of a task that robots can handle in healthcare?

Ans. Surgical robots can assist surgeons with precision and minimally invasive procedures.

Q.34. How can robots assist in agriculture?

Ans. They can automate planting, harvesting, and monitoring crops.

Q.35. How can society address the social acceptance of robots in various roles?

Ans. By fostering understanding of the benefits and potential of robotics while addressing ethical concerns.

Q.36. How can HRC impact income inequality?

Ans. If not managed properly, it could exacerbate income disparities.

Q.37. What's the importance of shared control in autonomous systems?

Ans. It ensures that humans can intervene if a robot makes a critical error.

Q.38. What's the role of public policy in addressing HRC's ethical challenges?

Ans. Policy can set guidelines, regulate usage, and promote responsible development.

Q.39. What's the potential impact of robots on the construction industry?

Ans. Robots can assist in tasks such as bricklaying, welding, and heavy lifting.

Q.40. How can HRC impact the way we interact with machines?

Ans. It can make interactions more intuitive, efficient, and natural.

Q.41. How can robots be programmed to respect privacy in healthcare settings?

Ans. By strictly limiting the collection and sharing of patient data.

Q.42. What's a potential concern in using robots for elder care?

Ans. The ethical consideration of maintaining human connection and companionship.

Q.43. How can robots be designed to understand and respond to human emotions?

Ans. By integrating emotion recognition algorithms and appropriate responses.

Q.44. What's the role of continuous improvement in HRI?

Ans. Robots can learn from their interactions, making them more effective over time.

Q.45. What's an example of an ethical guideline for HRC?

Ans. Transparency in how robots reach decisions to build trust.

Q.46. What's the potential impact of robots on job roles in manufacturing?

Ans. Robots can handle repetitive and physically demanding tasks, reducing the need for human workers in these roles.

Q.47. What's a key principle to address job displacement in HRC?

Ans. Focusing on reskilling and upskilling workers for new roles.

Q.48. How can we ensure that robots are designed to adapt to different user preferences?

Ans. By creating customizable interfaces and behavior patterns.

Q.49. How does HRC impact the concept of work?

Ans. It may change the nature of jobs, requiring humans to focus on tasks that demand creativity and decision-making.

Q.50. What's the role of human oversight in autonomous robots?

Ans. Ensuring that humans have control and can intervene when necessary to avoid harmful outcomes.

Multiple Choice Questions and Answers

Q.1. What does HRC stand for in robotics?

(*a*) Highly Robotic Computing (*b*) Human-Robot Collaboration

(*c*) High-level Robot Control (*d*) Human-Robot Communication

Q.2. What is the primary goal of HRC?
(*a*) Replacing humans with robots (*b*) Enhancing robot autonomy
(*c*) Integrating human and robot capabilities (*d*) Creating fully autonomous robots

Q.3. What is a significant benefit of HRC in manufacturing?
(*a*) Decreased productivity (*b*) Increased physical workload for humans
(*c*) Improved safety and efficiency (*d*) Higher labor costs

Q.4. What is a major ethical concern regarding job displacement due to robotics?
(*a*) Technological advancement (*b*) Economic growth
(*c*) Income inequality (*d*) Job creation

Q.5. How can job displacement challenges be mitigated in HRC?
(*a*) Increasing automation to replace all jobs
(*b*) Providing advanced training and reskilling for affected workers
(*c*) Ignoring the impact on the workforce
(*d*) Reducing the use of robots in industries

Q.6. What is a privacy concern in HRC?
(*a*) Lack of robot autonomy (*b*) Overreliance on human decision-making
(*c*) Data collection by robots (*d*) Lack of communication between robots

Q.7. How can robots impact personal privacy?
(*a*) By promoting transparency (*b*) By reducing the need for personal data
(*c*) By increasing data security
(*d*) By collecting sensitive personal information

Q.8. What should be obtained from individuals to ethically use their personal data collected by robots?
(*a*) Passive agreement (*b*) Technical knowledge
(*c*) Informed consent (*d*) Approval from regulatory authorities

Q.9. What's a potential ethical dilemma in autonomous robots making decisions?
(*a*) Ensuring complete robot autonomy
(*b*) Dealing with cognitive biases in decision-making
(*c*) Eliminating human intervention
(*d*) Reducing robot communication capabilities

Q.10. Why is transparency in robotic decision-making important?
(*a*) To confuse users (*b*) To avoid human intervention
(*c*) To build trust and accountability (*d*) To prioritize robot autonomy

Q.11. What's an important safety measure in HRC?
(*a*) Giving robots more speed and power (*b*) Avoiding sensors on robots
(*c*) Implementing collision detection and avoidance
(*d*) Making robots invisible

Q.12. How can robots avoid harming humans in shared spaces?
(*a*) By increasing robot speed
(*b*) By limiting human access
(*c*) By slowing down robot movements
(*d*) By increasing robot autonomy

Q.13. What's an essential aspect of handling unexpected situations in HRC?
(*a*) Robot autonomy above all else
(*b*) Ensuring robot speed
(*c*) Flexibility in robot behavior
(*d*) Avoiding human intervention

Q.14. What's a critical element in human-robot interaction (HRI)?
(*a*) Robot domination
(*b*) Effective communication
(*c*) Speed of robots
(*d*) Complexity of robots

Q.15. What's the primary focus of collaborative robots (cobots)?
(*a*) Replacing human workers
(*b*) Working autonomously
(*c*) Safely working alongside humans
(*d*) Performing high-risk tasks

Q.16. How can user interfaces enhance HRI?
(*a*) By making them complex and confusing
(*b*) By making them appealing to robots
(*c*) By making them intuitive and user-friendly
(*d*) By making them invisible

Q.17. What's the potential impact of robots on job roles in the service industry?
(*a*) Increase in job opportunities
(*b*) Reduction in job roles
(*c*) Improvement in job satisfaction
(*d*) Expansion of job markets

Q.18. How can shared control benefit autonomous systems?
(*a*) By removing all human oversight
(*b*) By increasing robot autonomy
(*c*) By allowing humans to intervene
(*d*) By making robots fully independent

Q.19. What's a potential challenge in using robots for elder care?
(*a*) Inadequate robotic capabilities
(*b*) Ethical concerns regarding human interaction
(*c*) Incompatibility with human environments
(*d*) Lack of interest from the elderly

Q.20. How can robots be designed to adapt to different user preferences?
(*a*) By avoiding customization
(*b*) By enforcing standard behavior
(*c*) By creating customizable interfaces
(*d*) By limiting user input

Q.21. What's a significant social implication of widespread robot adoption in healthcare?
(*a*) Increased job opportunities for medical professionals
(*b*) Improved patient care and outcomes
(*c*) Reduced costs of healthcare services
(*d*) Decreased demand for medical technologies

Q.22. How can the ethical concerns related to robot autonomy be addressed?
(*a*) By eliminating all autonomous features in robots
(*b*) By designing robots with clear boundaries and human oversight

(*c*) By giving robots complete control over decision-making
(*d*) By preventing robots from interacting with humans

Q.23. What's the primary ethical consideration when using robots in elder care?
(*a*) Ensuring robots have the same emotional understanding as humans
(*b*) Safeguarding the dignity and well-being of the elderly
(*c*) Enhancing the efficiency of caregiving without regard to human emotions
(*d*) Eliminating human involvement in caregiving

Q.24. What's the potential impact of robots on job roles in construction?
(*a*) Increased demand for human labor
(*b*) Enhanced creativity in construction projects
(*c*) Automation of hazardous and repetitive tasks
(*d*) Reduced need for architectural design

Q.25. What's an important principle to address job displacement due to robots?
(*a*) Fostering income equality
(*b*) Embracing complete automation
(*c*) Increasing the number of low-skilled jobs
(*d*) Providing retraining opportunities for affected workers

Q.26. What's the role of transparency in human-robot interaction (HRI)?
(*a*) To make robots invisible
(*b*) To build trust between humans and robots
(*c*) To eliminate the need for robot communication
(*d*) To increase robot autonomy

Q.27. What's a potential concern in using robots for military applications?
(*a*) Promoting international collaboration
(*b*) Ensuring accountability for robot actions
(*c*) Reducing the efficiency of military operations
(*d*) Decreasing reliance on advanced technology

Q.28. How can we ensure robots are designed to adapt to different user preferences?
(*a*) By enforcing standardized behavior
(*b*) By creating customizable interfaces and behaviors
(*c*) By limiting user input to predefined options
(*d*) By making robots completely autonomous

Q.29. What's the potential impact of robots on job roles in the entertainment industry?
(*a*) Reduction in demand for creative content
(*b*) Automation of production and post-production tasks
(*c*) Decreased public interest in entertainment
(*d*) Increased reliance on traditional forms of entertainment

Q.30. How can society address the social acceptance of robots in various roles?

(*a*) By completely avoiding the use of robots (*b*) By reducing the complexity of robots

(*c*) By ignoring ethical concerns

(*d*) By fostering understanding of the benefits and addressing ethical concerns

Q.31. What's the potential impact of robots on job roles in agriculture?

(*a*) Decreased demand for manual labor (*b*) Reduced need for precision farming

(*c*) Limited impact on crop yields

(*d*) Increased reliance on traditional farming methods

Q.32. What's the significance of shared control in autonomous systems?

(*a*) To eliminate human involvement (*b*) To ensure complete robot autonomy

(*c*) To allow humans to intervene when necessary

(*d*) To make robots fully independent

Q.33. What's the potential impact of robots on job roles in education?

(*a*) Decreased demand for teachers

(*b*) Enhanced learning experiences through personalized instruction

(*c*) Reduced access to education

(*d*) Elimination of human involvement in curriculum design

Q.34. What's an essential consideration in designing robots for healthcare applications?

(*a*) Maximizing robot autonomy (*b*) Minimizing human interaction

(*c*) Prioritizing cost savings

(*d*) Safeguarding patient well-being and privacy

Q.35. What's a critical ethical challenge in using robots for search and rescue missions?

(*a*) Ensuring complete robot autonomy

(*b*) Addressing the ethical concerns related to robot emotions

(*c*) Balancing the potential benefits with potential risks to human life

(*d*) Avoiding human involvement in decision-making during emergencies

Q.36. How can robots be designed to understand and respond to human emotions?

(*a*) By ignoring emotional cues (*b*) By avoiding emotional interactions

(*c*) By integrating emotion recognition algorithms

(*d*) By making robots completely logical

Q.37. What's an essential aspect of designing robots for elder care?

(*a*) Eliminating the need for human caregivers

(*b*) Focusing on efficiency over human interaction

(*c*) Ensuring dignity, companionship, and respect for the elderly

(*d*) Reducing the availability of healthcare services

Q.38. What's the potential impact of robots on job roles in the transportation industry?

(*a*) Increased demand for human drivers (*b*) Enhanced driver safety

(c) Reduced need for logistics management
(d) Elimination of all human involvement in transportation

Q.39. What's the role of ethics in the development of robotic systems?
(a) To eliminate the need for human intervention
(b) To ensure complete robot autonomy
(c) To address potential risks, biases, and ethical dilemmas
(d) To reduce the complexity of robotic technologies

Q.40. How can society ensure the responsible use of robots in military and security applications?
(a) By minimizing transparency in decision-making
(b) By eliminating human oversight
(c) By adhering to strict ethical guidelines and regulations
(d) By increasing robot autonomy

Q.41. What's a critical factor in preventing robots from making biased decisions?
(a) Ensuring complete robot autonomy (b) Eliminating human involvement
(c) Addressing biases in algorithms and training data
(d) Increasing robot communication capabilities

Q.42. How can robots be designed to respect privacy in public spaces?
(a) By maximizing data collection
(b) By using facial recognition without consent
(c) By strictly limiting data collection and sharing
(d) By avoiding public spaces

Q.43. What's a potential challenge in using robots for environmental monitoring?
(a) Eliminating the need for human involvement
(b) Addressing the potential impact on biodiversity
(c) Maximizing robot autonomy
(d) Reducing transparency in decision-making

Q.44. What's a key consideration in designing robots for disaster response missions?
(a) Reducing robot communication capabilities
(b) Maximizing robot autonomy
(c) Ensuring human safety and minimizing risks
(d) Eliminating human involvement in rescue operations

Q.45. How can robots be programmed to adapt to different cultural norms and preferences?
(a) By ignoring cultural differences (b) By enforcing a universal robot behavior
(c) By incorporating customizable settings (d) By reducing robot autonomy

Q.46. What's the potential impact of robots on job roles in the financial industry?
(a) Reduced need for cybersecurity (b) Increased demand for financial analysts
(c) Automation of routine financial tasks
(d) Elimination of all human involvement in finance

Q.47. How can robots contribute to sustainable practices in agriculture?

(*a*) By increasing water consumption (*b*) By reducing the use of pesticides

(*c*) By eliminating the need for soil management

(*d*) By increasing fuel consumption

Q.48. What's an important consideration in using robots for law enforcement?

(*a*) Reducing transparency in decision-making

(*b*) Maximizing robot autonomy

(*c*) Ensuring accountability and ethical use of force

(*d*) Eliminating human involvement in policing

Q.49. What's a potential challenge in using robots for disaster recovery operations?

(*a*) Addressing the need for human intervention

(*b*) Maximizing robot autonomy

(*c*) Ensuring safe interactions with humans

(*d*) Reducing transparency in decision-making

Q.50. What's the significance of public engagement in shaping the ethical use of robotics?

(*a*) To avoid public opinion (*b*) To limit access to robotic technology

(*c*) To ensure diverse perspectives are considered in policies

(*d*) To eliminate the need for ethical considerations

Key to Multiple Choice Questions

1. (*b*)	**2.** (*c*)	**3.** (*c*)	**4.** (*c*)	**5.** (*b*)	**6.** (*c*)	**7.** (*d*)	**8.** (*c*)	**9.** (*b*)
10. (*c*)	**11.** (*c*)	**12.** (*c*)	**13.** (*c*)	**14.** (*b*)	**15.** (*c*)	**16.** (*c*)	**17.** (*b*)	**18.** (*c*)
19. (*b*)	**20.** (*c*)	**21.** (*b*)	**22.** (*b*)	**23.** (*b*)	**24.** (*c*)	**25.** (*d*)	**26.** (*b*)	**27.** (*b*)
28. (*b*)	**29.** (*b*)	**30.** (*d*)	**31.** (*a*)	**32.** (*c*)	**33.** (*b*)	**34.** (*d*)	**35.** (*c*)	**36.** (*c*)
37. (*c*)	**38.** (*b*)	**39.** (*c*)	**40.** (*c*)	**41.** (*c*)	**42.** (*c*)	**43.** (*b*)	**44.** (*c*)	**45.** (*c*)
46. (*c*)	**47.** (*b*)	**48.** (*c*)	**49.** (*c*)	**50.** (*c*)				

Project Works

1. **Ethical Design Framework for Collaborative Robots:** Develop an ethical design framework for collaborative robots (cobots) that ensures they prioritize safety, respect human dignity, and address potential biases in decision-making.

   Potential Answers/Directions:

   - Identify key ethical principles for cobot design (e.g., safety, transparency, accountability).
   - Create guidelines for handling sensitive tasks, such as human-robot interactions in healthcare or elder care.
   - Explore ways to prevent discrimination and bias in cobot algorithms.
   - Discuss mechanisms for clear communication of cobot capabilities and limitations to users.

2. **Impact of Cobots on Workforce and Job Displacement:** Analyze the social and economic impact of introducing cobots in a specific industry or workplace, considering potential job displacement, income inequality, and retraining opportunities.

   Potential Answers/Directions:

   - Study the before-and-after scenario in the chosen industry.
   - Calculate the potential cost savings and productivity gains from cobot implementation.
   - Investigate strategies for retraining and upskilling affected workers.
   - Analyze the broader societal implications, such as income distribution and overall job market changes.

3. **Privacy and Data Security in Human-Robot Interaction:** Examine the privacy concerns related to data collection in human-robot interactions. Develop strategies to protect user privacy while maintaining effective collaboration.

   Potential Answers/Directions:

   - Explore data minimization techniques to limit unnecessary data collection.
   - Investigate encryption and secure storage of collected data.
   - Discuss obtaining informed consent from users for data collection and usage.
   - Consider potential legal and regulatory requirements for data protection.

4. **Public Perception and Acceptance of Cobots:** Conduct a survey or focus group to assess public perception and acceptance of collaborative robots in various contexts (e.g., manufacturing, healthcare, customer service).

   Potential Answers/Directions:

   - Analyze the factors influencing public acceptance (trust, perceived benefits, concerns).
   - Identify potential areas where public education and awareness campaigns are needed.
   - Discuss how to address misconceptions or fears about cobot technology.
   - Propose strategies for improving the public perception and acceptance of cobots.

5. **Ethics in Autonomous Decision-Making of Robots:** Explore the ethical challenges in allowing robots, particularly autonomous systems, to make critical decisions without human intervention. Consider scenarios where robots must prioritize between different ethical principles.

   Potential Answers/Directions:

   - Analyze real-world scenarios, such as autonomous vehicles making life-and-death decisions.
   - Discuss the importance of transparency, accountability, and fairness in autonomous decision-making.
   - Propose mechanisms to handle cases where predefined ethical principles conflict.
   - Examine the role of human oversight and intervention in critical decision-making by robots.

3

Autonomous Vehicles and Drones: Part of the broader Field of Robotics and AI

3.1 INTRODUCTION

Drones and autonomous cars are innovative technologies that have the potential to completely change the logistics, transportation, and many other industries. Both of these fall under the umbrella of robotics and artificial intelligence (AI). Autonomous vehicles have the potential to decrease collisions, enhance traffic flow, boost accessibility for all modes of transportation, and offer mobility options for those who are unable to drive, such as the elderly or those with impairments. Regulatory, moral, and technical obstacles must all be overcome, nevertheless, in order to achieve success.

3.2 RELATED EXCITING TECHNOLOGIES

Let's dive into an introduction to each of the related exciting technologies:

1. **Autonomous Vehicles:** Also known as self-driving automobiles, autonomous vehicles are ones that can drive themselves and navigate without a driver. To assess their surroundings, make quick judgments, and navigate safely, they use on a combination of sensors, cameras, radar, lidar (light detection and ranging), GPS, and complex software algorithms.

   Characteristics of autonomous vehicles

   - **Sensors:** These tools gather information about the surroundings of the moving object, such as the presence of other moving objects, pedestrians, road signs, and barriers.
   - **Control Systems:** Cutting-edge AI systems analyze sensor data and decide how to accelerate, brake, steer, and plot a route.
   - **Connectivity:** For increased safety and efficiency, autonomous cars frequently rely on high-speed internet connections to communicate with other vehicles (V2V) and infrastructure (V2I).

2. **Unmanned Aerial Vehicles or Drones:** Aircraft called drones fly without a human pilot on board. They can be directed by pre-programmed flight patterns or remotely operated by a human operator. Drones exist in a variety of shapes and sizes, from tiny consumer drones used for entertainment to massive industrial drones built for jobs like aerial photography, surveying, agricultural, and package delivery.

Key features of drones:

- **Remote Control:** Drones may be operated from a distance, making them adaptable instruments for diverse tasks.
- **Payloads:** Drones can be equipped with cameras, sensors, and other payloads to collect data or carry out particular activities.
- **Automation:** Modern drones can be programmed to fly along predetermined paths, maintain steady altitudes, and even do tasks on their own.

Industries like agriculture, construction, search and rescue, environmental monitoring, and filmmaking all use drones. They may increase productivity, lower expenses, and open up previously inaccessible locations. Privacy, airspace management, and the secure integration of drones into existing air traffic are issues, though. Drones and autonomous cars are being developed and adopted at a quick rate; as a result, their future will be shaped by continued technological developments, governmental restrictions, and popular acceptance.

3.3 DEVELOPMENT OF SELF-DRIVING CARS

An intricate combination of mathematics, computer science, engineering, and data analysis goes into the creation of self-driving cars. The following important mathematical ideas and methods are essential to the development of self-driving cars:

3.3.1 Sensor Fusion

To understand their surroundings, self-driving cars use a number of sensors, including cameras, lidar, radar, and ultrasonic sensors. The process of merging data from these sensors is known as sensor fusion, and it aims to produce a complete and accurate picture of the environment. To deal with sensor uncertainties, noise, and data integration, advanced mathematical approaches are used, such as Kalman filters, Bayesian inference, and probabilistic methods.

A key element of autonomous systems, such as self-driving automobiles, is sensor fusion. It entails combining information from various sensors to produce a more precise and reliable perception of the environment. The system's capacity to comprehend its surroundings, make wise decisions, and move safely is improved via sensor fusion. Let's investigate the mathematical theories and tools used in sensor fusion:

Mathematical Concepts:

1. **Bayesian Inference:** Sensor fusion relies heavily on Bayesian techniques. They offer a probabilistic framework for updating the belief or probability distribution about the state of a system using sensor readings and prior information. A key idea is the Bayes theorem, which enables us to update our predictions and take into account new sensor data.
2. Recursive estimation algorithms such as Kalman Filters and Extended Kalman Filters (EKF) are used to estimate the state of a dynamic system with unknown observations. While EKF expands its application to non-linear systems, Kalman filters are particularly helpful when working with linear systems. They are frequently employed in tracking applications, such as determining the position and speed of a moving object, for sensor fusion.
3. **Probabilistic Data Association:** In multi-object tracking scenarios, this method is utilised to link sensor values to particular environmental items. It entails putting measurements on already-existing tracks or making new ones based on the likelihood of those measurements while taking sensor noise and uncertainties into account.

4. **Graph Theory and Graph-based Methods:** Graph theory is used to do inference and express relationships between data. Because they can simulate complicated sensor interactions, dependencies, and limitations, graph-based approaches, such the Factor Graphs, are helpful for resolving sensor fusion issues.
5. **Non-linear Optimization:** When a system's behavior is non-linear, estimation of the system's state can be achieved by minimizing the difference between expected and actual sensor readings using optimization techniques like gradient descent or nonlinear least squares.

Technologies:

1. **Sensors:** A variety of sensors, including cameras, lidar, radar, ultrasonic sensors, GPS, IMUs, and more, are used in sensor fusion. Each sensor offers distinct information about the surroundings, and combining their data results in a perception of the world as a whole.
2. **Advanced Signal Processing:** Filtering, noise reduction, and feature extraction are signal processing techniques that are essential for preprocessing sensor data prior to fusion. This guarantees that the data is in a format that fusion algorithms can use.
3. **Real-time Processing and High-performance Computing:** Processing massive amounts of data in real-time is frequently necessary for sensor fusion. The computational demands of sensor fusion techniques are met by high-performance computing, parallel processing, and specialized hardware (such as GPUs and FPGAs).
4. **Machine Learning and Deep Learning:** To discover complicated correlations between many sensor modalities and increase the accuracy of fusion, machine learning techniques, including deep learning models, can be employed in sensor fusion.
5. **Redundancy and Reliability:** By supplying redundancy, sensor fusion improves the system's reliability. The fusion system can continue to operate safely even if one sensor malfunctions or delivers erroneous data by using data from other sensors.
6. **Connectivity and Communication:** Sensor fusion systems in autonomous cars occasionally interact with one another or with a central processing unit to exchange data and enhance situational awareness.

These mathematical ideas and methods support sensor fusion, which is essential for autonomous systems to accurately perceive their environment and make judgments.

3.3.2 Computer Vision

For cameras to recognize and comprehend the visual data they record, computer vision is required. It uses machine learning methods for object detection, feature extraction, and picture processing. Convolutional neural networks (CNNs) and other methods are employed to detect objects, pedestrians, traffic signs, and lane markings.

Self-driving automobiles use computer vision as a fundamental technology to sense and comprehend their environment. In order to extract useful information from visual input, modern image processing methods and cameras are used. Let's examine the mathematical ideas and tools that self-driving car computer vision relies on:

Mathematical Concepts:

1. **Image Formation and Geometry:** It is crucial to comprehend how images are created, including ideas such as camera projection, perspective, and image distortion. By using geometry, it is possible to connect 2D points in the image with 3D points in the real world.

2. **Feature Detection and Matching:** For object detection and tracking, it is essential to recognise important features in images, such as corners, edges, or keypoints. Finding and matching features across frames is done using methods like the Harris corner detector, SIFT (Scale-Invariant Feature Transform), and ORB (Oriented FAST and Rotated BRIEF).
3. **Object Detection and Classification:** This step entails locating particular objects in the photos, such as people, cars, and traffic signs. Techniques like YOLO (You Only Look) and deep learning-based methods (such as Haar cascades and Histogram of Oriented Gradients (HOG).
4. **Image Segmentation:** By dividing an image into pertinent parts, image segmentation enables the system to recognise object boundaries and apply semantic labels. For segmentation tasks, approaches including watershed segmentation, graph-based segmentation, and deep learning-based algorithms (such U-Net, Mask R-CNN) are employed.
5. **Optical Flow and Motion Analysis:** Tracking and predicting an object's trajectory requires an understanding of how it moves in the scene. Optical flow algorithms calculate the apparent velocity of pixels between two frames, giving important details about the dynamics of the image.
6. **3D Reconstruction and Stereo Vision:** Understanding depth and spatial relationships requires the creation of 3D representations of the world from numerous images. Stereo vision reconstructs the scene's three-dimensional structure using the discrepancy between two camera perspectives.

Technologies:

1. **Cameras and Multi-camera Systems:** Self-driving cars employ a variety of cameras strategically positioned all around the car to get a full picture of the surroundings. These cameras offer the computer vision system useful visual data to work with.
2. **Image Sensors and Resolution:** Accurate object detection and recognition depend on detailed visual information captured by high-resolution image sensors. Furthermore, improvements in sensor technology, such as multispectral or low-light sensors, improve the system's capacity to detect the world under a variety of conditions.
3. **Real-time Processing:** Computer vision in self-driving cars needs to be able to process data in real-time so that judgments can be made right away based on the collected visual information. The computational requirements of computer vision algorithms are met by high-performance processors, GPUs, and specialized hardware accelerators.
4. **Sensor Fusion:** In a sensor fusion technique, data from various sensors, such as lidar and radar, are frequently merged with computer vision. This enables the system to make the most of each sensor modality's advantages and provide a more thorough perspective of the surroundings.
5. **Deep Learning and Neural Networks:** To analyze images, identify objects, and categories them, deep learning techniques, in particular convolutional neural networks (CNNs), are used. These deep learning models can accurately perform computer vision tasks by learning complex features straight from the data.
6. **Data Annotation and Training:** Large labeled datasets are needed to build and train computer vision models. Images are labeled with ground truth data using annotation tools and data pipelines, and these datasets are utilized to train and improve the fine-tune the computer vision algorithms.

A key component of self-driving automobiles is computer vision, which is supported by these mathematical ideas and technological advancements. Vehicles can "see" their surroundings thanks to this technology, which also enables them to analyze visual clues and make wise decisions. As a result, driving is safer and more autonomous.

3.3.3 Localization and Mapping

In order to navigate safely, self-driving cars need to know their exact location (localization) and have a map of their surroundings. For computing positions using GPS, inertial measurement units (IMUs), and landmarks, math from geometry and trigonometry is needed. Real-time maps are created and updated using Simultaneous Localization and Mapping (SLAM) methods. In order for self-driving cars to comprehend their location in the surroundings and produce accurate maps for navigation, localization and mapping are crucial components. To achieve precise localization and mapping, these technologies combine sensors, algorithms, and mathematical ideas. Let's explore the underlying mathematical ideas and tools:

Mathematical Concepts:

1. **Geometry and Trigonometry:** For localization and mapping, it is essential to understand geometric concepts like angles, distances, and coordinate transformations. When combined with GPS information, trigonometry is used to determine locations based on sensor measurements.
2. Kalman filters and particle filters are probabilistic estimating techniques used for state estimation in dynamic systems (Monte Carlo localization). They estimate the vehicle's position and orientation over time by combining sensor measurements, motion models, and uncertainty.
3. **Simultaneous Localization and Mapping (SLAM):** In situations without pre-existing maps, SLAM is a key idea for mapping and localization. SLAM algorithms use sensor data to concurrently map the environment and estimate the position of the vehicle.
4. **Feature Matching and Correspondence:** Feature matching algorithms, such as RANSAC (Random Sample Consensus) or point cloud registration methods, are used to find shared features between frames and compute the transformation between them when using visual data (for example, from cameras).
5. **Graph Theory and Graph-based SLAM:** The relationships between the poses (positions) of the robot and the observable landmarks are represented using graph theory. Poses and landmarks are arranged in the most likely way possible using graph-based SLAM techniques like pose-graph optimization.

Technologies:

1. **GPS and GNSS:** Based on signals from satellites, the Global Positioning System (GPS) and Global Navigation Satellite Systems (GNSS) give location data. Although they are helpful for coarse localization, they could have accuracy and reliability issues, especially in urban canyons or places where there are signal obstacles.
2. **Inertial Measurement Units (IMUs):** IMUs track the movement of a vehicle by measuring its acceleration and angular rates. IMUs can improve localization accuracy and fill in the gaps left by lost GPS signals when used in conjunction with other sensors.
3. **Lidar and Radar Sensors:** By calculating the distances between objects, these sensors provide a thorough awareness of the surroundings. While radar provides data on object velocity and relative motion, helpful for mapping and spotting obstructions, lidar produces point clouds.
4. **Visual Odometry:** Based on image sequences, visual odometry uses cameras to estimate the motion of the vehicle. Map accuracy can be increased and localization can be done using this information.

5. **High-Definition Mapping:** Real-time localization can be complemented by previously created high-definition (HD) maps of the roads. Vehicles can fine-tune their position by comparing sensor data with recognized landmarks on the map.
6. **Sensor Fusion:** Using sensor fusion techniques, data from many sensors, including GPS, IMUs, cameras, lidar, and radar, is combined to enable reliable and precise localization and mapping, even in difficult conditions.
7. **Real-time Processing:** To keep up with the motion of the vehicle, localization and mapping algorithms need to be capable of real-time processing. These computations are expedited by specialized hardware and high-performance computers.
8. **Data Storage and Cloud Services:** For mapping, self-driving cars produce a lot of data. Updated maps may be stored, processed, and distributed throughout the fleet of autonomous vehicles via cloud-based services.

Self-driving cars may achieve exact localization, produce accurate maps, and navigate confidently by utilizing these mathematical ideas and cutting-edge technologies, making them safer and more effective on the road.

3.3.4 Path Planning and Control

Choosing a safe and effective path for the vehicle to travel on includes path planning. It takes into account things like following traffic laws, avoiding obstacles, speed restrictions, and driving smoothly. The car maintains a safe following distance, adjusts to changing conditions, and stays on the targeted course thanks to the use of control theory, including PID controllers and model predictive control. For self-driving cars to navigate safely and effectively, path planning and control are essential elements. The trajectory of the vehicle is determined by these technologies, which use a combination of algorithms, sensors, mathematical models, and control theory to make adjustments in real-time.

Let's explore the mathematical concepts and technologies involved:

Mathematical Concepts:

1. **Kinematics and Dynamics:** Planning and controlling a path requires an understanding of the kinematics (motion without taking forces into account) and dynamics (motion taking forces into account) of the vehicle. Equations that characterize the position, velocity, acceleration, and orientation of the vehicle are used in these ideas.
2. **Trajectory Generation:** When designing a vehicle's trajectory, it's important to take into account the vehicle's dynamics as well as road conditions, barriers, and other elements. Smooth and practical trajectories are produced using polynomial curves (such as cubic splines), clothoids (Euler spirals), and optimal control methods.
3. **Optimization:** Finding the shortest path while avoiding barriers or optimizing for fuel efficiency are two common optimization difficulties that arise during path planning. These issues can be resolved using optimization approaches such as linear and nonlinear programming, evolutionary algorithms, and A* search.
4. **Behavior Planning:** The vehicle's high-level actions, such as lane changes, highway merges, and pedestrian yielding, are determined by behavior planning. Different driving behaviors are modeled using Finite State Machines (FSMs) and decision trees dependent on the present circumstance.

5. **Model Predictive Control (MPC):** MPC is a control approach that determines control inputs that optimize a certain target (such as keeping on the path while taking into account vehicle restrictions) using a dynamic model of the vehicle and predictions of its future behavior.
6. **Feedback Control:** Real-time control adjustments are made to the vehicle's controls using sensor measurements. To maintain the desired trajectory, complex control methods (such as LQR – Linear Quadratic Regulator), state-space control, and proportional-integral-derivative (PID) controllers are used.

Technologies:

1. **Sensors:** Cameras, lidar, radar, ultrasonic sensors, GPS, and IMUs are just a few of the sensors used by self-driving automobiles. Real-time information from these sensors regarding the environment around the vehicle is used for path planning and control.
2. **Digital Maps and Localization:** High-definition digital maps include crucial details about the layout of the roads, the lane markings, the traffic signs, and potential hazards. Vehicle localization aids the vehicle in comprehending its position on the map and frequently makes use of GPS, lidar, and map matching.
3. **Real-time Processing:** In order to react swiftly to changing situations, path planning and control require real-time processing capabilities. The computations are sped up using high-performance computers, GPUs, and specialized devices.
4. **Sensor Fusion:** Using input from several sensors to create a single, accurate perception (sensor fusion) is essential for making judgments about course planning and control.
5. **V2V and V2I Communication:** V2V and V2I communication systems enable vehicle-to-vehicle (V2V) and vehicle-to-infrastructure (V2I) information transmission. Through communication, cooperative behaviors like merging or avoiding collisions can be improved, as well as situational awareness.
6. **Lane Keeping Assist and Adaptive Cruise Control (ACC):** These driver assistance systems use control algorithms to keep a safe following distance and stay in the lane, acting as the foundation for more sophisticated autonomous control.
7. **Machine Learning and AI:** Machine learning models can be used to forecast future risks, predict the behavior of other road users, and improve decision-making in challenging situations.

Self-driving cars will eventually be able to calculate optimal routes, regulate their motions, and negotiate the road with safety and precision thanks to the combination of fundamental mathematical ideas and cutting-edge technologies.

3.3.5 Machine Learning and AI

The use of machine learning algorithms is crucial in self-driving cars. Neural networks and other deep learning models are utilized for tasks including object recognition, decision-making, and anticipating other drivers' behavior. Vehicles can be taught to employ reinforcement learning to make the best decisions possible in challenging and dynamic circumstances. Self-driving cars can detect their environment, make judgments, and drive themselves thanks in large part to machine learning (ML) and artificial intelligence (AI). These technologies accomplish the amount of intelligence required for autonomous driving by utilizing mathematical ideas, algorithms, and cutting-edge hardware. Let's examine the main ML and AI features in self-driving cars, as well as the underlying math and technology:

Mathematical Concepts:

1. **Statistical Learning and Probability:** Statistical techniques are frequently used in machine learning. In order to create models that can handle uncertainty in sensor data and decision-making, it is crucial to have a solid understanding of probability, distributions, and statistical inference.
2. **Linear Algebra:** The use of matrices, transformations, and high-dimensional data all depend on linear algebra. It is utilized in methods like as PCA (Principal Component Analysis), matrix factorization, and feature engineering transformations.
3. **Calculus and Optimization:** Machine learning models are trained using optimization algorithms. During model training, loss functions are minimized by using gradient descent, a crucial optimization approach.
4. **Neural Networks and Deep Learning:** Developing and training deep learning models, which are used for tasks like image recognition and decision-making, requires a thorough understanding of the mathematics of neural networks, including activation functions, back propagation, and weight changes.
5. **Probability Graphical Models:** For probabilistic reasoning, which is crucial for sensor fusion, risk assessment, and comprehending the uncertainty of predictions, Bayesian networks and Markov models are used.

Technologies:

1. **Deep Learning Frameworks:** Applications such as TensorFlow, PyTorch, and Keras offer robust APIs for creating and refining deep learning models. These frameworks let engineers and researchers design intricate neural network designs.
2. **Computer Vision:** For object detection, image segmentation, and lane recognition, AI-driven computer vision techniques like Convolutional Neural Networks (CNNs) are used. The vehicle can "see" and "interpret" its surroundings thanks to these technologies.
3. **Sensor Fusion:** To overcome sensor constraints and enhance object detection, ML algorithms are used to merge data from numerous sensors (lidar, radar, cameras, etc.) to build a more complete and accurate representation of the environment.
4. **Reinforcement Learning:** Reinforcement learning, a branch of machine learning, can be used to teach self-driving cars how to make the best choices in changing settings.
5. **Data Annotation and Labeling:** To train ML models, high-quality datasets with labels are required. For training and validation purposes, ground truth labels are created with the aid of tools and services for data annotation.
6. **High-Performance Computing (HPC):** Deep neural networks, in particular, are ML and AI models that demand a lot of computational power for training and inference. Accelerated processing is carried out using TPUs (Tensor Processing Units) and GPUs (Graphics Processing Units).
7. **Real-time Processing:** Self-driving cars need to process sensor data in real-time in order to perceive, make decisions, and control themselves. Timely replies are ensured through high-performance computation and specialized hardware accelerators.
8. **Edge Computing:** Rather than depending primarily on cloud-based processing, some processing is done directly on the vehicle (edge computing) to reduce latency and guarantee quick decision-making.

9. **Semantic Segmentation and Mapping:** Semantic segmentation uses ML models to recognize and categorize various things in the environment. These models aid in producing precise maps that are used for localization and navigation.

Self-driving cars may assess their surroundings, make wise decisions, and drive independently by fusing these mathematical ideas and technology, leading to safer and more effective transportation overall.

3.3.6 Probability and Risk Assessment

Self-driving systems must evaluate potential hazards and uncertainties probabilistically. To determine the likelihood of various outcomes, Bayesian networks and decision theory are applied, enabling the vehicle to make safer and more educated judgments. The decision-making mechanisms used by self-driving automobiles must consider probability and danger. These technologies analyze and reduce potential dangers in various driving circumstances using mathematical ideas, statistical techniques, and sophisticated simulations. Let's explore the fundamental mathematics and technological underpinnings of probability and risk assessment in self-driving cars.

Mathematical Concepts:

1. **Probability Theory:** When determining the likelihood of various outcomes in ambiguous circumstances, probability theory is essential. In order to measure uncertainty and make wise decisions based on sensor data, concepts such as conditional probability, Bayes' theorem, and probability distributions are useful.
2. **Statistics and Data Analysis:** Statistics are crucial to the analysis of substantial datasets gathered from sensors and actual driving situations. To glean useful insights from data, methods including regression analysis, Monte Carlo simulations, and hypothesis testing are used.
3. **Decision Theory:** In decision theory, the best choices are made when there is uncertainty. The optimum course of action is chosen by weighing the potential outcomes of many options using ideas from decision trees and utility theory.
4. **Risk Assessment:** Risk assessment entails locating potential risks and assessing their ramifications. Quantitative risk assessment evaluates the overall risk connected to particular events by combining probability and impact analysis.
5. **Stochastic Modeling:** Stochastic models, which take uncertainty and unpredictability into account, are used to simulate environmental conditions, sensor noise, and other drivers' behavior. For identifying potential threats, these models are critical.

Technologies:

1. **Sensor Data Fusion:** Using probabilistic approaches to combine data from numerous sensors (lidar, radar, cameras, etc.), self-driving cars can develop a more reliable and accurate understanding of their surroundings. Object recognition is improved and the risk of false positives and negatives is decreased through sensor fusion.
2. **Bayesian Networks:** These networks simulate the probability connections between various variables. Incorporating sensor measurements, reasoning under uncertainty, and updating the system's perception of the surrounding environment and potential threats are all done with their help.
3. **Monte Carlo Simulations:** Complex systems with uncertainty are modeled using Monte Carlo simulations. They provide insights on the system's behavior in numerous scenarios, including unusual or extreme events, and aid in determining the likelihood of various outcomes.

4. **Predictive Modeling:** To identify future dangers, forecast the behavior of other road users, and evaluate the risks involved with particular activities, predictive models based on historical data and real-time sensor measurements are used.
5. **Safety-Critical Systems Design:** To minimize hazards, self-driving car systems are built with redundancy and fail-safes. Techniques like fault tolerance, fail-operational design, and thorough testing make sure the system is capable of handling unforeseen circumstances.
6. **Vehicle-to-Vehicle and Vehicle-to-Infrastructure Communication:** Self-driving cars may communicate with other vehicles and the road infrastructure via V2V and V2I communication technologies. Vehicle cooperation, collision avoidance, and overall safety are all improved by this data sharing.
7. **Real-time Processing:** Probability and risk evaluations are frequently carried out in real-time to make quick decisions depending on altering circumstances.
8. **Machine Learning for Anomaly Detection:** Machine learning models can be trained to recognize circumstances that differ from expected patterns in sensor data, detecting abnormalities. This aids in identifying potential hazards, such as sensor failures or unexpected actions by other road users.

By integrating these mathematical concepts and cutting-edge technologies, self-driving cars can quantitatively assess potential risks, make proactive decisions to ensure safety, and ultimately contribute to a more reliable and secure autonomous driving experience.

3.3.7 Data Analysis and Optimization

To increase the effectiveness and security of self-driving systems, large datasets from sensors and actual driving situations are examined. Algorithms are adjusted using optimization approaches to increase effectiveness and reduce errors. The performance, safety, and efficiency of self-driving automobiles must be improved through data analysis and optimization. These technologies handle massive amounts of sensor data, optimize algorithms, and improve various elements of autonomous driving by utilizing mathematical ideas, algorithms, and cutting-edge computing techniques. Let's examine the fundamental mathematics and technological underpinnings of data analysis and optimization in self-driving cars.

Mathematical Concepts:

1. **Statistics and Probability:** Analyzing sensor data, seeing trends, and calculating uncertainties all depend heavily on statistics. To make defensible conclusions based on data, probability distributions, hypothesis testing, and confidence intervals are used.
2. **Regression Analysis:** Regression models are used to examine how the variables in sensor data relate to one another. They aid in making predictions, comprehending correlations, and locating variables that have an impact on how well the vehicle performs.
3. **Feature Engineering:** This technique involves choosing pertinent features from unprocessed sensor data and developing new features to improve the efficiency of machine learning algorithms. For this goal, methods from linear algebra, signal processing, and image processing are frequently applied.
4. **Optimization:** Algorithms are tuned, control variables are adjusted, and the self-driving system's behavior is optimized using optimization techniques. Several parts of autonomous driving are optimized using evolutionary algorithms, gradient descent, and linear and nonlinear programming.

5. **Anomaly Detection:** Anomalies in sensor data can be used to spot problems that could arise, like sensor breakdowns or unforeseen behaviors. Anomaly detection can be accomplished using statistical techniques like Z-scores, clustering, and machine learning algorithms.

Technologies:

1. **High-Performance Computing (HPC):** For effective processing of the massive amounts of data produced by self-driving cars, high-performance processors and specialized hardware accelerators (such as GPUs) are needed. Real-time sensor data analysis is made possible by HPC.
2. **Data Pipelines:** Raw sensor data is preprocessed, cleaned, and transformed into a format that can be analyzed using data pipelines. These pipelines frequently use methods from data wrangling; signal processing, and image processing.
3. **Data Storage and Databases:** For historical analysis and developing machine learning models, it is crucial to efficiently store and retrieve big datasets. It is usual to employ distributed databases, NoSQL databases, and cloud-based storage options.
4. **Real-Time Processing:** In order to make quick judgments, self-driving cars need to analyze data in real-time. The system's real-time processing capabilities ensure that it can react quickly to shifting conditions on the road.
5. **Machine Learning and AI:** Models for machine learning are used to analyze sensor data, forecast outcomes, and improve algorithms. Data-driven decision-making is aided by methods like reinforcement learning, unsupervised learning, and supervised learning.
6. Data analysis is utilized to assess the effectiveness of self-driving algorithms. Model validation and testing. The behavior of the system is assessed, and potential areas for improvement are identified, using simulation environments and real-world testing situations.
7. **Simulation and Virtual Testing:** Before deploying self-driving algorithms on actual roads, simulation is a potent technique for evaluating them in a controlled setting. Algorithm iterations and optimization can happen quickly thanks to virtual testing.
8. **A/B Testing and Experimentation:** In real-world situations, A/B testing is utilized to compare several versions of algorithms or control schemes. Experimentation aids in improving the system based on empirical results.

Self-driving cars may analyze data, optimize algorithms, and continuously improve their performance by combining fundamental mathematical ideas with cutting-edge technologies, leading to safer and more effective autonomous driving experiences.

Self-driving car development and deployment success require a multidisciplinary strategy that incorporates these mathematical ideas and methods. The mathematics underlying self-driving cars will continue to develop as technology progresses, enabling more advanced and trustworthy autonomous vehicles.

3.4 AUTONOMOUS DRONES

Unmanned aerial vehicles (UAVs) that can function independently thanks to the use of robotics, artificial intelligence, and a variety of sensors are known as autonomous drones. Let's examine the fundamental mathematical ideas and innovations that underpin autonomous drones:

3.4.1 Kinematics and Dynamics

For autonomous navigation, a drone's motion must be understood. Kinematic equations are used to represent ideas like position, velocity, acceleration, and orientation, while dynamics considerations aid in foretelling how the drone will respond to outside forces. Fundamental ideas in the study of the motion of autonomous drones include kinematics and dynamics. Understanding the drone's position, velocity, acceleration, and the factors influencing its movement is required for these ideas. Let's investigate the mathematical ideas and methods that underlie the dynamics and kinematics of autonomous drones:

Mathematical Concepts:

1. **Position, Velocity, and Acceleration:** Position identifies the location of the drone in a three-dimensional (3-D) coordinate system; velocity shows how quickly the drone is travelling in a particular direction; and acceleration shows how the drone's velocity changes over time. Vectors and calculus are used to explain these ideas.
2. **Euler Angles and Quaternion's:** Roll, pitch, and yaw of the drone are represented by Euler Angles and Quaternion. For preventing gimbals' lock and maintaining stable orientation representation, quaternions are extremely helpful.
3. **Trajectory Equations:** Using the drone's beginning position, velocity, acceleration, and time, trajectory equations predict the drone's future position. The planning and control of paths depends on these equations.
4. **Forces and Torques:** Predicting the drone's movement and stability requires a thorough understanding of the forces—such as thrust, lift, drag, and gravity—and torques—resulting from the spinning of the propellers.
5. **Newton's Laws:** These laws of motion lay the groundwork for comprehending how forces impact the drone's acceleration and, in turn, its motion. The behavior of the drone is modeled using these principles.

Technologies:

1. **Inertial Measurement Units (IMUs):** IMUs are made up of gyroscopes and accelerometers that can measure the drone's rotational velocity and linear acceleration. The motion and orientation of the drone must be ascertained using this data.
2. **Barometers and Altimeters:** These sensors gauge the drone's height by gauging air pressure. For vertical positioning and 3D space navigation, this information is crucial.
3. **GPS and GNSS:** For horizontal positioning, geolocation data is provided by the Global Positioning System (GPS) and Global Navigation Satellite Systems (GNSS). For long-range navigation, GPS is essential even if it is not as precise as IMUs for short-term motion.
4. **Flight Controllers:** To accomplish the desired motion and stability, autonomous drones are outfitted with flight controllers that process sensor data, apply control algorithms, and send instructions to the drone's motors.
5. **Motor Control and Propulsion Systems:** The propulsion system of the drone is controlled to produce thrust and torques. It typically consists of electric motors and propellers. These elements are precisely controlled to provide the correct motion and orientation.
6. **Real-time Processing:** To effectively adapt to changes in the drone's environment and carry out commands, kinematics and dynamics calculations frequently require real-time processing. Real-time operating systems and high performance processors are employed.

7. **Simulation Software:** Engineers can simulate the drone's dynamics and kinematics using simulation software. This is essential for developing controllers, predicting, and testing algorithms.

Autonomous drones are able to accurately control their motion, maintain stability, navigate across difficult surroundings, and carry out duties ranging from aerial photography to search and rescue operations by fusing fundamental mathematical ideas with cutting-edge technologies.

3.4.2 Control Systems

Designing algorithms that stabilize and regulate the movement of the drone requires a solid understanding of control theory. PID controllers, state-space control, and optimum control strategies guarantee that the drone accurately follows intended pathways. For autonomous drones to maintain stability, carry out maneuvers, and follow planned flight paths, control systems are crucial. To ensure accurate and secure drone operations, these systems rely on cutting-edge technologies and mathematical concepts from control theory. Let's examine the fundamental mathematical ideas and innovations that underlie the drones' autonomous control systems:

Mathematical Concepts:

1. **Control Theory:** The foundation for creating algorithms that control the behavior of autonomous drones is provided by control theory. Concepts include system modeling, stability analysis, control signals, and feedback loops.
2. **Proportional-Integral-Derivative (PID) Control:** PID controllers are frequently used to stabilize drones. Based on the error (the difference between the desired state and the present state) and its history, these controllers modify control inputs.
3. **State-Space Control:** State-space control treats the drone's state—its position, velocity, orientation, and other characteristics—as a collection of variables, enabling the use of more sophisticated control techniques and multivariable systems.
4. **Optimal Control:** Optimal control systems, such as Linear Quadratic Regulator (LQR) and Model Predictive Control (MPC), look for control inputs that best achieve a particular goal, like minimizing energy usage or adhering to a predetermined trajectory.
5. **Nonlinear Control:** Because drones are nonlinear systems, standard linear control techniques might not be enough. For more complicated jobs and extremely manoeuvrable drones, nonlinear control techniques like back stepping or sliding mode control are employed.

Technologies:

1. **Flight Controllers:** These onboard components analyze sensor data, calculate control signals, and modify the drone's motors as necessary. These controllers carry out autonomous manoeuvres and maintain steady flight.
2. **Proportional-Integral-Derivative (PID) Controllers:** The software of the flight controller incorporates PID controllers. They compute modifications to the drone's motors to achieve stability and the required orientation using the sensor data (such as IMU measurements).
3. **State Estimate:** For control systems, precise state estimate, including location, velocity, and orientation, is essential. IMU, GPS, and other sensor data are combined using sensor fusion techniques, such as Kalman filters or complementing filters, to estimate the drone's state.

4. **Actuators and Motors:** To provide thrust and torque, drones employ electric motors and propellers as actuators. To achieve the necessary movements, the flight controller modulates these actuators depending on control commands.
5. **Real-time Processing:** Stability and quick response to environmental changes necessitate real-time processing in control systems. Low-latency control is accomplished with real-time operating systems and high-performance computers.
6. **Redundancy and Fail-safes:** To ensure safe operation, autonomous drones frequently include redundancy in their control systems. In case of sensor malfunctions or communication breakdowns, fail-safe mechanisms must be in place, such as the ability to return to a safe position.
7. **Control System Tuning:** To maximize the performance of the drone, control parameters must be precisely adjusted. This entails modifying PID controller gains or other control systems' weights.

Control systems in autonomous drones enable steady flying, precision navigation, and execution of complex manoeuvres by integrating these mathematical ideas and cutting-edge technologies, contributing to safer and more dependable autonomous aerial operations.

3.4.3 Path Planning

Path planning entails determining the best routes for drones to take in order to navigate between two points while avoiding hazards. Path planning is done using algorithms like A* search, Rapidly Exploring Random Trees (RRT), and prospective fields. The ability of autonomous drones to navigate through surroundings, avoid hazards, and arrive at particular locations while optimizing for effectiveness and safety depends on path planning. To determine the best flying paths, path planning algorithms combine cutting-edge technologies with concepts from computational geometry and optimization. Let's examine the fundamental mathematical ideas and methods underlying the autonomous drone path planning:

Mathematical Concepts:

1. **Graph Theory:** The world is represented by graphs, which have nodes for locations and edges for potential routes between those sites. Path discovery frequently makes use of graph search methods like the A* search or Dijkstra's algorithm.
2. **Search Algorithms:** These algorithms move over the graph in search of the shortest, fastest, or most energy-efficient path. The best path finding methods are Dijkstra's, A* (A-star), and variations like D* (D-star).
3. **Heuristic Functions:** Heuristics calculate the cost to get from one place to another. They direct search algorithms to give the most priority to paths that are probably the best. In A* search and related algorithms, heuristic functions are crucial.
4. **Sampling-based Approaches:** For planning in high-dimensional domains, sampling-based techniques like Rapidly Exploring Random Trees (RRT) or Probabilistic Roadmaps (PRM) are helpful. They create a collection of viable routes by sampling the surrounding area.
5. **Collision Detection:** Collision detection techniques determine whether a given path crosses over environmental obstacles. For effective collision verification, methods include bounding boxes, polygon intersection tests, or distance fields.
6. **Trajectory Optimization:** Trajectory optimization techniques calculate a series of control inputs that result in a path that is both safe and effective in dynamic conditions. Techniques from numerical optimization and optimum control are applied.

Technologies:

1. **Sensors:** The drone's surroundings can be learned using sensors including lidar, cameras, and obstacle detection systems. Using this information, impediments are found and the intended path is immediately updated.
2. **Digital Maps:** High-definition digital maps offer useful details about the topography, roads, and potential hazards. These maps are crucial for navigation and long-range planning.
3. **Flight Controllers:** These devices analyse the intended path and produce control commands to direct the drone in that direction. For perfect execution, real-time communication between the flight controller and path planner is crucial.
4. **Localization:** Adhering to the intended path requires precise localization (position estimate). To keep the drone on course, sensors like GPS, IMUs, and optical odometry are employed.
5. **Real-time Processing:** When the environment is changing dynamically, path planning frequently necessitates real-time processing. To guarantee prompt answers, real-time operating systems and high-performance processors are required.
6. **Simulators:** Using drone simulators, path planning algorithms can be examined and improved in simulated settings. Developers can test path planning ideas using simulations without the dangers of field testing.

Autonomous drones are able to create safe, effective routes through challenging situations, avoid hazards, and arrive at their destinations while taking into account variables like energy consumption, collision avoidance, and real-time flexibility.

3.4.4 Localization and Mapping

Autonomous drones must be able to locate themselves and make maps of their surroundings. Maps are created in real-time when the drone's location is estimated using methods like Simultaneous Localization and Mapping (SLAM). For autonomous drones to comprehend their position in the surroundings and produce precise maps that allow navigation, localization and mapping are crucial. These technologies rely on a mix of sensors, sophisticated algorithms, and mathematical ideas. Let's look at the main mathematical ideas and methods used in autonomous drone localization and mapping:

Mathematical Concepts:

1. **Geometry and Trigonometry:** For localization and mapping, it is essential to understand geometric concepts like angles, distances, and coordinate transformations. When combined with GPS information, trigonometry is used to determine locations based on sensor measurements.
2. **Probabilistic Approaches:** To address uncertainty in sensor measurements and precisely predict the drone's position and orientation, probabilistic approaches, such as Bayesian filtering, are used.
3. **Probabilistic Estimation Methods:** Probabilistic estimating methods for state estimation in dynamic systems include Kalman filters and particle filters. To increase the precision of localization, motion models and sensor data are combined using Kalman filters and particle filters.
4. **Simultaneous Localization and Mapping (SLAM):** Drones employ SLAM algorithms to create maps of their surroundings while also determining their location within those maps. For drones to function in uncharted or constantly changing surroundings, SLAM is crucial.

5. **Feature Matching and Correspondence:** Feature matching methods are used to find shared features between frames and compute the transformation between them when using visual data (such as from cameras). This is crucial for producing precise maps.

Technologies:

1. **GPS and GNSS:** Drones may get preliminary location estimates using the Global Positioning System (GPS) and Global Navigation Satellite Systems (GNSS). However, especially in situations with signal obstacles, their accuracy may be limited.
2. **Inertial Measurement Units (IMUs):** IMUs track the movement of the drone by measuring its acceleration and angular rates. IMUs can improve localization accuracy and fill in the gaps left by lost GPS signals when used in conjunction with other sensors.
3. **Lidar and Cameras:** Lidar sensors produce point clouds for 3D mapping, and cameras supply visual information for feature extraction and image-based mapping. Combining these sensors increases the precision of both mapping and localization.
4. **SLAM Algorithms:** More sophisticated SLAM algorithms, such Graph-based SLAM or V-SLAM (Visual SLAM), let drones map their surroundings in great detail while locating themselves on these maps.
5. **Visual Odometry:** Based on image sequences, visual odometry uses cameras to calculate the drone's motion. In the absence or presence of GPS signals, visual odometry is essential.
6. **Sensor Fusion:** Using sensor fusion techniques to combine data from many sensors, including GPS, IMUs, cameras, lidar, and radar, enables reliable and accurate mapping even in difficult conditions.
7. **High-Definition Mapping:** To supplement real-time localization, the environment may already have high-definition (HD) maps. Drones may fine-tune their positioning by comparing sensor data with recognized landmarks on a map.
8. **Real-time Processing:** To keep up with the drone's movements, localization and mapping algorithms need to be able to process data in real-time. These computations are expedited by specialized hardware and high-performance computers.

Autonomous drones may achieve exact localization, produce accurate maps, and navigate confidently by combining these mathematical ideas and cutting-edge technologies, making them safer and more effective in a variety of applications.

3.4.5 Computer Vision

For object identification, tracking, and environmental comprehension, computer vision is essential. Interpreting the visual data taken in by the drone's sensors requires the use of image processing, feature extraction, and deep learning models. For autonomous drones to sense and comprehend their environment through visual data, computer vision is an essential technology. To extract useful information from images or video streams, it combines mathematical ideas with image processing methods and cutting-edge algorithms. An outline of the key mathematical ideas and innovations behind computer vision in autonomous drones is given below:

Mathematical Concepts:

1. **Image Formation:** It is essential to comprehend how images are created through the interplay of light and camera sensors. For accurate computer vision, ideas like distortion reduction, intrinsic/extrinsic parameters, and camera calibration are crucial.

2. **Image Processing:** To preprocess visual data, extract features, and increase the quality of images, image processing techniques such as filtering, edge detection, image enhancement, and morphological operations are used.
3. **Feature Extraction:** For object detection, tracking, and scene comprehension, it is essential to find and describe characteristics in images, such as key points, corners, or edges.
4. **Image Transformations:** Aligning images, performing registration, and enabling precise object detection and mapping all require the usage of transformations such image rotation, scaling, translation, and perspective transformations.
5. **Geometric Transformations:** To correct the perspective distortion in images and enable accurate mapping and navigation, homograph transformations and perspective projections are used.

Technologies:

1. **Cameras:** Visual data for computer vision is provided by cameras (RGB, multispectral, thermal). Drones can be equipped with cameras to record images or movies of their surroundings.
2. **Lidar and Depth Sensors:** Using lidar or other depth sensors in conjunction with visual data to construct 3D maps, identify obstructions, and enhance object recognition.
3. **Real-time Processing:** To make prompt decisions based on visual data, computer vision in drones frequently needs real-time processing. The utilization of powerful CPUs, GPUs, and specialized hardware accelerators.
4. **Deep Learning:** For tasks like object detection, image segmentation, and scene interpretation, deep learning models such as convolutional neural networks (CNNs) are used. They make it possible for drones to recognize and learn from enormous datasets of objects.
5. **Object Detection and Tracking:** To locate objects in pictures or video streams, object detection techniques like YOLO (You Only Look Once) or SSD (Single Shot Multi Box Detector) are utilized. Tracking methods make sure that recognized items remain continuous across time.
6. **Semantic Segmentation:** Semantic segmentation models assign each pixel in an image to a certain category (such as a road, building, or tree). Drones using this technology are better able to comprehend their surroundings and make wise decisions.
7. **Visual Odometry:** Based on image sequences, visual odometry algorithms estimate the drone's motion and position using visual cues. This is especially helpful when GPS signals are poor or nonexistent.
8. **3D Reconstruction:** By using computer vision to create 3D models of the surroundings from photos, drones can navigate, map, and understand spatial relationships.

Autonomous drones can perceive their surroundings, detect objects, navigate through complex environments, and carry out tasks like surveillance, mapping, inspection, or search and rescue missions with greater accuracy and autonomy by combining these mathematical concepts and cutting-edge technologies.

3.4.6 Probabilistic Methods

Drones frequently operate in hazy conditions. To tackle errors in sensor data, Kalman filters, particle filters, and Bayesian inference are employed, strengthening the drone's perception. Numerous autonomous drone functions, such as vision, decision-making, navigation, and uncertainty management, heavily rely on probabilistic approaches. To deal with the inherent uncertainties in drone operations, these solutions make use of probabilistic models, Bayesian reasoning, and statistical techniques. Let's examine the main mathematical methods, technology, and probabilistic ideas employed in autonomous drones:

Probabilistic Concepts:

1. **Bayesian Inference:** Probabilistic approaches are built on the principles of Bayesian reasoning. It enables drones to make judgments while taking uncertainty into account and revise views in light of fresh information. A key idea in Bayesian inference is the Bayes' theorem.
2. **Probability Distributions:** It's important to comprehend and operate with probability distributions (such as Gaussian, Poisson, and Bernoulli). Drones model sensor measurement, prediction, and decision outcome uncertainty using distributions.
3. **Probabilistic Sensor Fusion:** Especially when individual sensors have inherent noise or limitations, probabilistically combining data from numerous sensors (GPS, IMUs, cameras, lidar) enables drones to construct a more accurate and robust representation of the environment.
4. **Monte Carlo Methods:** In drone operations, probabilistic sampling, estimating distributions, and managing complex scenarios all include the use of Monte Carlo techniques, such as Markov Chain Monte Carlo (MCMC) and Monte Carlo simulation.
5. **Kalman and Particle Filters:** These filters are employed for tracking and state estimation. Particle filters handle non-linear and non-Gaussian situations, while Kalman filters perform effectively in linear systems with Gaussian noise.

Technologies:

1. **Sensor Fusion:** Autonomous drones use sensor fusion algorithms to probabilistically combine data from many sensors. This enhances the precision of perception and judgment in ambiguous situations.
2. **Localization and Mapping (SLAM):** Algorithms for Probabilistic Simultaneous Localization and Mapping (SLAM) enable drones to map their surroundings while also determining their position within those maps. For navigation in uncharted or shifting areas, SLAM is crucial.
3. **Bayesian Networks:** These networks simulate the probability connections between various variables. Drones employ Bayesian networks for probabilistic decision-making, merging sensor data, and reasoning under uncertainty.
4. **Real-time Probabilistic Filtering:** Using real-time probabilistic filtering techniques (such as Extended Kalman Filters and Unscented Kalman Filters), drones can respond to shifting environmental variables while continuously gauging states and uncertainties.
5. **Support for Decision-making:** Probabilistic techniques help drones make decisions in the face of ambiguity. Probabilistic reasoning aids drones in evaluating risks and selecting the best course of action, whether they are planning a route, avoiding obstacles, or planning a mission.
6. **Probabilistic Motion Planning:** In dynamic contexts, algorithms for probabilistic motion planning provide trajectories that take into account uncertainties in sensor data, foreseeing future collisions and averting dangerous circumstances.
7. **Anomaly Detection:** Using probabilistic models, such as statistical techniques or machine learning algorithms, abnormalities in sensor data are found, unexpected occurrences are recognized, and appropriate responses are triggered.

Autonomous drones can improve their overall reliability and autonomy by incorporating these probabilistic concepts and technologies, enabling them to navigate safely in ambiguous settings, make informed decisions, sustain steady flight, and adapt to changing conditions.

3.5 TECHNOLOGICAL DETAILS OF AUTONOMOUS DRONES

We have elaborately presented the technology of Autonomous Drones below clarifying the salient features.

3.5.1 Sensors

Various sensors, including GPS, IMUs, ultrasonic sensors, LiDAR, and cameras, are used by autonomous drones. These sensors give information for positioning, mapping, avoiding obstacles, and perceiving the surroundings. Various sensors are used by autonomous drones to perceive their surroundings, navigate, and decide on actions. These sensors give crucial information that enables the drone to keep its balance, avoid obstructions, and carry out duties precisely.

Here is the key sensor types used in autonomous drones, along with their technological details:

1. **GPS (Global Positioning System):**

   Technology: In order to give worldwide location, velocity, and time data, GPS depends on a network of satellites. Signals from many satellites are triangulated to make it operate.

   GPS used in Drones: GPS is used in drones because it is necessary for initial position estimates, waypoint navigation, and geolocation data. Accuracy issues could arise, particularly in crowded or urban situations.

2. **IMU (Inertial Measurement Unit):**

   Technology: IMUs contain accelerometers and gyroscopes that measure linear acceleration and angular rates, respectively. They provide information about the drone's motion and orientation.

   Use in Drones: IMUs are crucial for stabilizing drones, estimating attitude (roll, pitch, yaw), and compensating for external forces. They help maintain stability, especially during GPS signal loss.

3. **Cameras:**

   Technology: Cameras record visual data as still photos or moving video streams. They could be multispectral cameras, infrared cameras, or RGB cameras.

   Use in Drones: Cameras are utilised for computer vision tasks such scene interpretation, object detection, obstacle avoidance, mapping, and localization. They are essential to perception.

4. **Lidar (Light Detection and Ranging):**

   Lidar sensors use technology to emit laser beams and time how long it takes for the beam to return after hitting an item. Using this information, precise 3D point clouds of the surroundings are produced.

   Use in Drones: Lidar is crucial for producing high-resolution 3D maps, detecting obstacles, modeling the landscape, and accurately estimating altitude, particularly in settings where exact distance information is important.

5. **Ultrasonic and Range Sensors:**

   Technology: High-frequency sound waves are emitted by ultrasonic sensors, which track how long it takes for the waves to return. Different technologies, such as time-of-flight and infrared, are used by range sensors to measure distances.

Ultrasonic and range sensors are employed in drones for terrain following, close-range object identification, and height control. In order to conduct low-altitude operations, they offer useful distance information.

6. **Barometers:**

   Technology: Barometers use technology to gauge atmospheric pressure. Altitude variations can be predicted using changes in pressure.

   Application in Drones: In addition to other sensors like GPS and IMUs, barometers are employed in drones to measure height (relative to a reference point) and offer additional altitude data.

7. **Magnetic Compass (Magnetometer):**

   Technology: Magnetometers use technology to gauge the magnetic field of the planet. They offer information about heading in relation to magnetic north.

   Use in Drones: Magnetometers are utilized in drones to help with orientation and heading. They may, however, experience local magnetic interference, necessitating careful calibration.

8. **Optical and Infrared Sensors:**

   Technology: These sensors use technology to identify light in the visible or infrared range. They can be used for a variety of things, such identifying particular light wavelengths or detecting temperature variations.

   Use in Drones: Remote sensing, multispectral imaging, thermal imaging, and environmental monitoring are just a few of the uses for optical and infrared sensors.

9. **Gyroscopes (Rate Sensors):**

   Technology: Gyroscopes use technology to measure rotational angular rates. They can be separate sensors, although they are frequently integrated into IMUs.

   Application in Drones: Gyroscopes aid in stabilization and response to changes in orientation by assisting with orientation estimation and motion control.

10. **Proximity Sensors:**

    Technology: Proximity sensors, like capacitive or inductive ones, can determine if an object is present or absent inside a given area.

    Use in Drones: Proximity sensors can be used in drones for close-range obstacle identification and to start certain processes when the drone approaches particular structures or objects.

11. **Environmental Sensors (Temperature, Humidity, etc.):**

    Technology: A variety of environmental sensors are capable of measuring a range of variables, including temperature, humidity, air pressure, and more.

    Use in Drones: Environmental sensors are useful for drones because they can add to the data collected for environmental monitoring, scientific study, or other purposes.

12. **Onboard Communication Systems:**

    Technology: The drone can communicate with other devices or ground control stations via communication technologies like Wi-Fi, radio transceivers, and cellular modules.

    Application in Drones: Communication systems are critical for real-time communication between drones and operators, remote piloting, receiving mission updates, transmitting telemetry data, and more.

Autonomous drones can detect their surroundings, navigate through complicated situations, make intelligent judgments, and carry out tasks with accuracy by integrating various sensors and utilizing their technological capabilities, ultimately leading to higher autonomy and reliability.

3.5.2 Flight Controllers

Onboard components called flight controllers process sensor data and manage the drone's motors. To stabilize the drone and carry out autonomous flight plans, these controllers employ control algorithms, sensor feedback, and feedback from the drone. In order to process sensor input, carry out control algorithms, stabilize the drone, and enable autonomous flight, flight controllers are essential parts of autonomous drones. To ensure precise and safe drone operations, they combine hardware and software.

Here is the technological detail of flight controllers in autonomous drones:

Microcontrollers and Processors:

- Microcontrollers or processors built into flight controls conduct real-time processing of sensor data and carry out control algorithms.
- The flight controller can efficiently handle complicated calculations, communication, and sensor fusion thanks to high-performance computers.

Sensor Integration:

- Flight controllers communicate with a variety of sensors, including barometers, gyroscopes, GPS, IMUs, and GPS, to gather data on the drone's status (position, orientation, velocity, etc.).
- Algorithms for sensor fusion combine data from many sensors to improve accuracy and dependability, particularly in hazy settings.

Control Algorithms:

- PID (Proportional-Integral-Derivative) controllers, state-space control, or advanced nonlinear control strategies are a few examples of control algorithms that flight controllers employ.
- These algorithms generate control signals that adjust motor speeds or servo positions to stabilize the drone, maintain desired orientation, and follow a predetermined flight path.

Real-time Operating Systems (RTOS):

- To ensure that control duties are completed promptly and that sensor data is processed, many flight controllers use real-time operating systems.
- RTOS enables the flight controller to give priority to important tasks, cutting latency and assuring responsiveness under flying situations that are rapidly changing.

Communication Interfaces:

- The ability to remotely pilot an aircraft, provide mission updates, and receive telemetry data for monitoring and control are all made possible by the communication interfaces that flight controllers have to receive commands from ground control stations or other devices.

Motor Control:

- According to the drone's present state and the required behavior, flight controllers produce motor control signals to modify the speed of individual motors or servos, ensuring steady flight and accurate manoeuvres.

Redundancy and Fail-Safes:

- In order to increase reliability, many modern flight controllers use redundant hardware or software.

- Fail-safe systems are frequently used to assure safe operation in the event of sensor failures or communication breakdowns, such as Return-to-Home (RTH) or autonomous landing.

Software Frameworks:

- Software frameworks created expressly for controlling drones are frequently used by flight controllers. Libraries for sensor integration, control algorithms, communication protocols, and security features are all part of these frameworks.

Motor Drive Electronics:

- Motor drive electronics, which govern the power provided to the drone's motors, interface with flight controllers and translate control signals from the flight controller into the proper voltage and current levels to drive the motors.

Telemetry and Logging:

- For real-time monitoring, flight controllers give operators with telemetry data, such as altitude, GPS locations, battery status, and more.
- Logging features make it possible to capture flight information for analysis and troubleshooting.

Power Management:

- Power distribution is controlled by flight controls, ensuring effective use of the drone's battery and avoiding over-discharge, which could result in loss of control.

By including these technological components, flight controller's work as the autonomous drone's brain, enabling it to carry out difficult tasks, maintain stability, navigate, and communicate with users and its surroundings successfully.

3.5.3 Embedded Systems and Microcontrollers

Strong embedded systems are needed by autonomous drones in order to process data, run control algorithms, and make important decisions. The necessary computing power is provided by single-board computers and microcontrollers. Autonomous drones are built on embedded systems and microcontrollers, which offer the processing power and control required for a variety of tasks include analyzing sensor data, controlling flight, communicating, and making decisions. Let's examine embedded systems and microcontrollers in autonomous drones in more technical detail:

Microcontrollers and Processors:

- Microcontrollers or processors created specifically for embedded applications are used by autonomous drones. These chips are frequently real-time processors and low-power devices.
- To meet the computational needs of drones, multi-core processors or specialized hardware accelerators (for signal processing, floating-point calculations, etc.) are frequently utilized.

Real-time Operating Systems (RTOS):

- Real-time operating systems (RTOS) are used by many embedded devices in drones. For sensor fusion, control algorithms, and responsiveness to external events, RTOS enables predictable and timely job execution.
- RTOS enables the system to effectively manage resource allocation and prioritise crucial tasks.

Low-level Hardware Control:

- Sensors, motors, and communication interfaces are just a few examples of the hardware elements that embedded systems directly interact with. To precisely control these components, they use low-level control approaches.

- To connect to actuators such motors and servos, microcontroller pins, PWM (Pulse-Width Modulation) outputs, and hardware timers are used.

Sensor Data Processing:

- Sensor data from a variety of sources, including IMUs, GPS, cameras, lidar, and more, is handled by embedded systems. In order to filter, calibrate, and fuse sensor data for precise perception and control, these systems require signal processing techniques.
- Maintaining steady flight, navigation, obstacle avoidance, and environment mapping all depend on the processing of sensor data.

Control Algorithms:

- PID (Proportional-Integral-Derivative) controllers, state-space control, or more sophisticated nonlinear control approaches are just a few of the control algorithms that microcontrollers can implement. To produce the intended drone behaviour, these algorithms compute control signals that modify motor speeds, servo positions, or other actuator outputs.

Communication Interfaces:

- Drones with embedded systems can exchange data with sensors, communication modules, and other devices via communication interfaces like UART, I2C, SPI, CAN, or Ethernet.
- The drone can connect with GPS receivers, IMUs, cameras, remote control systems, and other external components thanks to these interfaces.

Power Management:

- In order to maximize flight time and ensure effective use of the drone's battery, embedded systems manage power consumption through the regulation of voltage levels, control of sleep modes, and battery condition monitoring.

Redundancy and Fail-Safes:

- In order to increase reliability, advanced drones may include redundant microcontrollers or CPUs. Redundancy makes guarantee that crucial operations keep running even if one component fails.
- Fail-safe devices are frequently used to guarantee secure functioning in the event of sensor malfunctions or communication breakdowns.

Software Frameworks:

- Software frameworks and libraries designed for flight control, sensor integration, communication, and safety features are used by embedded systems in drones. These frameworks aid programmers in creating dependable and effective drone applications.

Security Considerations:

- As networked autonomous drones proliferate, embedded systems must address security issues. Measures for data integrity, secure boot, and encryption are crucial to safeguarding drone functioning and preventing unauthorized access.

The critical computational power and control abilities needed for autonomous drones to navigate, see their environment, make decisions, and carry out complicated tasks with accuracy and autonomy are provided by embedded systems and microcontrollers by integrating these technical elements.

3.5.4 Communication and Remote Control

Drones that are autonomous can communicate with remote pilots or centralized command centers. Real-time monitoring, remote control, and sending the drone updated mission plans all depend on this connectivity. Operating autonomous drones requires communication and remote control systems that allow for real-time monitoring, command transmission, data interchange, and remote flying. These systems create dependable and secure interactions with the drone using a combination of wireless technologies, protocols, and interfaces. The specific technology components of remote control and communication in autonomous drones are as follows:

Wireless Communication Technologies:

1. **Radio Frequency (RF) Communication:**

   Technology: RF communication transmits data between the drone and the ground control station using radio waves. It comprises a variety of modulation techniques and frequency ranges, including 2.4 GHz and 5.8 GHz.

   Use in Drones: Remote control, telemetry data transmission, and real-time video streaming from the drone to the operator are all frequent uses for RF communication in drones.

2. **Long-Range Communication:**

   Technology: Autonomous drones may use satellite communication, cellular (4G, 5G), or long-range Lora (Long Range) technologies for lengthy communication distances.

   Use in Drones: Beyond Visual Line of Sight (BVLOS) operations, remote missions, and data transmission over large areas all require long-range communication.

3. **WiFi and WLAN (Wireless Local Area Network):**

   Technology: To establish local communication networks with ground control stations, drones can use WiFi or WLAN. This allows for real-time data sharing, mission planning, and remote flying.

   Use in Drones: In areas with WiFi infrastructure already in place, such as indoor applications or close-range operations, WiFi is frequently employed for short-range communication.

4. **Mesh Networks:**

   In terms of technology, mesh networks are made up of interconnected nodes that can relay data in order to increase the network's robustness and communication range.

   Mesh networks can improve communication resilience in places with poor connectivity, allowing drones to transmit information across other drones or ground-based nodes. Mesh networks are used in drones.

Communication Protocols:

1. **MAVLink (Micro Air Vehicle Link):**

   Technology: MAVLink is a drone-specific, lightweight communication protocol. It gives drones and ground control systems a standardised mechanism to transmit telemetry data, instructions, and mission data.

   MAVLink is frequently utilised by the drone industry for ground control software and autopilot connectivity.

2. **UDP and TCP/IP:**

   Technology: Data transfer via IP networks is made possible by these common internet protocols. Real-time data streaming is best suited for UDP (User Datagram Protocol), while dependable data transmission is ensured by TCP/IP (Transmission Control Protocol/Internet Protocol).

Use in Drones: While TCP/IP can be used for more crucial control orders and data sharing, UDP is frequently employed for broadcasting video and telemetry data.

3. **Encryption and Security:**

   Technology: To prevent unauthorized access and guarantee data integrity, communication between autonomous drones and ground control stations must be secure. There are secure authentication methods and encryption systems.

   Use in Drones: Safe remote control, protection from hacking, and operation of the drone depend on secure connection.

Remote Control Interfaces:

1. **Ground Control Stations (GCS):**

   Technology: GCS software gives users a user interface through which they can submit commands, plan missions, view telemetry data, and monitor the drone's status.

   GCS is used in drones because it is crucial for mission planning, remote piloting, and real-time behavior monitoring.

2. **Joysticks and Transmitters:**

   Technology: For manual remote control of drones, especially when accurate piloting is needed, operators use physical joysticks or transmitters.

   Joysticks are used in drones because they offer simple control for takeoff, landing, manoeuvring, and reacting to changing conditions.

3. **Mobile Devices:**

   Technology: By using specialized software to connect to the drone's communication network, smart phones and tablets can act as remote control interfaces.

   Use in Drones: Mobile devices offer a practical way to check on the state of the drone, make settings, and issue commands.

4. **Web-Based Interfaces:**

   Technology: Web-based interfaces let users fly and configure drones remotely from a variety of devices, and they work with conventional web browsers.

   Use with Drones: Web-based interfaces give remote control and monitoring options, especially when installing specialized software is not an option.

Autonomous drones can effectively communicate with operators, exchange real-time data, receive commands, and operate remotely or autonomously while ensuring security, responsiveness, and mission success by integrating these communication technologies, protocols, and remote control interfaces.

3.5.5 Wireless Technologies

Drones frequently employ wireless communication methods like Wi-Fi, radio, or cellular networks for data transmission, remote control, and mission update notifications. For autonomous drones to communicate, be controlled, and share data, wireless technologies are essential. Drones can interface with other devices and networks, broadcast telemetry data, stream video, receive commands, and communicate with ground control stations thanks to these technologies. In the context of autonomous drones, the following are the main wireless technologies and their technological details:

1. **Radio Frequency (RF) Communication:**

   Technology: Radio waves are used in RF communication to transmit data. It employs modulation techniques to encode information and operates in a variety of frequency ranges, such as 2.4 GHz or 5.8 GHz.

   Use in Drones: Remote control, telemetry data transmission, and real-time video streaming from the drone to the operator are all frequent uses for RF communication in drones.

2. **Long-Range Communication:**

   Technology: Autonomous drones may use satellite communication, cellular (4G, 5G), or long-range Lora (Long Range) technologies for lengthy communication distances.

   Use in Drones: Beyond Visual Line of Sight (BVLOS) operations, remote missions, and data transmission over large areas all require long-range communication.

3. **WiFi and WLAN (Wireless Local Area Network):**

   Technology: To establish local communication networks with ground control stations, drones can use WiFi or WLAN. This allows for real-time data sharing, mission planning, and remote flying.

   Use in Drones: In areas with WiFi infrastructure already in place, such as indoor applications or close-range operations, WiFi is frequently employed for short-range communication.

4. **Mesh Networks:**

   In terms of technology, mesh networks are made up of interconnected nodes that can relay data in order to increase the network's robustness and communication range.

   Mesh networks can improve communication resilience in places with poor connectivity, allowing drones to transmit information across other drones or ground-based nodes. Mesh networks are used in drones.

5. **Satellite Communication:**

   Technology: Using satellite communication technologies, drones may connect in remote or out-of-line-of-sight situations where traditional ground-based connection is challenging.

   Satellite communication is useful for drone missions that need connectivity but are outside the range of terrestrial communication infrastructure due to its global coverage.

6. **Bluetooth and BLE (Bluetooth Low Energy):**

   Technology: Data transport between devices that are close to one another usually makes use of short-range wireless communication technologies like Bluetooth and BLE.

   Application in Drones: Bluetooth or BLE connections allow for direct command and control of drones as well as updates and configuration.

7. **Cellular Connectivity (4G, 5G):**

   Technology: Drones equipped with cellular modules can link to 4G or 5G networks, enabling data transmission and remote control over cellular networks.

   Cellular connectivity is helpful for real-time data exchange, remote piloting, and command transmission in drones when operating in areas with cellular coverage.

8. **NFC (Near Field Communication):**

   Technology: Near-field communication (NFC) is a short-range wireless technology that enables contactless data transfer between devices.

NFC's Use in Drones: Compatible devices can be quickly paired with or authenticated via NFC, streamlining the connection procedure for particular activities.

9. **GNSS (Global Navigation Satellite System):**

   Technology: Accurate location and time information is provided via GNSS, which includes GPS, Glonass, and other satellite positioning systems.

   Use in Drones: GNSS is crucial for precise geolocation and autonomous navigation, giving drones exact position information.

10. **Antennas and Signal Processing:**

    Technology: Drones communicate through specialized antennae. Signal processing techniques like beam formation can be utilized to improve signal reception, range, and interference resilience.

    Concerning Drones: Advanced antenna and signal processing techniques raise the efficiency and dependability of communications.

11. **Encryption and Security:**

    Technology: To prevent unauthorized access, safeguard data integrity, and guarantee safe operation, secure communication is crucial.

    Use in Drones: Communication between the drone and ground control stations is protected by encryption methods and secure authentication systems.

Autonomous drones can operate successfully in remote or autonomous modes, exchange vital information, and improve their capabilities by utilizing these wireless technologies and comprehending their technological quirks. This will also ensure the success of their missions.

3.5.6 Battery and Power Management

To maximize available energy and enable extended flight periods, autonomous drones must control their power usage. Power management techniques and advanced battery technology are crucial. In order to maximize flying time, operational range, and overall efficiency, autonomous drones must carefully manage their batteries and power sources. High-performance batteries and advanced power management systems are essential to autonomous drones' ability to maximize energy use, maintain stable flight, and carry a variety of payloads. The most important technological information on battery and power management in autonomous drones is provided below:

1. **Battery Chemistry:**

   Technology: Because lithium-based batteries have a high energy density, low self-discharge rates, and are lightweight, they are frequently utilised in autonomous drones. Lithium-based batteries, in particular Lithium Polymer (LiPo) and Lithium-ion (Li-ion), are employed in these batteries.

   Benefits: By providing a good balance between energy capacity and weight, these battery chemistries are appropriate for drones that need extended flying periods while yet retaining a tolerable overall weight.

2. **Battery Capacity:**

   In terms of technology, drone batteries are commonly measured in mill ampere-hours (mAh). Longer flight periods are provided by larger capacity batteries, but they also add to the overall weight.

Optimization: To achieve the best flying times for particular applications, drone makers attempt to balance battery capacity with the drone's weight and power consumption.

3. **Battery Management System (BMS):**

   Technology: A crucial component that maintains and monitors each cell in the battery pack is the BMS. It maintains balanced charging, guards against overcharging or undercharging, and keeps track of the battery's condition.

   Safety: By eliminating hazardous situations and extending the battery's total lifespan, a well-designed BMS improves battery safety.

4. **Power Density and Discharge Rate:**

   Technology: Batteries with a high power density can provide enough power for quick acceleration and manoeuvres. How quickly the battery can discharge its energy is indicated by the discharge rate (C-rating).

   Performance: To handle abrupt power demands, drones with high power needs, such as racing drones or those with large payloads, require batteries with higher discharge rates.

5. **Swappable Batteries:**

   Technology: Some autonomous drones, especially those used for business or industry, include interchangeable battery systems. By quickly replacing exhausted batteries with fully charged ones, the operational time of the drone is increased.

   Effectiveness: Swappable battery systems enable continuous operations and reduce downtime between flights, making them appropriate for applications where uninterrupted operation is required.

6. **Battery Charging and Management:**

   Technology: Intelligent battery management systems and chargers make sure that drone batteries are properly charged and maintained, increasing their lifespan and preserving peak performance.

   Smart Charging: Some drones and battery management systems can provide data on battery health, charging cycles, and anticipated battery life, assisting users in making decisions about battery replacement and maintenance.

7. **Energy-Efficient Components:**

   Technology: To reduce energy consumption, autonomous drones frequently use energy-efficient parts such sensors, low-power CPUs, and communication modules.

   Power Budgeting: Power-aware software and hardware design aids drones in effectively allocating power resources, ensuring that crucial operations (such as flight control, navigation, and communication) are given precedence while sparing non-essential components energy.

8. **Low-Voltage Detection and Fail-Safes:**

   Technology: Drones have low-voltage detecting systems that, when the battery voltage reaches a critical point, cause a fail-safe action to be taken (such as returning to the drone's home or landing).

   Safety: Fail-safes stop the drone from losing power in midair, lowering the possibility of uncontrolled accidents.

9. **Thermal Management:**

   Technology: A battery's longevity and efficiency can both be maintained with proper temperature control. The drone design may use cooling techniques like passive cooling, fans, or heat sinks.

   Preventing Overheating: Heat accumulation can impair battery performance, thus drones may use techniques to effectively dissipate heat while operating.

10. **Battery Monitoring and Telemetry:**

    Technology: Drones continuously check the temperature, voltage, and current of the batteries. The ground control station receives telemetry information regarding the battery's condition and operation.

    Real-time Information: Operators utilize this information to determine how much flying time is left, the condition of the battery, and the appropriate moment to return the drone or swap out the battery.

Autonomous drones can achieve longer flight times, reliable operations, and optimized power consumption by utilizing these technological advancements in battery chemistry, management, efficiency, safety, and monitoring. As a result, they are better suited for a variety of applications, from aerial photography to search and rescue missions.

3.5.7 Simulators

With the use of drone simulators, developers can test and train drones in safe virtual environments while also optimizing algorithms, simulating different scenarios, and validating their behavior. Before deploying autonomous drones in the real world, simulations are effective tools for testing and validating their behavior. They offer a low-risk alternative to real flight testing for the development, training, and optimization of algorithms. The following are the main technological specifics of drone simulators:

1. **Physics-Based Simulation:**

   Technology: Simulators use physics-based models to simulate the dynamics of drones in the actual world, including aerodynamics, motor behavior, sensor characteristics, and environmental elements (such as gravity and wind).

   The simulator is a solid testing platform because physics-based simulation assures that the drone's behavior in the simulated environment closely resembles that of the drone in the actual world.

2. **Sensor Emulation:**

   Technology: Simulators reproduce the results of sensors, including GPS, IMUs, cameras, lidar, and others, that autonomous drones employ for perception.

   Realism: Developers can test how algorithms and control systems react to sensor input by using sensor emulation, which enables them to assess navigation, mapping, object detection, and other functionalities.

3. **Environmental Rendering:**

   Technology: Simulators provide environments that are realistic, incorporating terrains, buildings, barriers, lighting, weather, and other factors that influence drone operations.

   Scalability: Developers can evaluate the drone's performance under varied circumstances by using simulators to model diverse scenarios and environmental variables.

4. **Real-Time Simulation:**

   Technology: Real-time simulators enable designers to watch drone behaviour as it would occur during actual flight.

   Real-time simulation is crucial for testing and debugging navigational systems, control algorithms, and decision-making procedures.

5. **Integration with Autopilots:**

   Simulators can communicate with real autopilots or software stacks used in autonomous drones thanks to technology. This enables direct testing of the algorithms created in simulation on the actual autopilot.

   Development Cycle: This integration shortens the development cycle by allowing developers to switch between simulation and real-world testing more quickly.

6. **Customizable Scenarios and Missions:**

   Technology: Simulators let programmers construct unique missions and situations, complete with flight paths, waypoints, obstacle configurations, and dynamic events.

   Training: Customizable scenarios make it possible to train the drone's behaviour in particular circumstances, which enhance the drone's reaction to mission-critical occurrences.

7. **Hardware-in-the-Loop (HIL) Simulation:**

   Technology: HIL simulation combines the simulator and the drone's autopilot hardware, enabling the autopilot to communicate with the simulated world.

   Validation: HIL simulation makes sure that when the autopilot is linked with the control and navigation algorithms in the simulator, it behaves as predicted.

8. **Machine Learning and AI Testing:**

   Technology: Machine learning and AI models, such as object recognition, path planning, and decision-making algorithms, are tested in simulators before being deployed in autonomous drones.

   Training: Before implementation, AI models can be trained and improved via simulations using real-world circumstances.

9. **Data Collection and Analysis:**

   Technology: Simulators can gather a tone of information regarding the behavior of the drone, sensor inputs, control outputs, and mission results.

   Data-Driven Development: Analyzing this data enables programmers to locate performance stumbling blocks, enhance algorithms, and enhance drone behavior in general.

10. **Integration with Development Environments:**

    Technology: Simulators frequently interact with well-known development environments, including ROS (Robot Operating System), enabling seamless integration of simulation and code development.

    - Workflow efficiency is improved through integrated development environments, which also make it simpler to implement and test changes.

Simulators make use of these technical advancements to offer an invaluable testing, training, and validation environment for autonomous drones. This speeds up the development process, lowers risk, and improves the dependability and performance of autonomous drone systems.

3.5.8 Edge Computing

For faster transmission and real-time answers, some data processing can be done on the drone itself (edge computing). A key technology idea for autonomous drones is edge computing, which enables them to process data locally on the drone rather than relying just on remote servers. Real-time decision-making is made possible by edge computing, which also lowers latency, saves bandwidth, and increases drone autonomy. The main technological specifics of edge computing for autonomous drones are as follows:

1. **Onboard Processing:**

   Technology: Powerful computers on autonomous drones can handle real-time data processing operations. Complex algorithms for vision, navigation, control, and decision-making can be run on these computers.

   Benefits: By eliminating the need to send raw sensor data to a remote server for analysis, onboard processing enables quicker response times and greater autonomy.

2. **Sensor Data Fusion:**

   Technology: Edge computing enables autonomous drones to locally combine data from several sensors, including cameras, lidar, IMUs, and GPS.

   Advantages: By giving the drone a thorough understanding of its surroundings, sensor data fusion improves perception, obstacle avoidance, and localization precision.

3. **Real-time Control:**

   Technology: Real-time control algorithms may now be directly executed by the drone's onboard processors thanks to edge computing. In reaction to shifting circumstances, these algorithms modify motor speeds, flight routes, and other control parameters.

   Reduction of Communication Delay: Real-time control facilitates faster responses to unforeseen occurrences by reducing communication delay between the drone and ground control systems.

4. **Object Detection and Recognition:**

   Technology: Thanks to edge computing, real-time control algorithms may now be run directly by the drone's onboard processors. These algorithms alter motor speeds, flight paths, and other control parameters in response to changing conditions.

   Reduction of Communication Delay: By minimizing communication lag between the drone and ground control systems, real-time control enables quicker responses to unanticipated events.

5. **Path Planning:**

   Using edge computing technology, drones may calculate their routes locally while taking into account topography, potential hazards, and mission objectives.

   Autonomy: Local path planning improves the drone's autonomy by enabling it to adjust to shifting circumstances and carry out difficult manoeuvres without continuously relying on remote commands.

6. **Communication Optimization:**

   Technology: Edge computing minimises the quantity of information that must be transferred via the communication link between the drone and the ground control systems.

Bandwidth Conservation: Edge computing conserves transmission bandwidth, which is necessary for maintaining a dependable connection, by processing data locally and delivering only pertinent information (such as high-level mission updates, telemetry, and vital alarms).

7. **Redundancy and Fail-Safes:**

   Technology: On-board redundant systems or fail-safe procedures can be implemented using edge computing. In the event of sensor malfunctions or communication breakdowns, these systems can take over crucial tasks.

   Reliability: Local redundancy improves the drone's safety and dependability by enabling it to carry out crucial tasks even in the face of malfunctions.

8. **Machine Learning Inference:**

   Technology: Lightweight machine learning models for inference tasks, such as object recognition or anomaly detection, can be run on drones using edge computing.

   Real-time Decisions: Making decisions in real-time based on the output of AI models is made possible by running machine learning inference locally on the drone.

9. **Energy Efficiency:**

   Technology: Edge computing may be energy-efficiently optimized, enabling the drone to utilize its onboard processing power effectively while preserving battery life.

   Autonomous Endurance: By extending the drone's autonomous flying time using energy-efficient edge computing, lengthier missions can be completed without frequent battery changes or recharging.

Autonomous drones may analyze data locally, act quickly, adapt to changing environments, and operate more independently by using these edge computing technological advancements. This improves speed, lowers latency, increases reliability, and makes better use of available resources.

Autonomous drones may carry out a variety of duties, such as aerial photography, surveillance, search and rescue missions, deliveries, and environmental monitoring, with little to no assistance from humans by fusing mathematical concepts and technologies.

3.6 AUTONOMOUS VEHICLES AND DRONES: THEIR POTENTIAL TO TRANSFORM TRANSPORTATION

Drones and autonomous vehicles have the potential to drastically change transportation in a variety of ways. Their influence touches on issues like safety, accessibility, urban planning, logistics, and environmental sustainability in addition to convenience and efficiency. The following are some of the most significant ways that drones and autonomous cars could change the transportation industry:

1. **Improved Safety:**
   - Human error, a primary cause of traffic accidents, is anticipated to decrease as a result of autonomous vehicles. Road safety may be improved overall with the help of cutting-edge sensor systems, real-time data processing, and AI-driven decision-making.
   - Drones can be used for surveillance, traffic monitoring, and infrastructure safety inspections, assisting in the identification of potential dangers and assuring the security of travel routes.
2. **Increased Accessibility:**
   - Drones can offer last-mile delivery solutions to remote or underserved areas, enabling access to goods and services that were previously difficult to reach.

- Autonomous vehicles have the potential to improve transportation accessibility for people with disabilities, the elderly, and those who are unable to drive.

3. **Enhanced Traffic Efficiency:**
   - Because autonomous vehicles can communicate with one other and traffic control systems, there will be less traffic jams and more efficient use of the roads.
   - Drones can be used to monitor traffic, spot congested locations, and assist in vehicle routing for shorter trip distances and better traffic management overall.
4. **Urban Planning and Infrastructure:**
   - As autonomous vehicles become more prevalent, urban planning may change as the need for parking facilities may decline and road layouts may change to facilitate more effective traffic flow.
   - Data from drones can be used for urban planning, such as 3D mapping, land-use analysis, and determining the upkeep requirements for infrastructure.
5. **Logistics and Delivery:**
   - Drones have the potential to revolutionize delivery, especially for small, high-value items, by offering quick and affordable solutions for last-mile delivery.
   - Autonomous delivery vehicles can streamline the logistics industry by reducing delivery times, optimizing routes, and increasing the efficiency of freight transport.
6. **Environmental Impact:**
   - Why By streamlining driving habits and easing traffic congestion, autonomous vehicles can result in higher fuel efficiency and lower emissions.
   - Drones have the potential to reduce road traffic and its accompanying environmental impact by monitoring the environment, protecting species, and perhaps providing urban air mobility in the future.
7. **Shared Mobility and Transportation-as-a-Service (TaaS):**
   - The necessity for private auto ownership may decline as a result of the development of shared mobility and transportation-as-a-service models made possible by autonomous vehicles.
   - Drones can offer on-demand services for a variety of applications, such airborne taxis or emergency medical deliveries, helping to make better use of available resources.
8. **Economic Opportunities:**
   - The creation, production, and use of autonomous vehicles and drones may open up new business opportunities across a range of sectors, from infrastructure and services to manufacturing and technology.

Autonomous vehicles and drones have tremendous potential benefits, but there are also issues that need to be resolved, including as legislative frameworks, public acceptance, cyber security worries, ethical issues, and the necessity for trustworthy, robust technology. Autonomous cars and drones, on the other hand, have the potential to revolutionize transportation, making it safer, more effective, and more accessible for everyone, with further technological developments, ongoing research, and cooperation between the public and commercial sectors.

3.7 AUTONOMOUS VEHICLES AND DRONES AND THEIR POTENTIAL TO TRANSFORM LOGISTICS

The logistics sector could undergo a transformation thanks to autonomous cars and drones, which would increase productivity, save costs, and enhance the entire supply chain. These technologies provide ground-breaking options for delivery, inventory control, and warehousing operations.

Here are some key ways in which autonomous vehicles and drones can transform logistics:

1. **Last-Mile Delivery:**
   - Drones can quickly transport small products to consumers' doorsteps, lowering delivery times and costs, especially in congested urban areas.
   - Autonomous delivery vehicles and drones can handle last-mile delivery, which is frequently the most expensive and time-consuming aspect of the supply chain.
2. **Faster Delivery Times:**
   - As opposed to conventional human-driven vehicles, autonomous vehicles and drones may operate around the clock, enabling quicker delivery times. This is especially useful for time-sensitive deliveries, such as those of medical supplies, perishable commodities, or critical spare parts.
3. **Cost Savings:**
   - By reducing fuel consumption, minimizing idle time, and optimizing delivery routes, autonomous trucks can save logistics firms money.
   - By doing away with the need for human drivers and lowering the cost of vehicle maintenance, drones can drastically reduce delivery costs.
4. **Scalability and Flexibility:**
   - In order to satisfy variable demand during peak seasons or unexpected increases in delivery requirements, autonomous vehicles and drones may be quickly expanded. This scalability enables logistics organizations to respond to changing market conditions and handle surges in demand more effectively.
5. **Warehouse Automation:**
   - Material handling, sorting, and inventory management tasks at warehouses can be handled by autonomous robots, which lowers labor costs and improves accuracy. These robots can assist human workers, increasing productivity and reducing the possibility of mistakes.
6. **Inventory Tracking and Monitoring:**
   - Large warehouse inventory may be promptly scanned by drones fitted with RFID or other tracking technology, increasing accuracy and minimising manual labour.
   - Monitoring inventory levels in real-time lowers the risk of stockouts and overstock situations while streamlining supply chain management.
7. **Remote and Rural Deliveries:**
   - This technology can enable access to necessary goods and services in areas with limited infrastructure, making drones particularly useful for transporting commodities to remote or rural areas where conventional distribution methods may be difficult.
8. **Reducing Traffic Congestion:**
   - Especially in urban areas where delivery trucks contribute to road congestion, autonomous vehicles and drones can help ease traffic congestion by optimising routes.As a result,

transportation networks become more effective, which is good for both logistics businesses and the general public.

9. **Safety Improvements:**
   - It is anticipated that autonomous vehicles will increase road safety by lowering the likelihood of accidents and assuring the safe transit of cargo.
   - Without putting human operators in risk, drones can operate in dangerous conditions or inaccessible places, such as disaster-stricken areas.
10. **Environmental Impact:**
    - Autonomous cars and drones can have a beneficial environmental impact by cutting emissions and minimizing the carbon footprint of logistical operations by optimizing routes and consuming less fuel.

The incorporation of autonomous vehicles and drones into the logistics industry holds great potential to transform how goods are transported, improving efficiency, lowering costs, and offering more convenient and dependable services to customers. However, there are regulatory, technical, and operational challenges to address.

Numerical Problems with Solutions

Autonomous Vehicles

Fuel Efficiency Improvement

Problem 1. An autonomous electric vehicle has a range of 300 miles on a full charge. If its energy consumption is 0.25 kWh per mile, how much energy does it consume for a 100-mile journey?

Solution: Energy consumed = Energy per mile * Distance = 0.25 kWh/mile * 100 miles = 25 kWh

Time Savings with Autonomous Driving

Problem 2. An autonomous vehicle reduces the average commute time by 20 minutes per day for a person. How much time does this save in a year?

Solution: Time saved per year = Time saved per day * Number of commuting days (assuming 250 workdays per year) = 20 minutes/day * 250 days = 5,000 minutes = 83.33 hours

Sensor Data Processing Rate

Problem 3. An autonomous vehicle's sensors generate data at a rate of 2 gigabytes (GB) per minute. How much data does it generate during a 1-hour drive?

Solution: Data generated = Data rate * Time = 2 GB/minute * 60 minutes = 120 GB

Battery Charging Time

Problem 4. An autonomous vehicle has a battery capacity of 75 kWh. If the charging station delivers power at a rate of 7.5 kW, how long does it take to fully charge the vehicle from 10% to 100%?

Solution: Energy needed for charging = Battery capacity * (100% – 10%) = 75 kWh * 90% = 67.5 kWh. Charging time = Energy needed/Charging rate = 67.5 kWh/7.5 kW = 9 hours.

Collision Avoidance Reaction Time

Problem 5. An autonomous vehicle's collision avoidance system can react to obstacles within 200 milliseconds. How far does the vehicle travel during this reaction time at a speed of 60 mph?
Solution: Distance traveled = Speed * Time = (60 miles/hour) * (200 milliseconds/1000) hours = 12 feet

Drones

Maximum Flight Time

Problem 6. A drone has a battery with a capacity of 20,000 milliampere-hours (mAh). If the drone consumes an average current of 2 amperes (A) during flight, how long can it fly before the battery is depleted?
Solution: Flight time = Battery capacity/Current = (20,000 mAh/1000)/2 A = 10 hours

Altitude Gain

Problem 7. A drone starts at an altitude of 100 meters above ground level (AGL) and climbs to an altitude of 300 meters AGL. What is the total altitude gain?
Solution: Altitude gain = Final altitude – Initial altitude = 300 meters – 100 meters = 200 meters

Drone Speed

Problem 8. A drone travels a distance of 1,500 meters in 2.5 minutes. What is its average speed in meters per second?
Solution: Speed = Distance/Time = 1,500 meters/(2.5 minutes * 60 seconds/minute) = 10 meters/second

Payload Capacity

Problem 9. A drone has a maximum payload capacity of 2 kilograms (kg). If the drone itself weighs 0.5 kg, how much additional weight can it carry?
Solution: Additional weight = Payload capacity – Drone weight = 2 kg – 0.5 kg = 1.5 kg

Flight Range

Problem 10. A drone has a flight range of 5 kilometers. If it flies a distance of 3.8 kilometers, how much remaining range does it have?
Solution: Remaining range = Total range – Distance flown = 5 km – 3.8 km = 1.2 km

Short Questions and Answers

Q.1. What is an autonomous vehicle?

Ans. An autonomous vehicle, also known as a self-driving vehicle, is a vehicle capable of navigating and operating without human intervention.

Q.2. What are the levels of autonomy in vehicles?

Ans. The Society of Automotive Engineers (SAE) defines five levels of autonomy, ranging from Level 0 (no automation) to Level 5 (full automation).

Q.3. How do autonomous vehicles perceive their environment?

Ans. Autonomous vehicles use a combination of sensors, such as cameras, lidar, radar, and ultrasonic sensors, to perceive their surroundings.

Q.4. What are the potential benefits of autonomous vehicles?

Ans. Benefits include improved safety, reduced traffic congestion, increased accessibility, enhanced energy efficiency, and new business models.

Q.5. Can autonomous vehicles operate in all weather conditions?

Ans. Current autonomous vehicles may face challenges in extreme weather conditions like heavy rain, snowstorms, or fog, which can affect sensor performance.

Q.6. What are the key applications of autonomous vehicles?

Ans. Applications include ride-sharing services, delivery logistics, public transportation, and autonomous shuttles in controlled environments.

Q.7. What is a drone?

Ans. A drone is an unmanned aerial vehicle (UAV) used for various purposes, including surveillance, photography, delivery, and remote sensing.

Q.8. How are drones powered?

Ans. Most drones are powered by rechargeable batteries, but some may use alternative power sources like gas or solar energy.

Q.9. What are the primary uses of drones in industry?

Ans. Drones are used for aerial surveying, mapping, agriculture, infrastructure inspection, and environmental monitoring.

Q.10. How do drones communicate with ground control stations?

Ans. Drones communicate using wireless technologies such as radio frequency (RF), WiFi, or cellular networks.

Q.11. What is the range of a typical drone flight?

Ans. The range of a drone depends on its design, but commercial drones typically have ranges between a few kilometers to tens of kilometers.

Q.12. What is the role of edge computing in autonomous vehicles?

Ans. Edge computing enables real-time data processing onboard autonomous vehicles, reducing latency and enhancing decision-making.

Q.13. What's the significance of sensor fusion in autonomous vehicles?

Ans. Sensor fusion combines data from multiple sensors to provide a comprehensive and accurate perception of the vehicle's environment.

Q.14. How can drones improve search and rescue operations?

Ans. Drones equipped with cameras and thermal sensors can quickly locate missing persons or assess disaster-stricken areas.

Q.15. How can autonomous vehicles impact urban planning?

Ans. Autonomous vehicles may lead to reduced parking needs, changed road designs, and more efficient use of urban space.

Q.16. Are fully autonomous vehicles available for purchase?

Ans. As of my knowledge cutoff in 2021, some semi-autonomous features are available, but fully autonomous consumer vehicles are not yet widely available.

Q.17. What is the role of AI in autonomous vehicles?

Ans. AI is essential for perception, decision-making, path planning, and adapting to dynamic environments in autonomous vehicles.

Q.18. What is the role of LiDAR in autonomous vehicles?

Ans. LiDAR (Light Detection and Ranging) is a remote sensing technology used in autonomous vehicles to create detailed 3D maps of the environment.

Q.19. Can drones be used for package delivery?

Ans. Yes, drones are being developed for last-mile package delivery, especially in urban areas.

Q.20. How do drones navigate without GPS indoors?

Ans. Indoor drones may use alternative methods such as visual odometry, ultrasonic sensors, or laser-based systems for indoor navigation.

Q.21. What challenges do autonomous vehicles face in complex urban environments?

Ans. Autonomous vehicles must handle unpredictable human drivers, pedestrians, cyclists, and complex traffic scenarios.

Q.22. What is the role of simulation in testing autonomous vehicles?

Ans. Simulation allows developers to test autonomous vehicle algorithms in virtual environments, reducing the need for extensive real-world testing.

Q.23. Can drones be used for agricultural purposes?

Ans. Yes, drones can be used for precision agriculture, monitoring crops, assessing soil health, and optimizing farming practices.

Q.24. How do autonomous drones avoid obstacles during flight?

Ans. Autonomous drones use obstacle detection and avoidance systems based on sensors like lidar, cameras, or ultrasonic sensors.

Q.25. Can drones be used for wildlife conservation?

Ans. Yes, drones can monitor wildlife populations, track animal movements, detect poaching activities, and protect endangered species.

Q.26. Are autonomous vehicles legal on public roads?

Ans. Regulations vary by region. Some places allow limited testing of autonomous vehicles on public roads under specific conditions.

Q.27. What role can drones play in disaster response?

Ans. Drones can quickly assess disaster damage, deliver emergency supplies, and help search and rescue teams locate survivors.

Q.28. Are drones allowed to fly near airports?

Ans. In most cases, drones are not allowed to fly near airports due to safety regulations.

Q.29. What challenges do drones face in terms of battery life?

Ans. Limited battery life restricts the flight duration of drones, which can impact their operational range and endurance.

Q.30. How are autonomous drones programmed to follow a specific path?

Ans. Autonomous drones use GPS waypoints or pre-defined paths programmed into their flight control systems.

Q.31. How can drones be used for environmental monitoring?

Ans. Drones can collect data on air quality, water quality, deforestation, and other environmental factors.

Q.32. What is the role of computer vision in drones?

Ans. Computer vision allows drones to interpret visual data from onboard cameras, enabling object detection, tracking, and mapping.

Q.33. Can drones be used for infrastructure inspection?

Ans. Yes, drones can inspect bridges, power lines, pipelines, and other infrastructure, reducing the need for manual inspections.

Q.34. How do autonomous vehicles handle complex intersections?

Ans. Advanced algorithms and communication between vehicles can enable safe and efficient passage through complex intersections.

Q.35. What are the primary challenges for widespread adoption of autonomous vehicles?

Ans. Legal and regulatory issues, technological challenges, public acceptance, and safety concerns are among the primary challenges.

Q.36. How do drones deal with windy conditions during flight?

Ans. Autonomous drones use advanced stabilization systems and flight control algorithms to handle gusty or windy conditions.

Q.37. What is the concept of swarm drones?

Ans. Swarm drones are a group of coordinated drones that can work together to achieve tasks such as search, surveillance, or mapping.

Q.38. What is the role of GPS in autonomous vehicles?

Ans. GPS provides precise location data, aiding in navigation, route planning, and overall vehicle positioning.

Q.39. How can autonomous vehicles enhance transportation for individuals with disabilities?

Ans. Autonomous vehicles can provide accessible transportation options for individuals who may have difficulty driving or using public transit.

Q.40. What is the role of remote sensing in drones?

Ans. Remote sensing capabilities, such as infrared or multispectral imaging, allow drones to collect valuable data for various applications.

Q.41. How do drones communicate with each other in a swarm?

Ans. Drones in a swarm can communicate using wireless protocols to share information and coordinate actions.

Q.42. What is the significance of redundancy in autonomous vehicles?

Ans. Redundancy, having multiple sensors or systems for critical functions, enhances the safety and reliability of autonomous vehicles.

Q.43. Can autonomous vehicles operate in mixed traffic with human-driven vehicles?

Ans. Yes, autonomous vehicles are being designed to safely interact with human drivers in mixed traffic environments.

Q.44. What role can drones play in monitoring and protecting natural resources?

Ans. Drones can monitor deforestation, wildlife habitats, water resources, and illegal activities in remote natural areas.

Q.45. How can autonomous vehicles reduce greenhouse gas emissions?

Ans. Autonomous vehicles can optimize driving patterns, leading to reduced fuel consumption and lower emissions.

Q.46. What are the challenges of integrating autonomous vehicles into existing transportation infrastructure?

Ans. Ensuring compatibility with current road systems, addressing legal liability, and establishing standard regulations are significant challenges.

Q.47. Can drones be used for 3D mapping?

Ans. Yes, drones equipped with lidar or photogrammetry systems can create detailed 3D maps for various applications.

Q.48. How do autonomous vehicles handle emergency situations?

Ans. Autonomous vehicles are programmed to detect emergencies and respond appropriately, which may involve pulling over, notifying authorities, or taking other necessary actions.

Q.49. What are the ethical considerations regarding autonomous vehicles and drones?

Ans. Ethical concerns include issues of safety, privacy, job displacement, decision-making algorithms, and potential misuse of technology.

Q.50. How can the public adapt to the presence of autonomous vehicles and drones on the roads and in the skies?

Ans. Education, public awareness campaigns, and clear communication about the benefits and limitations of these technologies are essential for public acceptance and adaptation.

Multiple Choice Questions and Answers

Q.1. What is the primary technology used for autonomous vehicles' perception of the environment?

(*a*) Radar (*b*) Sonar (*c*) GPS (*d*) Sensors

Q.2. What is the highest level of autonomy according to the SAE standards for autonomous vehicles?

(*a*) Level 3 (*b*) Level 4 (*c*) Level 5

(*d*) Full Automation

Q.3. Which of the following is a critical component for drones' precise location and navigation?

(*a*) LIDAR (*b*) Radar (*c*) GPS

(*d*) Inertial Measurement Unit (IMU)

Q.4. What is the primary use of autonomous drones in disaster response?

(*a*) Aerial photography (*b*) Search and rescue

(*c*) Agricultural monitoring (*d*) Package delivery

Q.5. How do autonomous vehicles handle lane-keeping and adaptive cruise control?

(*a*) AI-based decision-making (*b*) Human remote control

(*c*) Sensor-based systems (*d*) Satellite communication

Q.6. What is the primary sensor used in drones for altitude and distance measurement?

(*a*) Ultrasonic sensor (*b*) Infrared sensor (*c*) GPS (*d*) LIDAR

Q.7. Which of the following is an advantage of using autonomous drones in agriculture?

(*a*) Limited flight range (*b*) High labor cost

(*c*) Reduced water consumption (*d*) Inefficient energy use

Q.8. What role does machine learning play in autonomous vehicles' behavior?

(*a*) Providing real-time GPS updates (*b*) Enhancing battery life

(*c*) Decision-making and adapting to dynamic situations

(*d*) Remote control

Q.9. What type of drone is primarily used for capturing aerial imagery and videos?

(*a*) Fixed-wing drone (*b*) Quadcopter

(*c*) Racing drone (*d*) Tethered drone

Q.10. Which technology allows autonomous vehicles to communicate with each other to improve traffic flow and safety?

(*a*) Telepathy (*b*) AI decision-making

(*c*) Vehicle-to-vehicle (V2V) communication (*d*) GPS navigation

Q.11. What is the primary use of drones in surveying and mapping?

(*a*) Delivering packages (*b*) Aerial photography

(*c*) Traffic management (*d*) Creating accurate 3D maps

Q.12. What is the main challenge of using autonomous drones for long-range missions?

(*a*) High cost (*b*) Limited battery life

(*c*) Restricted airspace (*d*) Weather conditions

Q.13. Which technology is critical for drones' autonomous flight in confined indoor spaces?

(*a*) GPS (*b*) Sonar (*c*) Visual odometry (*d*) Radar

Q.14. How do autonomous vehicles ensure safety in emergency situations when the human driver needs to take control?

(*a*) By automatically contacting emergency services

(*b*) By slowing down and stopping gradually
(*c*) By providing clear visual and auditory warnings
(*d*) By requesting the human driver to take over

Q.15. What type of data is essential for drones in precision agriculture?
(*a*) Flight speed (*b*) Local weather forecast
(*c*) Soil moisture levels (*d*) All of the above

Q.16. What is the primary purpose of autonomous vehicles in logistics?
(*a*) Aerial photography (*b*) Last-mile delivery
(*c*) Human transportation (*d*) Weather monitoring

Q.17. What is the main advantage of using edge computing in autonomous vehicles?
(*a*) Longer battery life (*b*) Faster data transfer
(*c*) Reduced latency (*d*) Higher altitude

Q.18. What is the primary sensor used for obstacle detection in autonomous vehicles?
(*a*) LIDAR (*b*) Radar (*c*) GPS
(*d*) Ultrasonic sensors

Q.19. How can drones assist in wildlife conservation?
(*a*) By capturing wildlife photos for social media
(*b*) By participating in wildlife documentaries
(*c*) By monitoring animal populations, tracking poachers, and protecting habitats
(*d*) By creating virtual reality experiences for zoos

Q.20. What is the primary use of drones in construction and infrastructure inspection?
(*a*) Delivering building materials (*b*) Aerial surveying and monitoring
(*c*) Remote building design (*d*) Automated construction

Q.21. What type of communication technology is essential for real-time control of drones?
(*a*) Smoke signals (*b*) Bluetooth
(*c*) Wireless RF communication (*d*) Morse code

Q.22. Which technology helps autonomous drones maintain a stable hover?
(*a*) Wind prediction systems (*b*) GPS
(*c*) Inertial Measurement Unit (IMU) (*d*) Gyroscopes

Q.23. How do drones navigate autonomously to a specific location?
(*a*) Using manual joystick control (*b*) Following a predetermined flight path
(*c*) Contacting air traffic control (*d*) Autonomous decision-making

Q.24. What is the primary role of AI in autonomous drones?
(*a*) Piloting the drone (*b*) Capturing high-quality images
(*c*) Decision-making and object recognition (*d*) Charging the batteries

Q.25. What is the primary use of drones in the energy sector?
(*a*) Replacing traditional power sources (*b*) Powering remote areas
(*c*) Wind energy generation (*d*) Inspecting power lines and infrastructure

Q.26. Which sensor technology allows drones to create detailed 3D maps of the environment?

(*a*) Sonar
(*b*) GPS
(*c*) LIDAR
(*d*) Ultrasonic sensors

Q.27. What role can autonomous vehicles play in reducing traffic congestion?

(*a*) By increasing the number of vehicles on the road
(*b*) By optimizing routes and reducing stop-and-go traffic
(*c*) By enforcing stricter speed limits
(*d*) By eliminating the need for traffic lights

Q.28. How do drones avoid collisions with other objects during flight?

(*a*) By using radar to detect obstacles
(*b*) By sending out warning signals
(*c*) By using obstacle detection systems
(*d*) By relying on visual observation

Q.29. What is the primary advantage of using autonomous vehicles for public transportation?

(*a*) Faster travel times
(*b*) Reduced energy consumption
(*c*) Improved accessibility
(*d*) Lower maintenance costs

Q.30. What is the key challenge for drones operating in urban environments?

(*a*) Limited flight range
(*b*) Air traffic control restrictions
(*c*) Obstacle avoidance
(*d*) Battery charging locations

Q.31. What is the primary function of GPS in drones?

(*a*) Long-range communication
(*b*) Stabilizing flight
(*c*) Providing precise location data
(*d*) Collision avoidance

Q.32. What is the primary role of drones in the entertainment industry?

(*a*) Replacing traditional cameras
(*b*) Capturing aerial footage
(*c*) Providing live music
(*d*) Conducting live events

Q.33. How do drones improve the efficiency of search and rescue operations?

(*a*) By providing aerial views for news coverage
(*b*) By creating 3D models of the search area
(*c*) By quickly locating missing persons or disaster survivors
(*d*) By delivering food and water to search teams

Q.34. What is the primary technology used for autonomous vehicles' decision-making?

(*a*) Human intervention
(*b*) Onboard sensors and AI algorithms
(*c*) Satellite communication
(*d*) Remote control

Q.35. What is the primary use of drones in environmental monitoring?

(*a*) Measuring atmospheric pressure
(*b*) Tracking space weather
(*c*) Monitoring air quality, water quality, and wildlife habitats
(*d*) Forecasting earthquakes

Q.36. What is the role of edge computing in drones?

(*a*) Real-time data processing onboard the drone

(*b*) Long-range communication with ground control
(*c*) Charging the drone's batteries (*d*) Remote pilot training

Q.37. Which technology enables drones to detect obstacles and avoid collisions during flight?
(*a*) Satellite communication (*b*) Sonar
(*c*) LIDAR (*d*) All of the above

Q.38. What is the primary challenge for autonomous vehicles in handling unpredictable human drivers?
(*a*) Inability to navigate in unfamiliar environments
(*b*) Difficulty in identifying traffic signs
(*c*) Handling unpredictable behaviors of other drivers
(*d*) Lack of real-time weather updates

Q.39. What is the primary use of autonomous vehicles in ride-sharing services?
(*a*) Personal transportation (*b*) Delivering groceries
(*c*) Reducing traffic congestion (*d*) Providing on-demand transportation

Q.40. How do drones avoid collisions with birds and other flying objects?
(*a*) By flying at higher altitudes (*b*) By emitting ultrasonic signals
(*c*) By using collision avoidance algorithms (*d*) By relying on visual observation

Q.41. What is the primary role of machine learning in autonomous vehicles?
(*a*) Charging the vehicle's battery (*b*) Capturing high-definition videos
(*c*) Enhancing decision-making and adapting to different driving scenarios
(*d*) Sending real-time telemetry data

Q.42. What is the primary sensor used in drones for capturing images and videos?
(*a*) LIDAR (*b*) Infrared camera (*c*) GPS (*d*) Cameras

Q.43. What is the main challenge for autonomous drones in maintaining stable flight in turbulent weather conditions?
(*a*) Limited battery life (*b*) GPS signal interference
(*c*) Maintaining stability in windy conditions (*d*) Avoiding birds

Q.44. What technology allows drones to create 3D models of structures and terrain?
(*a*) GPS (*b*) Lidar
(*c*) Thermal imaging (*d*) Photogrammetry

Q.45. How can autonomous vehicles improve transportation for individuals with disabilities?
(*a*) By providing luxury features (*b*) By offering chauffeur services
(*c*) By increasing accessibility through self-driving features
(*d*) By offering fast travel

Q.46. What is the primary role of drones in monitoring and protecting natural resources?
(*a*) Monitoring endangered species (*b*) Measuring cosmic radiation
(*c*) Protecting water sources
(*d*) Monitoring deforestation and wildlife habitats

Q.47. What is the main advantage of using drones for aerial surveying and mapping?

(*a*) Reduced battery consumption
(*b*) High-speed flight
(*c*) Ability to capture detailed 3D maps
(*d*) Reduced cost compared to traditional surveying methods

Q.48. What type of communication technology is essential for remote control of autonomous drones?

(*a*) Satellite communication
(*b*) Morse code
(*c*) Wireless RF communication
(*d*) Infrared signals

Q.49. How do drones maintain stability and control during flight?

(*a*) Using remote control by a human operator
(*b*) Using onboard stabilization systems and flight control algorithms
(*c*) By following predefined flight paths
(*d*) By relying solely on GPS navigation

Q.50. What is the primary use of AI in drones?

(*a*) Capturing high-quality images
(*b*) Providing long-range communication
(*c*) Enhancing decision-making and object recognition
(*d*) Charging the batteries

Key to Multiple Choice Questions

1. (*d*)	**2.** (*b*)	**3.** (*c*)	**4.** (*b*)	**5.** (*c*)	**6.** (*d*)	**7.** (*c*)	**8.** (*c*)	**9.** (*b*)
10. (*c*)	**11.** (*d*)	**12.** (*b*)	**13.** (*c*)	**14.** (*d*)	**15.** (*d*)	**16.** (*b*)	**17.** (*c*)	**18.** (*d*)
19. (*c*)	**20.** (*b*)	**21.** (*c*)	**22.** (*d*)	**23.** (*b*)	**24.** (*c*)	**25.** (*d*)	**26.** (*c*)	**27.** (*b*)
28. (*c*)	**29.** (*c*)	**30.** (*c*)	**31.** (*c*)	**32.** (*b*)	**33.** (*c*)	**34.** (*b*)	**35.** (*c*)	**36.** (*a*)
37. (*d*)	**38.** (*c*)	**39.** (*d*)	**40.** (*c*)	**41.** (*c*)	**42.** (*d*)	**43.** (*c*)	**44.** (*d*)	**45.** (*c*)
46. (*d*)	**47.** (*c*)	**48.** (*c*)	**49.** (*b*)	**50.** (*c*)				

Project Works

1. **Autonomous Vehicle Simulation:**

   Project Idea: Develop a realistic simulation environment for testing autonomous vehicle algorithms. Use popular simulation tools (e.g., CARLA, Gazebo) and create challenging scenarios to evaluate the vehicle's perception, navigation, and decision-making capabilities.

   Outcome: A comprehensive simulation platform to assess the performance of autonomous vehicle control algorithms under various conditions.

2. **Obstacle Avoidance Drone:**

   Project Idea: Build a drone with obstacle detection and avoidance capabilities. Implement sensors (e.g., ultrasonic, Lidar) for detecting obstacles and develop algorithms to adjust the drone's flight path to avoid collisions.

   Outcome: A drone that can autonomously navigate through a predefined path while avoiding obstacles in real-time.

3. **Autonomous Vehicle Path Planning:**

   Project Idea: Design a path planning algorithm for autonomous vehicles in urban environments. Consider traffic, intersections, pedestrians, and other vehicles. Optimize the vehicle's route to minimize travel time and improve traffic flow.

   Outcome: An autonomous vehicle system capable of efficient and safe path planning in complex urban scenarios.

4. **Drone-based Remote Sensing:**

   Project Idea: Create a drone system to collect data for environmental monitoring. Use sensors (e.g., multispectral camera) to capture data on vegetation health, water quality, or pollution levels in a specific area.

   Outcome: A drone-based solution for environmental data collection, with potential applications in agriculture, conservation, or pollution assessment.

5. **Autonomous Delivery Robot:**

   Project Idea: Develop a small autonomous ground robot capable of navigating sidewalks and delivering packages to specified locations. Implement path planning, obstacle avoidance, and package handling mechanisms. **Outcome:** An autonomous robot that can perform last-mile package delivery, demonstrating efficiency and reliability.

6. **Traffic Management with Drones:**

   Project Idea: Design a drone-based system for monitoring traffic congestion in a city. Deploy a fleet of drones equipped with cameras to capture real-time traffic data and analyze it to identify congestion hotspots.

   Outcome: A traffic management solution that provides accurate and up-to-date traffic information to help optimize urban transportation.

7. **Autonomous Vehicle Safety Assessment:**

   Project Idea: Develop a safety assessment framework for autonomous vehicles. Define metrics to evaluate the vehicle's ability to detect and respond to potential hazards, emergency braking, and collision avoidance.

   Outcome: A standardized safety evaluation method for autonomous vehicle manufacturers to demonstrate their vehicles' safety capabilities.

8. **Drone-based Search and Rescue:**

   Project Idea: Build a drone system for search and rescue operations. Design algorithms to navigate through complex environments and identify missing persons or objects using thermal imaging or other sensors.

   Outcome: A drone system capable of assisting search and rescue teams by locating missing individuals or providing real-time situational awareness in challenging terrains.

9. **Autonomous Vehicle Data Analysis:**

   Project Idea: Analyze real-world autonomous vehicle data to understand driving patterns, behavior, and traffic interactions. Identify trends, safety metrics, and areas for improvement in autonomous vehicle operations.

   Outcome: Insights and recommendations for enhancing autonomous vehicle algorithms and optimizing traffic management.

10. Drone-based Aerial Mapping:

Project Idea: Create a drone-based mapping solution for generating high-resolution 3D maps of areas of interest. Integrate photogrammetry techniques, GPS, and gimbal-stabilized cameras to capture detailed aerial maps.

Outcome: A mapping system capable of producing accurate and up-to-date 3D models of landscapes, urban areas, or construction sites, with applications in urban planning, land surveying, or construction monitoring.

4 Robotics in Healthcare

4.1 INTRODUCTION

The use of robotic technology to several facets of medical care, diagnosis, treatment, and rehabilitation constitutes the fast evolving discipline of robotics in healthcare. By increasing the accuracy, effectiveness, and safety of medical treatments, better patient outcomes, and reducing the workload on healthcare personnel, it has the potential to revolutionize healthcare. There are several uses for using robots in the medical field, each of which addresses certain opportunities and challenges. Striking a balance between the capabilities of robotic technology and the human expertise and empathy that are necessary in providing complete healthcare services is crucial as the field of robotics in healthcare continues to develop. Realizing the full potential of robotics to improve healthcare outcomes would require close cooperation between medical practitioners, engineers, researchers, and politicians.

4.2 KEY AREAS WHERE ROBOTICS MAKING SIGNIFICANT CONTRIBUTIONS

Some key areas where robotics is making significant contributions include:

1. **Surgery:** With improved precision and less trauma to the patient, minimally invasive operations are made possible by surgical robots, which are frequently operated by qualified surgeons. These robots can offer better dexterity and visualization, improving patient outcomes and speeding up rehabilitation.
2. **Medical Imaging and Diagnosis:** Robots can help with medical imaging procedures, helping radiologists take high-quality pictures and with the interpretation of complex data. This may result in quicker and more precise diagnoses.
3. **Rehabilitation:** To assist patients in regaining mobility and strength following accidents or operations, robotic devices are employed in physical and occupational rehabilitation. These tools can deliver repeating exercises that are personalized, which is beneficial for rehabilitation.
4. **Drug Delivery:** In a medical setting, autonomous robots can dispense medications, eliminating the requirement for manual transportation and lowering the possibility of mistakes.

5. **Telemedicine:** Even when the patient and the healthcare professional are in separate locations, robots can enable remote consultations and medical procedures by providing a physical presence for the provider.
6. **Support for the Elderly and Disabled:** Robots can help the elderly or those with restricted mobility by helping with everyday duties, keeping an eye on vital signs, and providing company.
7. **Laboratory Automation:** By managing repetitive duties in laboratories including handling dangerous materials, preparing samples, and running experiments, robots can boost efficiency and accuracy.

Although the potential applications of robotics in healthcare are exciting, there are still a number of obstacles that must be overcome, such as governmental permission, integration with current healthcare systems, patient privacy issues, and the requirement for adequate training for healthcare personnel.

4.3 SURGICAL ROBOTS

Sophisticated medical tools called surgical robots are made to help surgeons with a variety of surgical procedures. They improve precision, dexterity, and visualization, enabling surgeons to carry out minimally invasive procedures more precisely and with better patient outcomes. The da Vinci Surgical System is a prominent illustration of a surgical robot.

4.3.1 Technical Aspects and Principles behind Surgical Robots Design

Surgical robots typically consist of several key components:

Robotic Arms

These are mechanical arms with several joints that are fitted with equipment for surgery. They mimic the surgeon's hand movements, but with less trembling and a wider range of motion. There can be a range in the number of arms, and some systems also have a camera arm for visualization.

Robotic Arm Design:

1. **Degrees of Freedom (DOF):** The amount of autonomous movements that a robotic arm is capable of executing is referred to as its degrees of freedom (DOF). To mimic the dexterity of a human hand, surgical robotic arms frequently include several degrees of freedom (DOFs). For instance, a typical system might include 6 DOFs, enabling rotations around each axis (roll, pitch, and yaw) as well as movements in three dimensions (x, y, and z).
2. **Joints and Links:** A combination of joints and links is used to produce each DOF. Revolute joints, which permit rotational motion, and prismatic joints, which permit linear motion, are examples of common joint types. To achieve the appropriate range of motion, these joints are joined together by rigid or flexible linkages.
3. **End Effectors:** The robotic arm's end effectors are the components that make contact with the surgical site. It keeps the cameras, instruments, and other supplies required for the surgery. Depending on the particular surgical application, the end effector's design will vary.

Mathematics and Kinematics:

1. **Kinematics:** Kinematics is the area of mathematics that studies how objects move. Kinematics is crucial in the setting of robotic arms for comprehending the position and orientation of the arm in relation to the joint angles. Homogeneous coordinates, rotation matrices, and transformation matrices are used to represent the kinematics of a robotic arm mathematically.

2. **Forward Kinematics:** Using the joint angles, this method establishes the location and orientation of the end effector. Matrix-based transformations of coordinates from one frame of reference, such as the robot's base, to another, such as the end effector, are used in forward kinematics equations.
3. **Inverse Kinematics:** This technique determines the joint angles necessary to create a particular end effector position and orientation. It is a more challenging issue than forward kinematics and frequently calls for the solution of systems of equations. Inverse kinematics problems must be solved in surgical robotics in order to control the robot's movement precisely.
4. **Path Planning:** Algorithms for path planning determine the best route for the robotic arm to take between two positions while avoiding hazards and guaranteeing safety. These algorithms frequently draw on ideas from computer science, robotics, and optimization.

Advanced control algorithms, real-time feedback, and sensor data are used by surgical robots to enable precise and secure movement of the robotic arms during surgery. These arms' design, control, and operation are greatly influenced by the mathematical underpinnings of robotics, which results in better surgical results and less invasive treatments.

Console

A console that gives the surgeon a 3D picture of the surgical site and lets him or her control the robotic arms is used to control the robot. The console of a surgical robot is an essential part that gives the surgeon control over the robotic arms and allows for surgery. Through its user interface, the surgeon can view the surgical site, control the robotic arms, and access a number of tools and features. Let's explore the console's technical aspects, including its design ideas, important components, and the use of mathematics.

Console Design:

1. The console has an intuitive user interface that enables the surgeon to communicate with the robotic device. It has screens for viewing the surgical site in real time, controllers for moving the robotic arms, and buttons or touchscreens for accessing other features.
2. The console is ergonomically comfortable for the surgeon to utilise for extended periods of time. To reduce fatigue, armrests, adjustable height, and comfortable seats are crucial.
3. **Controllers:** To mimic the movements of the robotic arms, the console may have haptic or tactile controllers. This might involve joysticks, pedals, or other tools that let the surgeon manipulate the robotic instruments naturally.
4. **Display:** The surgeon can see the surgical site clearly and in detail thanks to high definition monitors. For displaying various viewpoints, such as the endoscopic camera feed, 3D models, and vital signs, the console may include many screens.

Mathematics and Control:

1. **Coordinate Systems:** Surgical robotics primarily depends on coordinate system-related mathematical ideas. These ideas are used by the console to convert the surgeon's instructions into robotic arm movements. For precise control, it is crucial to comprehend the connections between several coordinate systems, such as those of the console and the robotic arm.
2. **Inverse Kinematics:** As was already noted, inverse kinematics is essential for controlling robotic arms. The console must figure out the joint angles necessary to manoeuvre the robotic arms inside the patient's body at the proper position and orientation. To gain accurate control, this involves solving challenging mathematical problems.

3. **Visualization:** Visual data on the console's displays is processed and rendered using mathematics. Filtering, interpolation, and 3D rendering are just a few of the image processing methods utilized to improve the endoscopic camera feed and give the surgeon a clear perspective.
4. **Sensors and Input:** The robotic system frequently provides the console with real-time input, including position data, force feedback, and other sensor data. In addition to giving the surgeon a sensation of touch (haptic feedback) and spatial awareness, mathematics is employed to analyze this data.

Endoscopic Camera

This high-definition camera offers live pictures of the operating room. It is fixed to one of the robotic arms and offers the surgeon a close-up, magnified view. An essential part of surgical robots is the endoscopic camera, which gives the surgeon real-time visual input during minimally invasive treatments. It enables precise navigation and instrument manipulation by enabling the surgeon to look into the patient's body without the need for big incisions.

Endoscopic Camera Design:

1. **Optics:** To take crisp, clear pictures of the surgery site, the endoscopic camera uses high-quality optical components, such as lenses and image sensors. The optics guarantee low distortion, excellent resolution, and suitable illumination.
2. **Light Source:** Clear visualization requires adequate illumination. To illuminate the surgical region, a light source is built within the camera, frequently using fiber-optic wires. This gives the surgeon a bright view even inside the body's small spaces.
3. **Sterilization:** To adhere to surgical standards, the camera and related equipment must be sterilizable. To avoid infection and guarantee patient safety, this is crucial.
4. **Small Form Factor:** Depending on the individual surgical technique, the camera's size is intended to fit through small incisions that are typically between a few millimeters and a centimeter in diameter.

Mathematics and Image Processing:

1. **Image Acquisition:** The endoscopic camera captures a series of images or frames during the procedure. The concept of frame rate (frames per second) is important, as it determines the smoothness of the visual feed and the ability to capture fast movements.
2. **Image Transformation:** The captured images are often transformed to correct for lens distortion, adjust brightness and contrast, and align with other data sources, such as 3D models or tracking information. These transformations involve mathematical operations.
3. **Spatial Calibration:** The endoscopic camera's image needs to be spatially calibrated in order to appropriately reflect the operative location. Mathematical methods are used in this calibration to translate the 2D image to the 3D surgical space.
4. **Image Stitching:** To create a panoramic view, often many cameras or photos need to be stitched together. To align and merge the photos seamlessly, complicated mathematical procedures are used in this process.
5. **Image Enhancement:** To improve the quality of the collected photographs, image processing techniques including filters, edge detection, and noise reduction are applied. Algorithms and mathematical processes may be used in these strategies.

6. **Visual Overlay:** The endoscopic camera may overlay extra information such as surgical instrument tracking, notes, or vital signs onto the video feed. The precise positioning of these overlays requires spatial calculations.

With the help of the endoscopic camera's technical features, sophisticated image processing, and spatial calibration, the surgeon is able to perform precise and less invasive surgeries while having access to vital visual information.

Surgical Instruments

The surgical instruments on the robotic arms are designed for precise movements including suturing, cauterizing, and cutting tissue.

Robotic Control: The surgeon operates the surgical robots from a console. The robotic arms transform the movements of the surgeon into exact actions. Safety safeguards are also included into the system to stop any jerky or rapid motions that would endanger the patient.

***Mathematics*:** Mathematics plays a crucial role in the design and operation of surgical robots:

(*a*) **Kinematics:** This area of mathematics studies how objects move. Kinematics is a branch of mechanics that studies how the motions of the hands of the surgeon at the control console are converted into movements of the robotic arms within the patient's body.

(*b*) **Coordinate Systems:** To track the location and orientation of the robotic arms and instruments in respect to the anatomy of the patient, surgical robots use coordinate systems.

(*c*) **Image Processing:** To process the visual data from the endoscopic camera and give the surgeon a good picture of the surgical site, image analysis techniques are frequently based on mathematical algorithms.

(*d*) **Path Planning:** Path planning is the process of determining the best route for robotic tools to take while taking into account the anatomy of the patient and the specific surgical procedure.

Safety, accuracy, and the capability to carry out difficult operations are all highly prioritized throughout the creation of surgical robots. They are utilized in a range of operations, including orthopedic surgery, heart surgery, and minimally invasive surgeries (like laparoscopic procedures). Robotics and medical knowledge are increasingly being combined, leading to better surgical methods and patient care.

4.4 REHABILITATION ROBOTICS

Rehabilitation robotics is a specialized subject that focuses on using robotic equipment and technologies to help people regain or enhance their physical capabilities after suffering an injury, undergoing surgery, or experiencing neurological problems. These tools can deliver focused, repeating exercises, help with mobility, and give patients and therapists real-time feedback.

4.4.1 Components of Rehabilitation Robotics

***Robotic Exoskeletons*:** These are limb-moving aids for patients that are wearable robotic devices. Exoskeletons can be created for the upper or lower limbs and offer aid, resistance, or support during particular actions. Robotic exoskeletons are wearable gadgets made to help and improve people's physical capabilities, especially in the context of rehabilitation. They can be used to increase mobility, assist or supplement limb movements, and offer therapeutic exercises. Let's look at some of the technological features of robotic exoskeletons and associated mathematical ideas.

Components of Robotic Exoskeletons:

1. **Mechanical Structure:** The mechanical structure of the exoskeleton consists of rigid frames, joints, and linkages that resemble human joints and support the user's limbs.
2. **Actuators:** These are in charge of producing the required torques and forces to support or enhance the user's movements. Pneumatic, hydraulic, electric motors, and even intelligent materials like shape memory alloys can be used as actuators.
3. **Sensors:** Sensors are frequently used in exoskeletons to track user movements, joint angles, and forces. For the exoskeleton to be able to be controlled and customised to the demands of the user, real-time feedback is essential.
4. **Control Algorithms:** Exoskeletons rely heavily on control systems. They decide when and how to support the user's motions using sensor data. This calls for quite complex algorithms.

Mathematics in Robotic Exoskeletons:

1. **Kinematics:** Complex mechanical systems with numerous degrees of freedom (DOF) are used in exoskeletons. The link between joint angles and the resulting end-effector positions can be modeled using kinematics. Different coordinate frames are transformed using mathematical equations.
2. **Dynamics:** For safety and helpful support, it's essential to comprehend the exoskeleton's dynamics. Dynamic equations aid in foreseeing how the movements of the exoskeleton will impact the user and vice versa. Understanding of Newtonian mechanics and the laws of motion are necessary for this.
3. **Control Theory:** Based on sensor data, the exoskeleton's behaviors are modified using proportional-integral-derivative (PID) controllers, adaptive control, and other control strategies. The forces that the actuators apply are controlled by these controllers using mathematical principles.
4. **Signal Processing:** Sensor data about joint angles, accelerations, and forces are continuously available. The data is processed using signal processing techniques, including as Fourier analysis, filtering, and feature extraction, to extract pertinent data for the control algorithms.
5. **Trajectory Planning:** Trajectory planning ensures smooth and effective routes while directing the user's limb via a certain movement. The desired trajectories can be produced using interpolation techniques such as cubic splines or other algorithms.
6. **Inverse Kinematics:** Equations for inverse kinematics can be used to determine the joint angles necessary to get the end-effector position that is sought. This is necessary in order to manage the exoskeleton's motions.
7. **Optimization:** In some situations, optimization techniques are utilized to identify the best actuator forces (control inputs) that accomplish a particular objective, such as minimizing

Robotic exoskeletons can effectively aid and rehabilitate people with mobility impairments or those undergoing physical therapy thanks to a mix of mechanical design, sensors, actuators, and control algorithms that are all based on mathematical concepts.

4.4.2 Robotic Prosthetics

Robotic limbs give people movement and the capacity to carry out daily tasks by replacing missing or dysfunctional limbs.

1. **Functional Electrical Stimulation (FES):** FES includes stimulating muscles with electrical currents to assist people regain their muscle control and strength.
2. **Sensor Systems:** Rehabilitation tools frequently have sensors built in to track patient motion, muscle activation, and other physiological data. These sensors offer insightful feedback that helps the rehabilitation program be adjusted.
3. **Control Algorithms:** These are crucial for synchronizing the robotic device's activities with the patient's movements, assuring efficiency and safety.

Mathematics in Rehabilitation Robotics:

1. **Kinematics and Dynamics:** Rehabilitation robots, like surgical robots, need to understand mathematical concepts related to kinematics, which is the study of motion, and dynamics, which is the study of forces and torques. These ideas support the design of the robotic devices' range of motion and mobility.
2. **Biomechanics:** For rehabilitation, it is essential to comprehend the biomechanics of human movement. Robotic systems that replicate natural movements are created using mathematical models of the human musculoskeletal system.
3. **Feedback Control:** To adapt to the user's motions and offer the proper support or resistance, rehabilitation robots frequently use feedback control systems. To provide smooth and precise control, control theory, particularly proportional-integral-derivative (PID) controllers, is used.
4. **Signal Processing:** Fourier analysis, digital filtering, and feature extraction are examples of signal processing techniques that are used to process data from sensors (such as accelerometers, gyroscopes, and electromyography).
5. **Path Planning:** Path planning algorithms for robotic exoskeletons assist in determining the best trajectory for moving the patient's limb while avoiding impediments or going beyond joint restrictions.
6. **Machine Learning:** In some circumstances, machine learning algorithms can be used to modify the rehab program in accordance with the patient's development and specific requirements.

Physical therapy and neuro rehabilitation are being advanced by the use of rehabilitation robots, which offers patients more focused and intensive therapy, which may result in a quicker and more complete recovery.

4.5 TELEMEDICINE

The practice of telemedicine involves providing medical care remotely through the use of communication technology, frequently the internet. It enables medical professionals to diagnose, treat, and keep an eye on patients remotely. Even while it might not necessitate complex mathematical calculations, telemedicine has a number of technical features, such as the technologies employed, data security, and service quality. Let's explore these aspects in more detail.

4.5.1 Components of Telemedicine

1. **Real-Time Conference:** Real-time video conferencing between the patient and the healthcare professional is one of the main telemedicine techniques. Accurate visual assessment requires high-quality video streams.
2. **Data Transmission:** Reliable and secure data transmission over the internet is essential to telemedicine. This covers video data, patient information, and other health data. To guarantee a seamless connection, data compression and effective transmission protocols can be crucial.

3. **Audio:** For telemedicine consultations, crystal-clear audio communication is essential. To improve audio quality, algorithms for noise cancellation and echo cancellation can be used.
4. **Medical Devices:** In some circumstances, telemedicine entails the use of distant medical tools like wearable health monitors or electronic stethoscopes or otoscopes. The healthcare provider receives medical data from these gadgets.
5. **Electronic Health Records (EHRs):** It is essential to have access to patient medical records while consulting via telemedicine. To enable the practitioner to review the patient's history, EHR systems and the telemedicine platform need to be smoothly connected.
6. **Security:** The confidentiality and security of patient data are essential. It is crucial to use encryption, secure authentication, and to abide by laws governing the safeguarding of medical data, such as HIPAA in the US.

Mathematics in Telemedicine: While mathematics may not be the primary focus of telemedicine, there are several areas where mathematical concepts are relevant:

1. **Bandwidth and Data Transfer:** It's essential to check that the internet connection has enough bandwidth to support the data transfer needs of video and audio. Providers must take into account the data rates for video compression as well as the available bandwidth.
2. **Quality of Service (QoS):** QoS measurements including delay (latency), jitter, and packet loss can affect how well telemedicine consultations are conducted. To optimize network settings for real-time communication, network engineers utilize mathematical models.
3. **Encryption:** Contemporary encryption methods, such RSA or AES, are founded on mathematical ideas. The security of patient data during transmission depends on your understanding of encryption.
4. **Medical Data Analysis:** Statistical analysis and algorithms may be utilized to find trends or anomalies in situations when telemedicine calls for the analysis of medical data (such as remote monitoring of vital signs).
5. **Resource Allocation:** To optimally distribute resources across numerous concurrent consultations in a telemedicine system with constrained resources (such as server capacity), mathematical optimization methods may be applied.

Even though telemedicine mostly relies on technological know-how and medical competence, some technical decisions—such as those regarding network optimization, data security, and resource allocation—are supported by mathematical principles.

4.6 CONCEPT OF HUMAN-ROBOT COLLABORATION

The integration of humans and robots in a shared workspace where they cooperate to accomplish shared objectives is known as human-robot collaboration (HRC). This idea goes beyond conventional robotics, which frequently involves robots working alone, and instead emphasises developing a synergy between human abilities, knowledge, and flexibility and the accuracy, sturdiness, and effectiveness of robots. A growing number of industries, including manufacturing, healthcare, agriculture, logistics, and others are exploring HRC. Let's look at the main components of human-robot cooperation:

1. **Types of Human-Robot Collaboration:**
   (*a*) **Collaboration:** Humans and robots coexist, frequently carrying out different but complementary jobs. For instance, a robot might undertake heavy lifting or repetitive jobs in manufacturing while a person might handle complex assembly tasks.

(*b*) **Coordination:** In order to accomplish a common goal, humans and robots coordinate their operations. Real-time planning and communication are frequently involved in this. For example, humans and robots can work together in a warehouse to streamline the order picking and packing procedures.

(*c*) Cobots (Collaborative Robots) are robots that have been especially created to work closely and safely with people in a shared office. To detect human presence and lower the risk of injury, they frequently contain features like force-sensing technology. Cobots are employed in jobs that require human supervision or skill.

2. **Benefits of Human-Robot Collaboration:**

(*a*) **Enhanced Productivity:** Robots are highly efficient at repetitive and precise operations. Humans can concentrate on making complex decisions, solving problems, and being creative.

(*b*) **Safety:** Because collaborative robots can handle physically taxing or dangerous activities, the risk of harm to human workers is reduced. Robots can work with people in a safe environment thanks to safety features like force restriction.

(*c*) **Flexibility:** Human workers are flexible and provide context awareness, critical thinking, and flexibility to jobs. Robots may be designed to do a variety of jobs, which makes it simpler to respond to shifting demands.

(*d*) **Skill Development:** Both humans and robots can be designed to learn from the activities of others. Both human workers and robotic systems can benefit from this mutual learning in terms of their skills.

3. **Challenges and Considerations:**

(*a*) **Safety:** Protecting human employees from harm is of utmost importance. Robots used in collaboration must be built to reduce the possibility of crashes, recognise human presence, and act appropriately.

(*b*) **User Interface:** Human-robot interactions should be simple and straightforward. Humans should be able to interact with and programme robots with ease thanks to interfaces.

(*c*) **Ethical and Social Implications:** The introduction of robots into workplaces raises ethical concerns regarding job displacement, data privacy, and the possible effects on human civilization.

4. **Applications of Human-Robot Collaboration:**

(*a*) **Manufacturing:** In assembly lines, handling materials, welding, and other tasks, robots operate alongside human workers.

(*b*) **Healthcare:** In hospitals, robots dispense medications, do physical rehabilitation, and help surgeons with delicate operations.

(*c*) **Logistics and Warehousing:** Collaborative robots improve inventory control, order fulfillment, and package handling.

(*d*) In agriculture, robots help in activities like planting, harvesting, and crop health monitoring.

Innovations in robotics, artificial intelligence, and human-machine interaction are all driving changes in the idea of human-robot collaboration. We may anticipate more sophisticated and seamless human robot collaboration across a range of fields as technology advances.

4.7 ROBOTS WORKING ALONGSIDE HUMANS IN VARIOUS DOMAINS

Robots may collaborate with people in a variety of fields by using their specialized abilities to supplement human abilities and increase overall efficiency. The safety, task distribution, communication, and adaptability aspects of this collaboration must be carefully taken into account. Here are a few instances of how robots can successfully collaborate with people in many fields:

1. **Manufacturing:**

   Cooperative Assembly: Humans can concentrate on jobs that call for dexterity, decision-making, and quality control while robots undertake repetitive work, precision assembly, and heavy lifting.

   Collaborative Welding: Human welders perform inspection and intricate, non-standard welds while robots can execute automated welding.

2. **Healthcare:**

   Surgical Support: Robots can support surgeons during minimally invasive procedures, improving accuracy and minimizing invasiveness. Human surgeons maintain discretion and autonomy.

   Rehabilitation: Robotic equipment can help with physical treatment and rehabilitation by offering dependable, individualized workouts under human supervision.

3. **Logistics and Warehousing:**

   Order Fulfillment: While people manage inventory, handle exceptions, and do quality checks, robots can explore warehouses to pick and pack things.

   Material Handling: While human workers supervise logistics operations and handle difficult jobs, robots can deliver big loads or dangerous goods.

4. **Agriculture:**

   Harvesting: Robots can do harvesting duties like picking fruits or vegetables, freeing up human personnel to oversee operations and deal with specific problems.

   Monitoring: Drones and robotic systems can keep an eye on crop health and spot anomalies, giving human farmers important information they may use to make wise choices.

5. **Construction:**

   Precise Activities: While people supervise project planning and handle difficult or specialized activities, robots may conduct precise construction operations in controlled surroundings, such as bricklaying or welding.

   Risky Construction Sites: Robots can be used at risky construction sites to lower worker dangers.

6. **Customer Service:**

   Chatbots: Automated systems with the ability to understand natural language may answer common customer questions and give rudimentary information, freeing up human agents to tackle more sophisticated conversations.

   Data Analysis: To support human marketers and customer service teams, robots can process enormous volumes of customer data to find trends and preferences.

7. **Education and Research:**

   Assistive Teaching: Robots can support instructors by delivering individualised lesson plans, assisting students with practise exercises, and keeping track of advancement.

Assistance in the Laboratory: Robots can automate tedious lab procedures, freeing researchers to concentrate on the design of experiments and data processing.

Successful human-robot collaboration in each discipline depends on elements like precise task demarcation, effective human-robot communication, safety procedures, user-friendly interfaces, on-going training, and monitoring. The possibility for humans and robots to collaborate in a variety of industries is growing as technology progresses, which will raise productivity, increase safety, and enhance human capabilities.

4.8 ROBOTS ENHANCING PRODUCTIVITY

By doing jobs more quickly, precisely, and consistently than human workers alone, robots have the potential to greatly increase productivity in a variety of industries. The following are some ways that robots can increase productivity:

1. **Repetitive and Labor-Intensive Tasks:** Robots are excellent at completing repetitive jobs without becoming tired or losing accuracy. Robots, for instance, may conduct assembly, welding, and material handling activities with consistency in the manufacturing industry, increasing output rates and lowering human error.
2. **High-Speed Operations:** Robots can move quickly, which makes them ideal for activities that call for quick and accurate movements. Robots can considerably increase productivity in sectors like electronics production, where accuracy is essential and cycle durations are short.
3. **24/7 Operations:** Robots can work continually, unlike humans who need pauses and rest. This enables continuous production, which can be useful in markets with strong demand and constrained deadlines.
4. **Reduction of Manual Labor:** Robots free up human workers to concentrate on more difficult and valuable jobs by automating manual labour. Human workers may experience increased job satisfaction and productivity as a result of this.
5. **Consistent Quality:** Robots can routinely do jobs with great accuracy, improving the quality of the final product and reducing faults. Thus, the need for rework is diminished, and customer satisfaction is increased.
6. **Dangerous or Hazardous Environments:** Robots can be used in situations when humans are at risk, such as while handling harmful chemicals, operating in intense heat, or negotiating hazardous terrain. By doing this, productivity is maintained under difficult circumstances while ensuring human safety.
7. **Data Analysis and Decision Support:** Artificially intelligent (AI)-enhanced robots can quickly process enormous volumes of data and produce insights that can aid in decision-making. AI-driven analytics can enhance productivity in industries like finance, marketing, and logistics.
8. **Flexibility and Scalability:** To adjust to shifting production demands, robots can have their programming or configuration changed. This adaptability enables producers to quickly modify production processes, launch new goods, or meet changing customer demands.
9. **Precision and Accuracy:** Robots can perform better than humans at highly precise jobs like microsurgery. In fields like aerospace, electronics, and medical devices, this precision is crucial.
10. **Collaborative Work with Humans:** Cobots (collaborative robots) can aid human workers in jobs that call for strength or dexterity while increasing their abilities. Overall productivity may increase as a result of this teamwork.

By utilizing robotic technology in this way, businesses may increase output, save costs, improve safety, and boost competition, all of which contribute to increased overall productivity.

4.9 HUMAN-ROBOT COLLABORATION AND SAFETY

Safety must come first in human-robot collaboration (HRC) in order to guarantee the welfare of both people and robots in co-work spaces. Implementing safety precautions, design considerations, and processes that reduce the chance of mishaps or injuries is crucial when people and robots collaborate. Key elements of safety in human-robot collaboration are listed below:

1. **Risk Assessment:** Conduct a thorough risk analysis to determine potential dangers and weigh the dangers of human-robot interactions prior to deploying HRC. The tasks carried out, the surroundings, the pace and force of the robotic motions and the presence of human workers should all be taken into account during this evaluation.
2. **Safe Physical Interaction:** Utilize collaborative robots (cobots), which are made to work safely alongside people. Cobots have sensors for force and torque, which enable them to recognize human presence and lessen their force in the event of contact. The possibility of harm is reduced by this "soft" encounter.
3. **Physical Barriers and Zones:** Create physical barriers or safety zones to keep people and robots apart when direct connection is not necessary. These barriers maintain a safe distance and stop inadvertent touch.
4. **Emergency Stop:** Make sure that emergency stop buttons that immediately cease robotic operations are accessible to both human workers and robots. In the event of unforeseen situations, this feature enables prompt intervention.
5. **Training and Education:** Provide enough training to human employees on how to utilise robots safely, including knowledge of the robot's capabilities, constraints, and emergency protocols. The safe handling of encounters with robots should be covered in training.
6. **Risk Mitigation:** Use risk-reduction techniques including slowing down robotic motions when people are nearby, wearing safety gear (such safety gloves), and creating task-specific safety protocols.
7. **Regular Inspections and Maintenance:** Make sure robotic systems, including their sensors, actuators, and safety precautions, are functioning properly on a regular basis. To avoid unplanned problems, routine maintenance and calibration are crucial.
8. **Compliance with Regulations:** Follow all applicable safety rules and guidelines that are relevant to collaborating between humans and robots, such as ISO 10218-1 and ISO 10218-2 for industrial robots and ISO/TS 15066 for collaborative robots.
9. **Risk Communication:** Employ tactics for effective communication to tell human workers about the presence of robots, the duties they are supposed to perform, and any potential risks. The presence of robots can be detected by signs and other visual clues.
10. **Continuous Monitoring and Improvement:** Regularly evaluate the effectiveness of safety precautions, get employee feedback, and make the required adjustments to increase safety in the context of human-robot collaboration.

When humans and robots work together, putting safety first guarantees a safe and productive workplace and builds trust between them as they cooperate to accomplish shared objectives.

4.10 REVOLUTIONARY IMPACT OF ROBOTICS IN THE HEALTHCARE SECTOR

The integration of robotic technology has significantly advanced medical treatment, surgery, diagnostics, rehabilitation, and patient assistance, having a revolutionary impact on the healthcare sector. Here are some crucial areas in healthcare where robotics has had a profound impact:

1. **Minimally Invasive Surgery:** With the use of robotic surgery, such as the da Vinci Surgical System, surgeons may complete intricate procedures with more accuracy, less invasiveness, and smaller incisions. As a result, recuperation times are shortened, pain is decreased, and the risk of complications is decreased.
2. **Precision Medicine:** Targeted medicine delivery, individualized therapy, and treatment regimens based on unique patient data are all made possible by robotics. This strategy lessens side effects while improving therapeutic results.
3. **Telemedicine and Remote Monitoring:** Remote consultations are made possible by robots and telemedicine platforms, enabling medical experts to treat patients in far-off locations or in times of emergency. Tracking vital signs and sending real-time information to healthcare professionals is made possible by remote monitoring systems, which frequently use robots.
4. **Rehabilitation and Assistive Devices:** Robotic exoskeletons and other assistive technology can help people with mobility issues regain their strength, mobility, and independence. In physical therapy and rehabilitation programs, several gadgets are employed.
5. **Laboratory Automation:** In laboratories, robotics automates labor-intensive and repetitive operations like sample handling, testing, and high-throughput analysis. This expedites analysis, enhances precision, and frees up scientists to work on more imaginative projects.
6. **Drug Discovery and Development:** Robots can automate many stages of the drug development process, including chemical production and high-throughput screening. By accelerating the drug development process, it becomes more effective and economical.
7. **Elderly Care:** Social robots and robotic companions help seniors keep their independence and lessen the load on careers by offering support and companionship, monitoring their health, and helping with everyday tasks.
8. **Diagnostics and Imaging:** Diagnostic imaging methods are improved by robotic systems, such as robotic-assisted ultrasonography or automated image processing in medicine. This results in earlier disease identification and more precise diagnostics.
9. **Mental Health Support:** In order to assist mental health, resources, counseling, and early mental health issue detection are offered via robots and AI-driven chatbots.
10. **Drug Dispensing and Pharmacy Automation:** Robotic systems are effective at managing the delivery of medications, cutting down on mistakes, ensuring accurate dosages, and streamlining pharmacy processes.

Robotics' transformational effects on healthcare go beyond just bettering patient outcomes. It also covers issues with healthcare process optimization, access to care in underprivileged areas, and issues with ageing populations. The combination of robotics and AI will probably lead to more innovation and breakthroughs in the healthcare industry as technology develops.

Numerical Problems with Solutions

Calculating Surgical Time Savings

Problem 1. A hospital introduced a robotic-assisted surgical system for a specific procedure. On average, the traditional open surgery for this procedure takes 4 hours. The robotic-assisted surgery for the same procedure takes 2.5 hours. Calculate the time saved per surgery using the robotic system.

Solution: Time saved per surgery = Traditional surgery time – Robotic surgery time Time saved per surgery = 4 hours – 2.5 hours = 1.5 hours

Result: The robotic-assisted surgery saves 1.5 hours per procedure compared to traditional open surgery.

Cost Comparison of Robotic Surgery

Problem 2. Hospital A and Hospital B are considering adopting robotic-assisted surgery systems. Hospital A's system costs $1.2 million, while Hospital B's system costs $1.8 million. Both hospitals expect to perform 250 surgeries annually using the robotic system. Calculate the cost per surgery for each hospital.

Solution: Cost per surgery = Total system cost/Number of surgeries Cost per surgery (Hospital A = $1,200,000/250 = $4,800 Cost per surgery (Hospital B = $1,800,000/250 = $7,200

Result: The cost per surgery is $4,800 for Hospital A and $7,200 for Hospital B.

Rehabilitative Robot Usage

Problem 3. A rehabilitation center has a robotic exoskeleton for assisting patients with mobility impairments. The exoskeleton is used for 3 hours per day, and it operates at an average power consumption of 1000 Watts. Calculate the total energy consumed by the exoskeleton in kilowatt-hours (kWh) over a month.

Solution: Energy consumed (kWh) = Power (kW) × Time (hours) Energy consumed (kWh) = (1000 W/1000) kW × 3 hours/day × 30 days

Result: The exoskeleton consumes 90 kWh of energy in a month.

Telemedicine Consultation Time

Problem 4. A telemedicine robot is used for remote consultations. On average, a consultation using the robot lasts 20 minutes. Calculate the total time saved by the doctor if they conduct 15 telemedicine consultations in a week compared to traditional in-person consultations that last 30 minutes each.

Solution: Time saved per consultation = Traditional consultation time – Telemedicine consultation time Time saved per consultation = 30 minutes – 20 minutes = 10 minutes

Total time saved in a week = Time saved per consultation × Number of consultations Total time saved in a week = 10 minutes × 15 consultations = 150 minutes

Result: The doctor saves 150 minutes (2.5 hours) of consultation time in a week using the telemedicine robot.

Robotic Pharmacy Dispensing

Problem 5. A robotic pharmacy system dispenses medications. The system fills an average of 500 prescriptions per day, and it operates for 10 hours a day. Calculate the rate at which the robotic system dispenses prescriptions.

Solution: Dispensing rate = Number of prescriptions dispensed/Operating hours Dispensing rate = 500 prescriptions/10 hours = 50 prescriptions per hour

Result: The robotic pharmacy system dispenses prescriptions at a rate of 50 prescriptions per hour.

Short Questions and Answers

Q.1. What is healthcare robotics?

Ans. Healthcare robotics refers to the use of robotic technology in medical applications, including surgery, diagnostics, rehabilitation, and patient care.

Q.2. How are robots used in surgery?

Ans. Robots assist surgeons in performing precise and minimally invasive surgeries, enhancing dexterity and visualization.

Q.3. What is the da Vinci Surgical System?

Ans. The da Vinci system is a robotic surgical platform widely used for minimally invasive procedures.

Q.4. How do robotic exoskeletons benefit patients?

Ans. Robotic exoskeletons help patients regain mobility and strength, particularly in rehabilitation after injuries or surgeries.

Q.5. What is telemedicine?

Ans. Telemedicine is the remote delivery of medical services using communication technology, allowing patients and healthcare providers to connect without physical presence.

Q.6. How can robots be used in remote patient monitoring?

Ans. Robots equipped with sensors can monitor patients' vital signs and transmit real-time data to healthcare providers for remote monitoring.

Q.7. What role do robots play in drug discovery?

Ans. Robots automate various steps in drug discovery, including high-throughput screening and compound synthesis, accelerating the process.

Q.8. What are assistive robotic devices used for in healthcare?

Ans. Assistive robots help individuals with disabilities in daily activities, improving their quality of life and independence.

Q.9. How do robots contribute to laboratory automation?

Ans. Robots automate repetitive tasks in labs, such as sample handling and analysis, increasing efficiency and reducing human error.

Q.10. What's the impact of robotics on elderly care?

Ans. Robots provide companionship, monitoring, and assistance to the elderly, helping them stay independent and easing caregiver burden.

Q.11. How are robots used in mental health support?

Ans. Robots and AI-driven chatbots can provide resources, counseling, and early detection of mental health issues.

Q.12. What's the significance of robotics in diagnostic imaging?

Ans. Robotics enhances diagnostic imaging by improving accuracy and early disease detection.

Q.13. How can robots help in drug dispensing?

Ans. Robots in pharmacy automation ensure proper dosages, reduce errors, and streamline medication dispensing.

Q.14. What challenges does robotics address in healthcare?

Ans. Robotics addresses challenges related to precision, access to care, automation of tasks, and improving patient outcomes.

Q.15. What is the role of robotics in drug delivery?

Ans. Robots can assist in precise drug delivery, targeting specific areas in the body.

Q.16. How do surgical robots enhance precision?

Ans. Surgical robots provide more precise movements and better visualization, reducing the risk of human error.

Q.17. What types of surgeries can robotic systems perform?

Ans. Robotic systems can perform a wide range of surgeries, including urological, gynecological, and cardiothoracic procedures.

Q.18. How do robotic devices impact physical therapy?

Ans. Robotic devices in physical therapy assist patients in regaining strength and mobility after injuries or surgeries.

Q.19. How can robotics improve patient rehabilitation?

Ans. Robotics can provide repetitive and targeted exercises, speeding up the rehabilitation process.

Q.20. What's the future of robotics in healthcare?

Ans. The future involves more advanced robotic systems, AI-driven diagnostics, and personalized treatment options, revolutionizing healthcare.

Q.21. What is the potential of AI in healthcare robotics?

Ans. AI can enhance robotic decision-making, analyze medical data, and provide predictive insights for better patient care.

Q.22. How do collaborative robots (cobots) work in healthcare?

Ans. Cobots work safely alongside humans, assisting in tasks that require strength or precision, such as lifting patients.

Q.23. How does robotics improve surgical outcomes?

Ans. Robotics enhance surgical precision, reducing complications, speeding up recovery, and minimizing scarring.

Q.24. What are the benefits of telemedicine in rural areas?

Ans. Telemedicine brings medical expertise to remote areas, improving access to healthcare services.

Q.25. How do surgical robots improve patient safety?

Ans. Surgical robots reduce the risk of human errors, enhance visualization, and provide steady hands, leading to safer surgeries.

Q.26. What's the role of robots in physical rehabilitation?

Ans. Robots assist patients in regaining muscle strength, coordination, and mobility, leading to faster recovery.

Q.27. How can robotics enhance diagnostic accuracy?

Ans. Robotic systems can analyze medical images with high precision, helping doctors make accurate diagnoses.

Q.28. What's the impact of robotics on healthcare workforce demands?

Ans. Robotics can alleviate the shortage of skilled healthcare workers by automating repetitive tasks and allowing professionals to focus on complex cases.

Q.29. How do robotic exoskeletons work?

Ans. Robotic exoskeletons are wearable devices that provide support to limbs, enhancing movement and strength for individuals with mobility impairments.

Q.30. How can robots be used in medical research?

Ans. Robots automate experiments, handle samples, and assist researchers in high-throughput data collection, accelerating medical research.

Q.31. What challenges does the integration of robotics in healthcare face?

Ans. Challenges include regulatory compliance, initial costs, acceptance by medical professionals, and addressing ethical concerns.

Q.32. How do robotic devices in surgery reduce the risk of infections?

Ans. Minimally invasive robotic surgeries require smaller incisions, reducing the risk of infection compared to traditional open surgeries.

Q.33. How can robotic telemedicine be used during emergencies?

Ans. Robotic telemedicine platforms allow remote consultation and guidance during emergencies, enabling quick medical response.

Q.34. What's the impact of robotics on healthcare cost-efficiency?

Ans. Robotics can reduce healthcare costs by minimizing errors, optimizing resource allocation, and streamlining procedures.

Q.35. How are robotic systems used in neurosurgery?

Ans. In neurosurgery, robots assist with precision, navigation, and tumor removal, reducing risks in delicate procedures.

Q.36. How do robots help in laparoscopic surgery?

Ans. Robotic systems offer improved dexterity in laparoscopic surgery, enabling intricate movements and reducing surgeon fatigue.

Q.37. What's the potential for robots in geriatric care?

Ans. Robots can provide companionship, monitor health parameters, and assist in daily activities for elderly individuals, enabling aging in place.

Q.38. How can robots improve the accuracy of drug dosing?

Ans. Robots ensure precise drug dosing, reducing the risk of medication errors and adverse effects.

Q.39. What's the role of robotics in physical disability rehabilitation?

Ans. Robotic devices aid in retraining muscles and improving coordination, enabling individuals with physical disabilities to regain functionality.

Q.40. How can robots handle hazardous materials in healthcare?

Ans. Robots can be used in environments involving hazardous materials, reducing the risk to human workers in tasks such as chemical handling.

Q.41. How are robots used in orthopedic surgeries?

Ans. Robots assist in orthopedic surgeries, such as joint replacements, by providing precise measurements and guidance during surgery.

Q.42. What's the significance of AI-driven robotics in disease diagnosis?

Ans. AI-driven robotics can analyze large medical datasets, identify patterns, and provide early disease detection, enhancing diagnostic accuracy.

Q.43. How do robots assist in wound care?

Ans. Robots can automate wound dressing, reducing infection risks and ensuring proper wound management.

Q.44. What's the impact of robots on hospital logistics?

Ans. Robots in hospital logistics can handle material transportation, assist with inventory management, and streamline supply chain operations.

Q.45. How can robotics be used in pediatric care?

Ans. Robots can be designed to engage with children, providing distraction, companionship, or support during medical procedures.

Q.46. How do robots help in surgical training?

Ans. Robots can simulate surgical scenarios, providing a safe environment for surgical training, practice, and skill development.

Q.47. What's the role of robotics in prosthetics?

Ans. Robotics plays a significant role in the design and development of advanced prosthetic limbs, providing amputees with improved mobility and functionality.

Q.48. How can robots assist in endoscopy?

Ans. Robots can provide stability and precision in endoscopic procedures, enhancing the surgeon's control and improving visualization.

Q.49. What's the potential for robotics in managing chronic diseases?

Ans. Robotics can assist in managing chronic diseases by providing continuous monitoring, reminders for medication, and data analysis for personalized treatment plans.

Q.50. How do robots contribute to faster patient recovery?

Ans. Robotic assistance in surgery and rehabilitation leads to faster recovery times and reduced post-operative complications, allowing patients to resume normal activities sooner.

Multiple Choice Questions and Answers

Q.1. What is the primary benefit of robotic-assisted surgery in the healthcare sector?

(*a*) Faster surgeries (*b*) Reduced need for human surgeons
(*c*) Minimally invasive procedures (*d*) Lower costs

Q.2. Which surgical system is commonly used for robotic-assisted surgeries?

(*a*) UltraSurgery System (*b*) MEDibot
(*c*) da Vinci Surgical System (*d*) RoboDoc

Q.3. What's the advantage of telemedicine in the healthcare sector?

(*a*) In-person consultations (*b*) Remote delivery of medical services
(*c*) Only used in emergencies (*d*) Limited to rural areas

Q.4. How do robotic exoskeletons benefit patients in rehabilitation?

(*a*) They provide mental support (*b*) They enable patients to fly
(*c*) They assist in regaining mobility and strength
(*d*) They reduce the need for physical therapy

Q.5. What is the role of robotic devices in laboratory automation?

(*a*) Providing entertainment for lab workers
(*b*) Handling repetitive tasks like coffee making
(*c*) Increasing the number of coffee breaks
(*d*) Automating lab processes such as sample handling and analysis

Q.6. How do collaborative robots (cobots) differ from traditional robots?

(*a*) They work alone
(*b*) They collaborate with humans in shared workspaces
(*c*) They are used only in large factories
(*d*) They are too expensive for most industries

Q.7. Which area of healthcare can benefit from robots providing companionship and monitoring for the elderly?

(*a*) Surgery (*b*) Radiology (*c*) Cardiology (*d*) Geriatric care

Q.8. How can robots assist in mental health support?

(*a*) They prescribe medication (*b*) They provide emotional support
(*c*) They perform psychotherapy (*d*) They build furniture

Q.9. In which field can robotics enhance diagnostic accuracy through precise analysis of medical images?

(*a*) Dentistry (*b*) Radiology (*c*) Orthopedics (*d*) Psychiatry

Q.10. How can robots improve patient safety in healthcare?

(*a*) By taking over all medical procedures
(*b*) By reducing the need for human healthcare providers

(*c*) By providing companionship

(*d*) By enhancing surgical precision and reducing human errors

Q.11. What type of surgery involves robotic systems for precise movements and visualization?

(*a*) Cosmetic surgery (*b*) Orthopedic surgery

(*c*) Minimally invasive surgery (*d*) Traditional open surgery

Q.12. What is the primary purpose of robotic exoskeletons in healthcare?

(*a*) Enhancing athletic performance (*b*) Providing physical therapy for children

(*c*) Assisting astronauts in space

(*d*) Aiding in rehabilitation for individuals with mobility impairments

Q.13. What is the primary benefit of telemedicine?

(*a*) Decreased access to healthcare (*b*) Reduced quality of care

(*c*) Convenience and remote access to medical services

(*d*) Increased in-person consultations

Q.14. Which robotic system is commonly used for surgical procedures?

(*a*) RoboNurse (*b*) RoboCleaner

(*c*) RoboAssistant (*d*) da Vinci Surgical System

Q.15. What role can robotic exoskeletons play in the healthcare sector?

(*a*) Enhancing robotic surgeries (*b*) Providing remote telemedicine services

(*c*) Assisting in physical rehabilitation (*d*) Managing patient records

Q.16. How do robots contribute to laboratory automation?

(*a*) By handling tasks like coffee making

(*b*) By providing entertainment for lab workers

(*c*) By automating repetitive lab processes like sample handling and analysis

(*d*) By designing experiments

Q.17. What's the main difference between collaborative robots (cobots) and traditional robots?

(*a*) Cobots only work in isolation (*b*) Cobots are much larger

(*c*) Cobots collaborate with humans in shared workspaces

(*d*) Cobots are cheaper

Q.18. In which area of healthcare can robots provide companionship and monitoring for the elderly?

(*a*) Cardiology (*b*) Surgery (*c*) Geriatric care (*d*) Radiology

Q.19. How can robots assist in mental health support?

(*a*) By providing emotional support (*b*) By performing surgery

(*c*) By prescribing medication (*d*) By building furniture

Q.20. Which field of healthcare can benefit from robotics for precise analysis of medical images?

(*a*) Dentistry (*b*) Radiology (*c*) Cardiology (*d*) Psychiatry

Q.21. How can robots enhance patient safety in healthcare?
(*a*) By taking over all medical procedures
(*b*) By reducing the need for human healthcare providers
(*c*) By providing companionship
(*d*) By enhancing surgical precision and reducing human errors

Q.22. What type of surgery involves robotic systems for precise movements and visualization?
(*a*) Cosmetic surgery (*b*) Orthopedic surgery
(*c*) Minimally invasive surgery (*d*) Traditional open surgery

Q.23. What is the primary purpose of robotic exoskeletons in healthcare?
(*a*) Enhancing athletic performance (*b*) Providing physical therapy for children
(*c*) Assisting astronauts in space
(*d*) Aiding in rehabilitation for individuals with mobility impairments

Q.24. What is the primary benefit of telemedicine?
(*a*) Decreased access to healthcare (*b*) Reduced quality of care
(*c*) Convenience and remote access to medical services
(*d*) Increased in-person consultations

Q.25. Which robotic system is commonly used for surgical procedures?
(*a*) RoboNurse (*b*) RoboCleaner
(*c*) RoboAssistant (*d*) da Vinci Surgical System

Q.26. What role can robotic exoskeletons play in the healthcare sector?
(*a*) Enhancing robotic surgeries (*b*) Providing remote telemedicine services
(*c*) Assisting in physical rehabilitation (*d*) Managing patient records

Q.27. How do robots contribute to laboratory automation?
(*a*) By handling tasks like coffee making
(*b*) By providing entertainment for lab workers
(*c*) By automating repetitive lab processes like sample handling and analysis
(*d*) By designing experiments

Q.28. What's the main difference between collaborative robots (cobots) and traditional robots?
(*a*) Cobots only work in isolation (*b*) Cobots are much larger
(*c*) Cobots collaborate with humans in shared workspaces
(*d*) Cobots are cheaper

Q.29. In which area of healthcare can robots provide companionship and monitoring for the elderly?
(*a*) Cardiology (*b*) Surgery (*c*) Geriatric care (*d*) Radiology

Q.30. How can robots assist in mental health support?
(*a*) By providing emotional support (*b*) By performing surgery
(*c*) By prescribing medication (*d*) By building furniture

Q.31. Which field of healthcare can benefit from robotics for precise analysis of medical images?

(*a*) Dentistry (*b*) Radiology (*c*) Cardiology (*d*) Psychiatry

Q.32. How can robots enhance patient safety in healthcare?

(*a*) By taking over all medical procedures

(*b*) By reducing the need for human healthcare providers

(*c*) By providing companionship

(*d*) By enhancing surgical precision and reducing human errors

Q.33. What type of surgery involves robotic systems for precise movements and visualization?

(*a*) Cosmetic surgery (*b*) Orthopedic surgery

(*c*) Minimally invasive surgery (*d*) Traditional open surgery

Q.34. What is the primary purpose of robotic exoskeletons in healthcare?

(*a*) Enhancing athletic performance (*b*) Providing physical therapy for children

(*c*) Assisting astronauts in space

(*d*) Aiding in rehabilitation for individuals with mobility impairments

Q.35. What is the primary benefit of telemedicine?

(*a*) Decreased access to healthcare (*b*) Reduced quality of care

(*c*) Convenience and remote access to medical services

(*d*) Increased in-person consultations

Q.36. Which robotic system is commonly used for surgical procedures?

(*a*) RoboNurse (*b*) RoboCleaner

(*c*) RoboAssistant (*d*) da Vinci Surgical System

Q.37. What role can robotic exoskeletons play in the healthcare sector?

(*a*) Enhancing robotic surgeries (*b*) Providing remote telemedicine services

(*c*) Assisting in physical rehabilitation (*d*) Managing patient records

Q.38. How do robots contribute to laboratory automation?

(*a*) By handling tasks like coffee making

(*b*) By providing entertainment for lab workers

(*c*) By automating repetitive lab processes like sample handling and analysis

(*d*) By designing experiments

Q.39. What's the main difference between collaborative robots (cobots) and traditional robots?

(*a*) Cobots only work in isolation (*b*) Cobots are much larger

(*c*) Cobots collaborate with humans in shared workspaces

(*d*) Cobots are cheaper

Q.40. In which area of healthcare can robots provide companionship and monitoring for the elderly?

(*a*) Cardiology (*b*) Surgery (*c*) Geriatric care (*d*) Radiology

Q.41. How can robots assist in mental health support?

(*a*) By providing emotional support
(*b*) By performing surgery
(*c*) By prescribing medication
(*d*) By building furniture

Q.42. Which field of healthcare can benefit from robotics for precise analysis of medical images?

(*a*) Dentistry
(*b*) Radiology
(*c*) Cardiology
(*d*) Psychiatry

Q.43. How can robots enhance patient safety in healthcare?

(*a*) By taking over all medical procedures
(*b*) By reducing the need for human healthcare providers
(*c*) By providing companionship
(*d*) By enhancing surgical precision and reducing human errors

Q.44. What type of surgery involves robotic systems for precise movements and visualization?

(*a*) Cosmetic surgery
(*b*) Orthopedic surgery
(*c*) Minimally invasive surgery
(*d*) Traditional open surgery

Q.45. What is the primary purpose of robotic exoskeletons in healthcare?

(*a*) Enhancing athletic performance
(*b*) Providing physical therapy for children
(*c*) Assisting astronauts in space
(*d*) Aiding in rehabilitation for individuals with mobility impairments

Q.46. What is the primary benefit of telemedicine?

(*a*) Decreased access to healthcare
(*b*) Reduced quality of care
(*c*) Convenience and remote access to medical services
(*d*) Increased in-person consultations

Q.47. Which robotic system is commonly used for surgical procedures?

(*a*) RoboNurse
(*b*) RoboCleaner
(*c*) RoboAssistant
(*d*) da Vinci Surgical System

Q.48. What role can robotic exoskeletons play in the healthcare sector?

(*a*) Enhancing robotic surgeries
(*b*) Providing remote telemedicine services
(*c*) Assisting in physical rehabilitation
(*d*) Managing patient records

Q.49. How do robots contribute to laboratory automation?

(*a*) By handling tasks like coffee making
(*b*) By providing entertainment for lab workers
(*c*) By automating repetitive lab processes like sample handling and analysis
(*d*) By designing experiments

Q.50. What's the main difference between collaborative robots (cobots) and traditional robots?

(*a*) Cobots only work in isolation
(*b*) Cobots are much larger
(*c*) Cobots collaborate with humans in shared workspaces
(*d*) Cobots are cheaper

Key to Multiple Choice Questions

1. (*c*)	**2.** (*c*)	**3.** (*b*)	**4.** (*c*)	**5.** (*d*)	**6.** (*b*)	**7.** (*d*)	**8.** (*b*)	**9.** (*b*)
10. (*d*)	**11.** (*c*)	**12.** (*d*)	**13.** (*c*)	**14.** (*d*)	**15.** (*c*)	**16.** (*c*)	**17.** (*c*)	**18.** (*c*)
19. (*a*)	**20.** (*b*)	**21.** (*d*)	**22.** (*c*)	**23.** (*d*)	**24.** (*c*)	**25.** (*d*)	**26.** (*c*)	**27.** (*c*)
28. (*c*)	**29.** (*c*)	**30.** (*a*)	**31.** (*b*)	**32.** (*d*)	**33.** (*c*)	**34.** (*d*)	**35.** (*c*)	**36.** (*d*)
37. (*c*)	**38.** (*c*)	**39.** (*c*)	**40.** (*c*)	**41.** (*a*)	**42.** (*b*)	**43.** (*d*)	**44.** (*c*)	**45.** (*d*)
46. (*c*)	**47.** (*d*)	**48.** (*c*)	**49.** (*c*)	**50.** (*c*)				

Projects Ideas with Hints

1. **Robotic Surgical Assistant:** Develop a robotic system that assists surgeons during minimally invasive surgeries, providing steady hands, precise movements, and enhanced visualization. This could lead to safer and more efficient surgeries.
2. **Robotic Rehabilitation Device:** Create a wearable robotic exoskeleton that assists individuals with mobility impairments in their rehabilitation process, helping them regain strength and movement.
3. **Telemedicine Robot:** Build a mobile robot equipped with cameras and communication tools to enable remote consultations between doctors and patients, particularly in rural or remote areas.
4. **Robotic Laboratory Automation:** Design a robotic system that automates repetitive tasks in medical laboratories, such as sample handling, testing, and analysis, improving efficiency and accuracy.
5. **AI-driven Diagnostics:** Develop an AI-powered platform that analyzes medical images (e.g., X-rays, MRIs) to assist doctors in early disease detection and accurate diagnoses.
6. **Elderly Care Robot:** Create a social robot that provides companionship to the elderly, monitors their health parameters, reminds them to take medication, and alerts caregivers in case of emergencies.
7. **Robotic Drug Dispensing System:** Build an automated robot for accurately dispensing medication in hospitals or pharmacies, reducing human errors in drug dosing.
8. **Robotic Physical Therapy Assistant:** Develop a robot that guides patients through physical therapy exercises, ensuring proper form and providing real-time feedback, which could enhance the effectiveness of therapy.
9. **Remote Monitoring of Patients:** Create a network of wearable sensors that monitor patients' vital signs and transmit data to healthcare providers in real-time, allowing remote monitoring and timely interventions.
10. **Robotic Mental Health Support:** Build a chatbot or robot that provides resources, offers counseling, and assists in identifying early signs of mental health issues, helping individuals access mental health support.
11. **Robotic Medication Management:** Develop a robotic system that helps patients manage their medication, ensuring timely doses and reducing the risk of missed doses.
12. **Robotic Ultrasound Assistant:** Design a robotic arm that holds and moves the ultrasound probe during procedures, improving stability and accuracy in ultrasound imaging.

13. **Robotic Prosthetic Limbs:** Create advanced robotic prosthetic limbs with natural movement and sensory feedback for amputees, enabling them to regain functional mobility.
14. **Robotic Patient Transport:** Build a robot designed to safely transport patients within hospitals, reducing strain on healthcare staff and improving efficiency.
15. **Robotic Data Analysis for Clinical Trials:** Develop an AI-based system that analyzes medical data from clinical trials, identifying trends and patterns to optimize trial designs and drug development.

These project ideas combine robotics, AI, and healthcare to address various aspects of patient care, diagnosis, rehabilitation, and support. Each project has the potential to make a significant impact on the healthcare sector.

5

Robotics in Agriculture

5.1 INTRODUCTION

For millennia, agriculture has been the foundation of human civilization, giving innumerable communities all over the world food and a means of subsistence. The problems facing agriculture are getting more complicated as the world's population keeps expanding. The challenge is to increase food production while utilizing finite resources and having as little negative environmental impact as possible. In this situation, robotics can be a game-changer. Imagine a world where farmers collaborate with intelligent machines to increase production, accuracy, and sustainability. Through the adoption of robotics and automation in the agriculture sector, this is no longer just a pipe dream.

Robotics in agriculture has a wide range of uses, including crop planting, crop monitoring, harvesting, and even fine-tuning pesticide and fertilizer application. These robots can move through fields, detect plants, evaluate the state of the soil, and carry out duties with surprising accuracy because they are fitted with a range of sensors, cameras, and actuators. One of the main advantages of deploying robots in agriculture is their 24/7 availability, regardless of the weather. As a result, tasks can be completed quickly, maximizing yield and minimizing losses brought on by bad weather. Robots also provide a high degree of uniformity, which eliminates human unpredictability and faults that can lower crop quality.

Precision is yet another essential component of agricultural robotics. Robots can distribute resources like water, fertilizer, and insecticides precisely where and when they are required by using data gathered from sensors and drones. This decreases waste and promotes more environmentally friendly farming methods, protecting our natural resources for future generations. We must be aware of the obstacles posed by this game-changing technology as we explore the promise of robotics in agriculture. It takes interdisciplinary cooperation to create robots that can move across difficult terrain, recognize different plant species, and adapt to changing surroundings. To create solutions that address actual problems, engineers, agronomists, computer scientists, and environmental experts must work together. The economic viability of implementing robotics in agriculture is another factor to take into account, particularly for small-scale farmers who may have limited resources. A promising route to a future that is more effective, robust, and sustainable is provided by the convergence of robotics and agriculture. We may attempt to meet the ever expanding demands of our globe while protecting our planet's resources by using technology to operate in harmony with nature.

5.2 AGRICULTURAL ROBOTICS

The use of robotics and automation technology to various farming and agricultural practises is known as "agricultural robotics" or "agribotics," and it is a growing field. The agriculture industry's efficiency, productivity, and sustainability are all intended to be increased through this integration. Here are some significant applications of robotics in agriculture:

5.2.1 Planting and Seeding

Robotic devices can carefully place seeds or seedlings at the right depth and spacing to ensure uniform growth and minimize waste. The creation of automated technologies that can precisely and effectively plant seeds or seedlings in a field is known as "planting and seeding" in agriculture. Precision in seed location, depth, and spacing is desired because these factors have a big impact on crop output.

Here's a detailed overview of the process, along with some mathematics involved:

1. **Seed Dispensing Mechanism:** Typically, robotic seed planting systems include a device that releases seeds at predetermined intervals. It is possible for this device to operate on a variety of theories, including hoover suction, pneumatic pressure, or mechanical scooping. The planting density is determined by the distance between seeds and can change depending on the crop needs.
2. **Mathematical Considerations:** There are various mathematical computations that must be performed in order to sow seeds accurately:

   Planting Density: The amount of seeds in a given area is known as planting density. The formula is used to compute it: Total Seeds/Area = Planting Density

   Distance between Rows: The distance between seed rows influences how easily the seeds can obtain sunlight, water, and nutrients. The formula is used to compute it: Field Width/(Number of Rows – 1) provides the spacing between rows.

   Distance between Seeds: The struggle for resources is determined by the distance between each seed in a row. The formula is used to compute it: Row Length/(Number of Seeds – 1) is the formula for spacing between seeds.
3. **Seed Depth and Soil Compaction:** For best germination, seeds must be sown in the soil at the proper depth. To achieve uniformity, robotic devices must regulate the depth of planting. The type of crop and the soil's characteristics determine the depth.
4. **Precision and Sensor Integration:** In order to precisely estimate locations and distances, robotic seed planting systems frequently incorporate sensors like cameras and LiDAR. The robot can make necessary adjustments to its motions and planting mechanism thanks to the real-time data provided by these sensors.
5. **Autonomous Navigation:** Robots need to be able to move through the field on their own to plant effectively. In order to do this, path planning algorithms must take into account the field's overall structure, turns, and obstructions.
6. **Variability Consideration:** Fields aren't always level and the soil quality can change. Robotic systems must modify planting depth and pressure in response to these changes to achieve consistent planting over the field.
7. **Case Study: Lettuce Seeding** Consider the sowing of lettuce as a case study. Let's say we have a rectangle with the following measurements:

The field is 20 meters wide.

40 meters is the field length.

The prerequisites for planting are as follows:

- Planting Density: 50,000 seeds per hectare
- Row Spacing: 30 cm
- Seed Spacing: 10 cm

Calculations:

1. **Calculate the Planting Density:** Planting Density = Total Seeds/Area. Area = Field Width * Field. Length = 20 m * 40 m = 800 m^2. Total Seeds = Planting Density * Area = 50,000 seeds/hectare * (800 m^2/10,000 m^2/hectare) = 4,000 seeds
2. **Calculate the Number of Rows:** Spacing between Rows = Field Width/(Number of Rows – 1). Number of Rows = (Field Width/Spacing between Rows) + 1 Number of Rows = (20 m/0.3 m) + 1 $\approx$ 67 rows
3. **Calculate the Number of Seeds per Row:** Spacing between Seeds = Row Length/(Number of Seeds – 1). Number of Seeds per Row = (Row Length/Spacing between Seeds) + 1 Row Length = Field Length = 40 m. Number of Seeds per Row = (40 m/0.1 m) + 1 = 401 seeds

These calculations show how important mathematical considerations are in the design and programming of robotic systems used in agriculture for planting and seeding. Optimal crop growth and yield are influenced by the accuracy of these estimations and the robotic system's execution.

5.2.2 Weeding and Pest Control

Robots using cameras and sensors can precisely locate and get rid of weeds, eliminating the need for chemical herbicides. Similar to this, robotic systems can use methods like precise spraying or even the release of beneficial insects to target pests. Important agricultural processes like weeding and pest control can considerably benefit from the incorporation of robotics. Robotic systems can assist in precisely locating and getting rid of weeds or pests, decreasing the need for chemical interventions and supporting organic agricultural methods.

Let's delve into the details of how robotics can be used for weeding and pest control, along with some mathematical considerations:

1. **Weed Detection:** Robotic systems can recognize and categories weeds based on their shape, colour, and appearance using cameras and sensors. Algorithms for machine learning can be used to discriminate between crops and weeds.
2. **Precision Spraying:** Once weeds have been identified, robotic systems with sprayers can target and apply herbicides only where weeds are present, using less chemicals and having a smaller negative environmental impact.
3. **Mechanical Weed Removal:** Robotic systems can be built to physically pull weeds out of the ground using mechanical means, such as revolving brushes or uprooting arms.
4. **Mathematical Considerations:** The design and operation of robotic weeding and pest control systems include a number of mathematical calculations:

   Precision Measures: Precision and recall measures can be used to assess the accuracy of weed detection. For instance, recall shows the proportion of correctly identified weeds among all weed occurrences that were actually detected, whereas precision shows the proportion of correctly identified weeds among all weed instances that were actually detected.

Spraying Accuracy: To reduce chemical usage, the robot must precisely target the weedy region. Calculations for exact placement and orientation are necessary.

Coverage Efficiency: The robotic system's capacity to thoroughly cover a field while successfully identifying and eliminating weeds. Calculating the best route for the robot to take through the field is required for this.

Resource Savings: The decreased use of herbicides or pesticides as a result of the accuracy of the robotic system. This can be quantified in terms of cost savings and environmental impact.

5. **Case Study: Precision Herbicide Application.** Let's take the example of a robotic system that is intended to spray pesticides on a field that is overrun with weeds. The following dimensions apply to the field:

   Field Width: 30 meters **Field Length:** 50 meters

   The spraying requirements are as follows:

   Herbicide Consumption Rate: 500 mL per square meter

   Calculations:

   1. **Calculate the Total Area:** Area = Field Width * Field Length = 30 m * 50 m = 1500 m^2.
   2. **Calculate the Herbicide Consumption:** Herbicide Consumption = Area * Herbicide Consumption Rate. Herbicide Consumption = 1500 m^2 * 500 mL/m^2 = 750,000 mL
   3. **Potential Savings with Precision Application:** Suppose the robotic system can identify and spray only the weed-infested areas, covering 70% of the field. Area with Weeds = 0.7 * Area = 0.7 * 1500 m^2 = 1050 m^2. Herbicide Consumption with Precision Application = Area with Weeds * Herbicide Consumption Rate. Herbicide Consumption with Precision Application = 1050 m^2 * 500 mL/m^2 = 525,000 mL

      Savings in Herbicide Consumption = Herbicide Consumption – Herbicide Consumption with Precision Application Savings in Herbicide Consumption = 750,000 mL – 525,000 mL = 225,000 mL

These calculations demonstrate how precisely locating and treating the weed-infested areas with robotics can result in significant chemical consumption reductions. In order to maximise the effectiveness of the robotic system, cut expenses, and lessen the negative effects of weed control methods on the environment, mathematical considerations are essential.

5.2.3 Harvesting

Robotics can be used to harvest crops including grains, fruits, and vegetables. Advanced vision systems are used by these harvesting robots to spot ripe food, and soft grasping mechanisms are used to prevent harm. Robotic harvesting in agriculture entails the creation of automated devices that can precisely and effectively gather crops including grains, fruits, and vegetables. These devices identify ripe produce, handle it gently, and assure little harm during the harvesting process using cutting-edge sensors, algorithms, and mechanical components.

Let's explore the details of how robotics can be used for harvesting, along with some mathematical considerations:

1. **Crop Maturity Detection:** Crop maturity detection is carried out by robotic harvesters using sensors, cameras, and spectroscopy. To evaluate whether a fruit or vegetable is ripe for harvest, they examine characteristics including color, size, texture, and even internal qualities like sugar concentration.

2. **Gentle Handling:** Crops are handled gently by robotic harvesting equipment to prevent bruises or damage. To simulate human touch, soft grasping mechanisms and sensitive actuators are used.
3. **Precision Cutting:** Robotic systems employ precise cutting instruments to ensure clean cuts without extra waste while harvesting products that require cutting, such as grains or vegetables.
4. **Mathematical Considerations:** When constructing and using robotic harvesting systems, the following mathematical calculations must be made:

   Detection of Ripeness Accuracy: Precision and recall measures can be used to assess the accuracy of ripe produce detection, much like weed detection. This aids in assessing the system's capacity to accurately detect and gather ripe crops.

   Yield Estimation: Robots can also calculate a crop's yield by figuring out how many harvested units there are per square foot of land. Farmers can manage resources and arrange logistics thanks to this.

   Time Efficiency: To maximize overall efficiency throughout the harvesting process, it is essential to calculate the time needed for the robot to harvest a specific area.

   Harm Reduction: Measuring the reduction in waste or harm brought on by the robotic system's accuracy. This can be measured in terms of economic value and a decrease in food loss.
5. **Case Study: Robotic Fruit Harvesting** Let's consider a case where a robotic system is designed to harvest apples from an orchard. The orchard has the following dimensions:

   Orchard Width: 40 meters **Orchard Length:** 60 meters

   The harvesting requirements are as follows:

   Apples per Tree: 100 apples **Time per Apple Harvest:** 10 seconds

   Calculations:

   1. **Calculate the Total Number of Apple Trees:** Number of Apple Trees = (Orchard Width/ Apple Tree Spacing) * (Orchard Length/Apple Tree Spacing). Apple Tree Spacing = 3 meters (both width and length). Number of Apple Trees = (40 m/3 m) * (60 m/3 m) = 800 trees.
   2. **Calculate the Total Number of Apples:** Total Apples = Number of Apple Trees * Apples per Tree. Total Apples = 800 trees * 100 apples/tree = 80,000 apples
   3. **Calculate the Total Harvest Time:** Total Harvest Time = Total Apples * Time per Apple Harvest. Total Harvest Time = 80,000 apples * 10 seconds/apple = 800,000 seconds.

   Convert to Hours: Total Harvest Time (hours) = Total Harvest Time (seconds)/3600 seconds/ hour Total Harvest Time (hours) = 800,000 seconds/3600 seconds/hour ≈ 222.22 hours

These calculations demonstrate how automating the operation and offering precise predictions of time, yield, and resource usage can dramatically improve harvesting efficiency. For the robotic system to operate at its best and harvest crops successfully, mathematical considerations are essential.

5.2.4 Monitoring and Data Collection

Crop health, soil moisture, and nutrient levels may all be monitored by drones and ground-based robots that are outfitted with sensors and cameras. Farmers can use this information to make well-informed decisions about disease control, fertilization, and irrigation. Robotic monitoring and data collecting play a major role in modern agriculture by giving farmers instantaneous insights into the health of their crops, the state of the soil, and other important characteristics. Robotic systems can gather data from fields and relay it to farmers, assisting them in making decisions that would maximize crop productivity and resource management.

Here's an overview of how robotics is used for monitoring and data collection in agriculture, along with relevant mathematical considerations:

1. **Sensor Integration:** To monitor various characteristics of the agricultural environment, robotic systems are fitted with a variety of sensors. Cameras, multispectral and hyperspectral sensors, temperature and humidity sensors, soil moisture sensors, and more are examples of these sensors.
2. **Data Gathering:** Robotic platforms move throughout the fields, gathering information from different sensors. Images, temperature readings, humidity levels, nutritional concentrations, and other information can all be included in the data collection.
3. **Data processing:** To extract useful information, the obtained data are organized, analyzed, and processed. In order to find patterns, anomalies, and trends, machine learning techniques are frequently used.
4. **Mathematical Considerations:** The use of robotics in agriculture for data collecting and monitoring involves a number of mathematical considerations:

   Spatial Sampling: Choosing the placement and spacing of data collection locations within a field to maximize resources and assure proper coverage.

   Data Fusion: Combining information from various sensors to produce a complete picture of the state of the field.

   Data Interpretation: Using mathematical models to analyze sensor data to calculate factors like nutrient concentrations, plant stress levels, and soil moisture content.

   Path planning: determining the best route for the robot to take while navigating the field, taking into account different types of terrain and obstacles.

   Yield Estimation: Using sensor data to calculate crop yield based on elements like plant density and health.
5. **Case Study: Soil Moisture Monitoring.** Let us look at a scenario where a robotic system with soil moisture sensors is employed to keep an eye on the moisture level of a rectangular field. The following dimensions apply to the field:

   Field Width: 30 meters **Field Length:** 50 meters

   The soil moisture measurements are taken at regular intervals of 2 meters both along the width and length of the field.

   Calculations:

   1. **Calculate the Number of Sampling Points along Width and Length:** Number of Sampling Points along Width = Field Width/Sampling Interval. Number of Sampling Points along Width = 30 m/2 m = 15 points

      Number of Sampling Points along Length = Field Length/Sampling Interval. Number of Sampling Points along Length = 50 m/2 m = 25 points
   2. **Calculate the Total Number of Sampling Points:** Total Number of Sampling Points = Number of Sampling Points along Width * Number of Sampling Points along Length. Total Number of Sampling Points = 15 points * 25 points = 375 points

These calculations show how mathematical factors, such as spatial sampling and data interpretation, are essential to the success of robotic data gathering in agriculture. Decisions regarding the timing of irrigation, nutrient management, and general crop health can subsequently be made using the data provided.

5.2.5 Precision Agriculture

Robots are capable of carrying out operations like variable-rate fertilization and planting while adjusting to particular crop and soil conditions. This increases yield and decreases resource waste. Utilizing cutting-edge technologies, such as robotics, precision agriculture aims to maximize crop output by adjusting methods to particular field circumstances. Robotics is essential to precision agriculture because it makes it possible to take precise, focused activities based on real-time data collection and analysis.

Let's investigate precision agriculture in depth, along with how robotics helps with its application and pertinent mathematical considerations:

1. **Data Gathering:** Robotic systems with sensors, cameras, and other data-gathering equipment acquire a variety of data about the field, such as soil moisture, temperature, nutrient levels, plant health, and more.
2. **Data Analysis:** To find spatial differences and trends within the field, the obtained data is examined. The data is processed and analyzed using mathematical algorithms, yielding information about crop health, nutritional deficits, irrigation requirements, and more.
3. **Application at Variable Rates:** Using the data analysis, robotic systems are able to apply inputs like water, fertilizer, and pesticides at various rates throughout the field. This focused strategy reduces waste and maximizes resource use.
4. **Precision Planting:** Depending on the soil's characteristics and the intended planting densities, robotic seed-planting devices can assure precision seed placement.
5. **Mathematical Considerations:** Robotic precision agriculture involves a number of mathematical components:

   Geospatial Analysis: Making use of Geographic Information Systems (GIS) to produce maps that display the variability of various field-level variables such soil types, elevation, and yield history.

   Interpolation: Using mathematical methods, values at unsampled places are estimated based on values at neighboring sampled points, assisting in the creation of precise and thorough field maps.

   Optimization Algorithms: Based on field data, these algorithms find the best way to distribute resources (such as water and fertilizers) in order to maximize crop yield and cut expenses.

   Yield Mapping: By calculating average yields in various fields using information gathered during harvest, farmers can pinpoint high- and low-performing areas.
6. **Case Study: Variable Rate Fertilization** Let's consider a case where a robotic system is used for precision fertilization in a rectangular field. The field has the following dimensions:

   Field Width: 40 meters **Field Length:** 60 meters

   Soil testing indicates varying nutrient levels across the field, and the following fertilization rates are recommended:

Low-Nutrient Areas: Apply 150 kg/ha of fertilizer

Medium-Nutrient Areas: Apply 250 kg/ha of fertilizer

High-Nutrient Areas: Apply 350 kg/ha of fertilizer

Calculations:

1. **Calculate the Field Area:** Area = Field Width * Field Length = 40 m * 60 m = 2400 m^2
2. **Calculate the Recommended Fertilizer Amounts:** Total Fertilizer for Low-Nutrient Areas = 150 kg/ha * (2400 m^2/10,000 m^2/ha) = 36 kg. Total Fertilizer for Medium-Nutrient Areas = 250 kg/ha * (2400 m^2/10,000 m^2/ha) = 60 kg. Total Fertilizer for High-Nutrient Areas = 350 kg/ha * (2400 m^2/10,000 m^2/ha) = 84 kg

These calculations demonstrate how robotic precision agriculture optimizes fertilizer application depending on changing nutrient levels in the field. The proper resource allocation for each area is determined in large part by mathematical considerations, which improves crop productivity and resource efficiency.

5.2.6 Greenhouse Management

Robots are capable of managing duties in greenhouse environments, including modifying the temperature, humidity, and lighting settings as well as shifting potted plants for the best possible exposure to sunshine.

Robotic systems are integrated to automate and optimize a variety of functions within a regulated greenhouse environment as part of greenhouse management in agriculture. These duties could involve managing the temperature, humidity, lighting, irrigation, and keeping an eye on the health of the plants.

Here's a detailed overview of how robotics is utilized for greenhouse management, along with relevant mathematical considerations:

1. **Climate Control:** To maintain the best possible growth conditions for crops, robotic systems can regulate the greenhouse's ventilation, heating, cooling, and humidity levels.
2. **Lighting Control:** Robots are capable of managing artificial lighting to deliver the ideal spectrum and intensity of light at various phases of plant development.
3. **Irrigation and Fertilization:** Robotic systems can automate the delivery of nutrients and water to plants, ensuring that they receive the proper amount of both.
4. **Control of Pests and Disease:** Robotic sensors can identify pests or diseases early on, enabling focused action and lowering the need for chemical treatments.
5. **Data Gathering and Analysis:** Robots with sensors can gather information on plant health, CO_2 levels, humidity, and temperature. By analyzing this data, greenhouse conditions can be improved.
6. **Mathematical Considerations:** Robotic greenhouse management involves a number of mathematical considerations:

   Control Algorithms: Creating control algorithms that modify parameters (such as temperature, humidity, and other variables) in accordance with sensor data and predetermined goal values.

   Optimization: Applying mathematical optimization techniques to determine the best environmental conditions for a certain crop's growth.

   Predictive Models: Constructing mathematical simulations that forecast how changes in the environment will impact plant development, enabling preemptive modifications.

Resource Efficiency: Determining the best way to distribute resources (such as water, nutrients, and energy) in order to maximize output while minimizing waste and expense.

7. **Case Study: Temperature Control** Let's consider a case where a robotic system is used to control the temperature in a greenhouse. The desired temperature range is 22°C to 26°C. The greenhouse's current temperature is 24°C, and the robot can adjust the temperature by activating fans or heaters.

   Calculations:

   1. **Calculate the Temperature Deviation:** Temperature Deviation = Current Temperature – Desired Temperature. Temperature Deviation = 24°C – 24°C = 0°C
   2. **Determine the Action:**

      If Temperature Deviation > 0: Activate Fans to cool the greenhouse.

      If Temperature Deviation < 0: Activate Heaters to warm the greenhouse.

      If Temperature Deviation is within an acceptable range, no action is needed.

   These calculations show how the robotic system employs mathematical reasoning to decide how to regulate the greenhouse's temperature. Robotic greenhouse management optimizes plant growth by continuously monitoring and modifying environmental factors, resulting in improved yields and resource efficiency.

5.2.7 Autonomous Tractors and Machinery

Without any direct human involvement, self-driving tractors and other machinery may carry out operations like tilling, cultivating, and ploughing. To correctly navigate fields, they can use GPS or pre-programmed paths. Modern agriculture has greatly advanced with the development of autonomous tractors and equipment, which automates many field operations chores. Without direct human involvement, these autonomous systems may carry out operations like planting, growing, tilling, and more.

Here is a thorough explanation of the operation of autonomous tractors and other equipment, along with pertinent mathematical points:

1. **GPS and Navigation:** Autonomous tractors utilize GPS to precisely locate themselves in the field and navigate there. To cover the entire area, they either utilize guiding algorithms or follow established paths.
2. **Path Planning and Obstacle Avoidance:** Algorithms for path planning choose the tractor's optimal course while avoiding hazards including rocks, trees, and other machines. These methods optimize the path and guarantee effective coverage by using mathematical models.
3. **Control of Implements:** Tractors that operate autonomously are fitted with a variety of implements, including ploughs, seeders, and sprayers. These tools are operated by robotic systems, which provide precise and reliable field operations.
4. **Mathematical Considerations:** Agricultural autonomous tractors and machines include a number of mathematical issues:

   Path Optimization: Using methods for optimization, choosing the path that crosses the entire field with the least amount of overlap and lag time.

   Guiding Algorithms: Creating algorithms for guiding the tractor based on GPS coordinates and heading, ensuring precise navigation.

Collision Avoidance: To use successful collision avoidance techniques, calculate the distance and trajectory of obstacles.

Coverage Efficiency: Calculating the remaining area to ensure complete coverage after estimating the area the tractor covered after each pass.

5. **Case Study: Autonomous Plowing.** Let's consider a case where an autonomous tractor is used for plowing a rectangular field. The field has the following dimensions:

   Field Width: 50 meters

   Field Length: 100 meters

   The plow's working width is 2 meters, and the tractor's average speed is 4 km/h.

   Calculations:

   1. **Calculate the Number of Passes:** Number of Passes along Width = Field Width/Plow Width. Number of Passes along Width = 50 m/2 m = 25 passes
   2. **Calculate the Total Distance Traveled:** Distance Traveled per Pass = Field Length. Total Distance Traveled = Distance Traveled per Pass * Number of Passes along Width. Total Distance Traveled = 100 m * 25 passes = 2500 meters
   3. **Calculate Time Taken for Plowing:** Time = Distance/Speed Time = 2500 m/(4 km/h * 1000 m/km) = 0.625 hours

   Convert to Minutes: Time (minutes) = Time (hours) * 60 minutes/hour Time (minutes) = 0.625 hours * 60 minutes/hour = 37.5 minutes

These calculations show how self-driving tractors can cover a field for ploughing effectively while maximizing path planning and travel time. In order to ensure correct navigation, resource efficiency, and overall operational performance, mathematical considerations are essential.

5.2.8 Sorting and Packaging

Produce can be gathered and sorted by quality, size, and ripeness using robots. They can also efficiently pack the produce into cartons or containers. Robotic sorting and packing in agriculture entails categorizing and packaging harvested crops according to numerous standards such size, shape, color, and quality. In post-harvest procedures, these robotic devices increase efficiency and lower labor costs.

Below is a detailed description of how sorting and packing robotics work, along with relevant mathematical points:

1. **Sorting Procedure:** Robotic devices collect information about each harvested item using sensors and cameras. The qualities of this data are examined, and the object is then categorized according to those attributes.
2. **Packaging Method:** After the products have been sorted, robotic arms or mechanisms can package them into containers, boxes, or bags in accordance with predetermined criteria, such as a specific quantity per package.
3. **Mathematical Considerations:** Robotic sorting and packaging in agriculture involves several mathematical aspects:

   Quality Measurements: Establishing quantitative measurements to gauge qualities like size, color, and flaws. These measurements support correct item classification.

   Sorting Algorithms: Creating algorithms that classify items into various bins according to predetermined criteria using quality metrics.

Packaging Optimization: determining the best placement of things inside a packaging to maximize use of available space and reduce waste.

Quantity Distribution: Ensuring that a predetermined number of items are divided equally among packages while taking into account the size of the packages and the total number of items available.

4. **Case Study: Fruit Sorting and Packaging**. Let's consider a case where a robotic system is used to sort and package apples based on size. The system is set to pack 10 apples into each box, and apples are categorized into three sizes: Small, Medium, and Large. The harvested apples include the following quantities:

Small Apples: 150

Medium Apples: 300

Large Apples: 200

Calculations:

1. **Calculate the Total Number of Boxes for Each Size:** Number of Boxes for Small Apples = Total Small Apples/Apples per Box. Number of Boxes for Small Apples = 150 apples/10 apples/box = 15 boxes

   Number of Boxes for Medium Apples = Total Medium Apples/Apples per Box. Number of Boxes for Medium Apples = 300 apples/10 apples/box = 30 boxes

   Number of Boxes for Large Apples = Total Large Apples/Apples per Box. Number of Boxes for Large Apples = 200 apples/10 apples/box = 20 boxes

2. **Calculate the Total Number of Boxes:** Total Number of Boxes = Number of Boxes for Small Apples + Number of Boxes for Medium Apples + Number of Boxes for Large Apples. Total Number of Boxes = 15 boxes + 30 boxes + 20 boxes = 65 boxes

These calculations illustrate the effective sorting and packaging of harvested crops according to predetermined parameters using robotics. In order to maximize resource efficiency, ensure accurate categorization, and optimize the packing process, mathematical considerations are essential.

5.2.9 Labor Shortages

By automating monotonous and physically taxing work, robotics might help the agriculture sector deal with labor shortages, especially when human labor is expensive or in limited supply. In many parts of the world, the lack of agricultural workers has become a serious problem. Robotics can be extremely helpful in solving this problem by automating many processes and decreasing the need for human labor.

Let's delve into how robotics can mitigate labor shortages in agriculture, along with relevant mathematical considerations:

1. **Task Automation:** A variety of jobs, including as planting, harvesting, weeding, spraying, and more, can be performed by robotic devices. When carried out manually, these procedures can be time-consuming and labor-intensive.
2. **Labor Savings:** By comparing the time and effort needed for manual labor with the time taken by robotic systems to execute the identical tasks, the labor savings brought about by robotics can be estimated.
3. **Return on Investment (ROI):** By taking into account the initial investment in robotic technology, operational costs, and the possible increase in productivity and yield, farmers can assess the financial benefits of adopting robotics.

4. **Mathematical Considerations:** The assessment of labor shortages and the effect of robotics in agriculture involve several mathematical aspects:

   Calculating Labor Time: This involves figuring out how long it will take manual labor to perform a given activity while taking labor efficiency and task complexity into account.

   Robotic Task Time: Calculating the time required by robotic systems to complete the same activities while taking into account their efficiency and speed.

   Cost Comparison: weighing the costs of labor (wages, benefits, etc.) against the startup and ongoing expenses of putting robotic technologies into place.

   ROI Calculation: Calculating the return on investment by comparing the original cost of robotics with the economic returns of higher productivity, lower labor costs, and improved yield.

5. **Case Study: Harvesting Labor Shortage.** Let's consider a case where a farm is facing a labor shortage for harvesting. The farm has a rectangular field of dimensions:

   Field Width: 40 meters **Field Length:** 80 meters

   The harvest requirements are as follows:

   Number of Workers Needed: 10 workers

   Average Harvesting Rate per Worker: 20 kg/hour

   Calculations:

   1. **Calculate the Total Area:** Area = Field Width * Field Length = 40 m * 80 m = 3200 m^2
   2. **Calculate the Time Needed for Manual Harvesting:** Total Harvesting Rate = Number of Workers * Harvesting Rate per Worker. Total Harvesting Rate = 10 workers * 20 kg/hour = 200 kg/hour

      Time Needed for Manual Harvesting = Total Area/Total Harvesting Rate. Time Needed for Manual Harvesting = 3200 m^2/200 kg/hour = 16 hours

   These figures show how long it will take for physical labor to do the harvesting job. Robotics may be used to speed up the harvesting process, shorten the time required, and effectively address labor shortages.

   In conclusion, by automating jobs and boosting operational efficiency, robotics provides a solution to the labor deficit in agriculture. Evaluation of the effects of robotics on labor savings, return on investment, and total farm productivity heavily relies on mathematical factors.

5.2.10 Livestock Management

Robotic systems can help with livestock management by automating and humanely milking cows, monitoring health problems, and feeding cattle. Robotics are used in agriculture's livestock management to monitor, feed, and care for animals more effectively and humanely. Robotic systems can aid in enhancing farm productivity overall, lowering labor costs, and improving animal care.

Here's a detailed overview of how robotics is used for livestock management, along with relevant mathematical considerations:

1. **Automation of Feeding:** Robotic systems can feed animals in accordance with predetermined schedules or in response to signals from sensors that track the amount of food being consumed. The animals will receive constant and proper diet as a result.

2. **Health Monitoring:** Animals can have sensors attached to them to track health indicators including body temperature, heart rate, and level of activity. Robotic devices are able to spot irregularities and warn farmers of potential health problems.
3. **Automated Milking:** Cows can be automatically recognized and milked using robotic milking equipment. Without human assistance, they find udders using sensors and milk cows.
4. **Garbage Management:** By cleaning animal living areas and correctly disposing of garbage, robotic systems can assist manage waste by fostering improved hygiene and minimizing environmental effect.
5. **Mathematical Considerations:** Robotic cattle management involves a number of mathematical considerations:

   Improving feeding schedules, calculating the amount of feed needed for each animal, and reducing waste are all aspects of feeding efficiency.

   Algorithms for Health Monitoring: Creating algorithms that examine sensor data to find anomalous health situations, compute health indices, and forecast potential illnesses.

   Milking Optimization: Making use of mathematical models to determine the best time and length of milking in order to maximize cow comfort and milk output.

   Resource Allocation: To optimize resource allocation and lessen environmental effect, resource usage such as feed consumption, water consumption, and waste generation are calculated.
6. **Case Study: Feeding Automation** Let's consider a case where a robotic feeding system is used to automate the feeding of pigs on a farm. The farm has 200 pigs, and each pig requires 2 kg of feed per day.

   Calculations:

   1. **Calculate the Total Daily Feed Requirement:** Total Feed Requirement per Day = Number of Pigs * Feed per Pig. Total Feed Requirement per Day = 200 pigs * 2 kg/pig = 400 kg/day.
   2. **Calculate the Feeding Interval:** Feeding Interval = 24 hours/Number of Feedings per Day. Suppose pigs are fed twice a day (morning and evening): Feeding Interval = 24 hours/2 = 12 hours

      Convert to Minutes: Feeding Interval (minutes) = Feeding Interval (hours) * 60 minutes/hour. Feeding Interval (minutes) = 12 hours * 60 minutes/hour = 720 minutes
   3. **Calculate the Feed Rate per Minute:** Feed Rate per Minute = Total Feed Requirement per Day/Feeding Interval (minutes). Feed Rate per Minute = 400 kg/720 minutes ≈ 0.5556 kg/minute

   These calculations show how a robotic feeding system can be set up to feed pigs at a predetermined rate to satisfy their daily nutritional requirements. Optimizing feeding effectiveness, ensuring animal health, and fostering sustainable livestock management all depend heavily on mathematical considerations.

5.3 IMPLEMENTING ROBOTICS IN AGRICULTURE

Benefits of using robotics in agriculture include:

- **Increased Productivity:** Robots' ability to work continuously and precisely lowers human error rates.

- **Resource Conservation:** Robot-driven precision agriculture techniques can reduce the usage of water, fertilizers, and pesticides, which has a less negative influence on the environment.
- **Labor Savings:** With a labor deficit in the agricultural sector, robots can fill in the gaps by carrying out labor-intensive or repetitive jobs.
- **Data-Driven Decision Making:** Robotic systems can gather and analyze data, giving farmers the knowledge they need to make knowledgeable crop management decisions.
- **Lessened Environmental Impact:** Robotics can reduce the harmful ecological effects of conventional farming practices by using resources in a targeted manner.

Despite these benefits, there are also challenges in implementing robotics in agriculture:

- **High Initial Costs:** Small farms with limited resources may face difficulties due to the high initial costs of robotic technology.
- Robots must be able to adapt to various terrains, weather patterns, and crop varieties because agricultural environments are complicated and varied.
- **Technical Difficulties:** It might be technically difficult to create robots that can do delicate jobs like fruit picking without harming the produce.
- **Maintenance and Repair:** It can be difficult and expensive to maintain and repair robots in isolated and challenging agricultural situations.
- **Integration with Traditional Practices:** Adapting current agricultural processes and practices to incorporate robotic systems can be necessary.

Despite these challenges, ongoing research and development in the field of agricultural robotics are steadily addressing these issues, making robotics an increasingly viable option for modernizing and improving the agricultural sector.

5.4 PRECISION FARMING IN AGRICULTURE

Precision farming is often referred to as precision agriculture, uses cutting-edge technologies, such as robotics, to optimize agricultural practices and resource utilization on a field-by-field basis. Through measures that are tailored to the particular requirements of each sector of the industry, it seeks to increase productivity, decrease waste, and promote sustainability.

Let's explore precision farming using robotics in detail, along with relevant mathematical considerations:

1. **Data Gathering and Analysis:** The first step in precision farming is gathering data from multiple sources, including sensors, drones, robotic devices, and satellite photography. This data contains details about the properties of the soil, the health of the crops, the weather, and other things.
2. **Spatial Analysis and Mapping:** Using Geographic Information Systems (GIS), the gathered data is converted into thorough maps that highlight the variation of many parameters throughout the field. This aids in locating locations with particular needs or problems.
3. **Variable Rate Technology (VRT):** Using insights from data analysis and mapping, VRT applies inputs including water, fertilizer, insecticides, and seeds at different rates across the field.
4. **Robotic Systems in Precision Farming:** Robotics is crucial for putting precision farming techniques into practice.

   Autonomous Navigation: Robotic vehicles go across the field on their own, gathering information and carrying out different activities in response to real-time input from sensors.

Actuation Systems: Based on particular demands found through data analysis, robotic arms or mechanisms can carry out operations like planting, spraying, or even carrying out targeted interventions.

Data Integration: Robotic systems combine data from numerous sensors to make judgments in real-time, assuring the best course of action in response to shifting circumstances.

5. **Mathematical Considerations:** Precision farming with robotics involves a number of mathematical considerations:

   Interpolation Techniques: To interpolate data points between sampled locations and produce thorough maps of field conditions, mathematical models are used.

   Yield Mapping: Using information gathered during harvest to determine average yields in various fields, helping farmers to make wise decisions.

   Optimization Algorithms: Using optimization algorithms to decide on the optimal resource allocation based on geographic information, crop needs, and economic considerations.

   Spatial Analysis: Using statistical and geo-statistical techniques, spatial patterns and trends in field data are analyzed.

6. **Case Study: Variable Rate Fertilization**. Let's consider a case where a robotic system is used for variable rate fertilization in a rectangular field. The field has the following dimensions:

   Field Width: 60 meters **Field Length:** 100 meters

   Soil testing indicates varying nutrient levels across the field, and the following fertilization rates are recommended:

   Low-Nutrient Areas: Apply 150 kg/ha of fertilizer

   Medium-Nutrient Areas: Apply 250 kg/ha of fertilizer

   High-Nutrient Areas: Apply 350 kg/ha of fertilizer

   Calculations:

   1. **Calculate the Total Area:** Area = Field Width * Field Length = 60 m * 100 m = 6000 m^2.
   2. **Calculate the Total Fertilizer Amounts for Each Area:** Total Fertilizer for Low-Nutrient Areas = 150 kg/ha * (6000 m^2/10,000 m^2/ha) = 90 kg. Total Fertilizer for Medium-Nutrient Areas = 250 kg/ha * (6000 m^2/10,000 m^2/ha) = 150 kg. Total Fertilizer for High-Nutrient Areas = 350 kg/ha * (6000 m^2/10,000 m^2/ha) = 210 kg

   These calculations show how robotic systems may apply different fertilizer rates depending on the available space, maximizing resource efficiency and fostering precision farming. For agricultural production to be effective and sustainable, mathematical considerations are crucial.

5.5 POTENTIAL BENEFITS FOR FOOD PRODUCTION

The use of robotics in agriculture has many potential advantages for the production of food. These advantages include better precision and efficiency as well as less resource waste and higher sustainability.

Let's examine these potential advantages in further detail and discuss pertinent mathematical issues.

1. **Greater Efficiency:** Compared to human labor, robotic systems are more productive since they can operate continuously without becoming tired. Metrics like job completion time and throughput can be used to measure this improved efficiency.

2. **Precision Farming:** Robotic technologies make it possible to perform precision tasks including accurate planting, spraying, and harvesting. This accuracy reduces the need for inputs like water, fertilizer, and pesticides, which saves money and has a smaller negative impact on the environment.
3. **Resource Optimization:** Sensors for soil moisture, nutrient levels, and other characteristics can be added to robotic systems. The best possible distribution of resources can then be determined via optimization algorithms, minimizing waste and maximizing yield.
4. Robotic automation decreases the demand for physical labor, particularly in repetitive and labor-intensive jobs. Based on the amount of hours performed by robots as opposed to human workers, the labor savings can be determined.
5. **Data-Driven Decision Making:** Robotic systems gather information on soil properties, meteorological conditions, and crop health. Mathematical models can be used to analyze this data and help you decide on planting, irrigation, and other practices.
6. **Decreased Chemical Usage:** Robotic precision application of pesticides and fertilizers lowers over application and consequently, chemical usage. Chemical quantity reductions can be measured in terms of financial savings and environmental impact.
7. **Output Improvement:** Robotic systems can help to improve plant health and increase output by taking care of issues like planting depth, spacing, and nutrient application. By comparing the average yield with and without robotic interventions, this increase can be estimated.
8. **Timely Harvesting:** By operating at the best time for each crop, robotic harvesting systems can guarantee that produce is harvested at the height of its nutritional value and quality.
9. **Environmental Sustainability:** By reducing resource use, chemical runoff, and soil erosion, robotic precision farming lessens its negative effects on the environment. Metrics like a smaller carbon footprint and water savings can be used to measure environmental advantages.
10. **Scalability:** Because robotic systems can be scaled to efficiently cover huge regions, farmers may manage larger farms with fewer resources.

Mathematical Considerations: Mathematical calculations can frequently be used to quantify the possible advantages of deploying robotics in agriculture:

- **Cost Savings:** Figuring out how much less money is spent on labor, resources, and waste disposal as a result of robotic interventions.
- **Yield Increase:** Calculating the percentage increase in crop yield brought about by robotic operations such as precise planting, fertilization, and other tasks.
- Measuring the decrease in chemical and water use as well as the improvement in resource utilization efficiency.
- **Return on Investment (ROI):** This metric compares the initial cost of robotic technology to the entire financial returns, such as higher yield and lower costs.
- **Environmental Influence:** Measuring how using fewer chemicals and managing resources more effectively has a positive influence on the environment, such as reducing greenhouse gas emissions.

These mathematical factors are essential for determining the concrete advantages that robotics can have for agricultural food production. They support the adoption of robotic technologies by farmers and stakeholders, enhancing agricultural practices and ensuring sustainable food production.

5.6 POTENTIAL BENEFITS FOR SUSTAINABILITY IN AGRICULTURE

Using robotics in agriculture could have a number of positive effects on sustainability, promoting eco-friendly practices and long-term resource preservation. These advantages include lessened resource use and environmental effect, enhanced soil health, and less waste.

Let's explore these potential benefits in detail, along with relevant mathematical considerations:

1. Robotics allows for accurate and focused resource deployment, reducing wastage and misuse. This reduces the need for pesticides, fertilizers, and water, saving money and protecting the environment.
2. **Precision Farming:** By performing activities with great accuracy, robotic devices can optimize planting, fertilizing, and harvesting. This accuracy determines the precise amounts required for each region and lowers the danger of under or overusing resources.
3. **Improving Soil Health:** Robotic technology can keep an eye on the state of the soil and provide nutrients only where they are required. This preserves soil health and encourages sustainable land use by preventing soil degradation and nutrient discharge.
4. **Energy Efficiency:** Robotic systems may run on renewable energy, including solar power, which lessens dependency on fossil fuels and lowers the carbon impact of agricultural activities.
5. **Lessened Chemical Impact:** Robotic pesticide and herbicide administration with precision reduces the environmental effects of chemical use, reducing pollution and safeguarding ecosystems.
6. **Soil Erosion Prevention:** Robotics can aid in preventing soil erosion by enhancing planting and cultivating techniques, which preserves healthy topsoil and upholds ecosystem equilibrium.
7. **Minimizing Waste caused by Early or Excessive Harvesting:** Precision harvesting utilizing robotics ensures that only ripe produce is collected.
8. **Biodiversity Preservation:** Robotics-enabled sustainable farming methods and reduced chemical use help protect regional ecosystems and biodiversity.
9. **Water Conservation:** Robotic systems can track soil moisture levels and only use irrigation when it's essential, conserving water resources and tracking the water that was saved.
10. **Reducing Carbon Footprint:** Robotic technologies optimize workflows, resulting in less travel, energy use, and emissions. Based on the energy savings made, it is possible to compute the reduction in carbon emissions.

Mathematical Considerations: Mathematical calculations can frequently be used to estimate the possible sustainability benefits of deploying robotics in agriculture:

- **Resource Savings:** Calculating the economic savings and environmental advantages of using less water, fertilizer, and pesticides.
- **Energy Efficiency:** Calculating the amount of energy saved when robotic operations are powered by renewable energy sources.
- **Carbon Footprint:** Calculating the reduction in carbon emissions brought on by less travel, less energy use, and less chemical use.
- **Improving Soil Health:** Measuring the decrease in soil erosion and nutrient runoff, which results in better soil health and long-term sustainability.
- **Biodiversity and Ecosystem Impact:** Monitoring the biodiversity and general well-being of the local fauna as well as the preservation of local ecosystems.

These mathematical factors are crucial for assessing the concrete sustainability advantages that robotics in agriculture can offer. They support decision-making by stakeholders about the adoption of robotic technologies to accomplish sustainable and ecologically friendly farming practices.

Numerical Problems with Solutions

Problem 1. A farmer wants to use a robotic tractor to plant crops in a field that is 100 meters long and 50 meters wide. The robotic tractor can plant 1 row of crops every 2 meters. How many rows can the tractor plant in one pass, and how many passes are needed to plant the entire field?

Solution: Number of rows planted in one pass = Width of the field/Spacing between rows. Number of rows planted in one pass = 50 meters/2 meters = 25 rows

Number of passes needed to plant the entire field = Length of the field/Width covered in one pass. Number of passes needed = 100 meters/50 meters = 2 passes

Problem 2. A drone is used to monitor a wheat field that is 500 meters long and 200 meters wide. The drone flies at an altitude of 50 meters and captures images with a resolution of 10 pixels per square meter. How many pixels are in one image, and how many images are needed to cover the entire field?

Solution: Number of pixels in one image = Field area * Resolution. Number of pixels in one image = (500 m * 200 m) * (10 pixels/m^2) = 1,000,000 pixels

Total number of images needed to cover the field = Field area/Area covered by one image. Total number of images = (500 m * 200 m)/(500 m * 50 m) = 4 images

Problem 3. A robotic weed control system operates in a field and can identify and remove weeds at a rate of 2 weeds per second. If there are 1,000 weeds in the field, how many seconds will it take for the system to remove all the weeds?

Solution: Number of seconds to remove all weeds = Total number of weeds/Removal rate. Number of seconds = 1,000 weeds/(2 weeds/second) = 500 seconds

Problem 4. A tractor is equipped with GPS guidance and autonomous steering. The tractor is programmed to plow a field in straight rows. If the field is 300 meters long, and the tractor can maintain a straight row accuracy of 5 centimeters, how much overlap will there be between adjacent rows?

Solution: Overlap between adjacent rows = Straight row accuracy/Width of the field Overlap = 0.05 meters/300 meters = 0.00167 or 0.167%

Problem 5. A soil sampling robot is programmed to collect soil samples from a square field that is 150 meters on each side. It collects a sample every 10 meters. How many samples will the robot collect in one pass?

Solution: Number of samples collected in one pass = Side length of the field/Spacing between samples. Number of samples = 150 meters/10 meters = 15 samples

Problem 6. A robotic harvester is tasked with picking apples from an orchard. It can pick 10 apples per minute. If there are 5000 apples in the orchard, how long will it take for the robot to pick all the apples?

Solution: Time = (Number of Apples)/(Rate of Picking). Time = 5000 apples/10 apples per minute. Time = 500 minutes

Problem 7. A drone is used for crop spraying in a field. It can cover an area of 2 acres in 20 minutes. If the entire field is 10 acres, how long will it take to spray the entire field?

Solution: Time = (Total Area)/(Area Covered per Minute). Time = 10 acres/2 acres per 20 minutes. Time = 50 minutes

Problem 8. A robotic weeder can operate for 8 hours on a single battery charge. If it consumes 0.5 kWh of energy per hour, how much energy does it consume during a full day of operation?

Solution: Energy Consumption = (Operating Hours) × (Energy Consumption per Hour). Energy Consumption = 8 hours × 0.5 kWh per hour. Energy Consumption = 4 kWh

Problem 9. A robotic tractor is used for plowing a field. It can plow at a speed of 5 miles per hour. If the field is 2 miles long, how long will it take to plow the entire length of the field?

Solution: Time = (Distance)/(Speed). Time = 2 miles/5 miles per hour. Time = 0.4 hours or 24 minutes

Problem 10. A drone is used to monitor crop health by capturing images of a 100-acre field. It can capture images at a rate of 5 acres per minute. How long will it take to capture images of the entire field?

Solution: Time = (Total Area)/(Area Captured per Minute). Time = 100 acres/5 acres per minute. Time = 20 minutes

Problem 11. A robotic fruit sorter can process 1000 fruits per hour. If there are 10,000 fruits to be sorted, how many hours will it take to complete the sorting?

Solution: Time = (Number of Fruits)/(Processing Rate per Hour). Time = 10,000 fruits/1000 fruits per hour. Time = 10 hours

Problem 12. A robotic irrigation system pumps water at a rate of 500 liters per minute. If it needs to irrigate a field with 5,000 liters of water, how long will it take to complete the irrigation?

Solution: Time = (Water Required)/(Pumping Rate per Minute). Time = 5,000 liters/500 liters per minute. Time = 10 minutes

Problem 13. A robotic planting machine can plant 200 seeds per minute. If there are 5,000 seeds to be planted, how long will it take to complete the planting?

Solution: Time = (Number of Seeds)/(Planting Rate per Minute). Time = 5,000 seeds/200 seeds per minute. Time = 25 minutes

Problem 14. A drone is used for crop scouting and can cover a distance of 10 kilometers in 15 minutes. If the scouting area is 40 square kilometers, how long will it take to scout the entire area?

Solution: Time = (Total Area)/(Area Covered per Minute). Time = 40 square kilometers/(10 kilometers/15 minutes). Time = 60 minutes or 1 hour

Problem 15. A robotic milking machine can milk 20 cows per hour. If there are 200 cows on a dairy farm, how many hours will it take to milk all the cows?

Solution: Time = (Number of Cows)/(Milking Rate per Hour). Time = 200 cows/20 cows per hour. Time = 10 hours.

Precision Planting

Problem 16. A robot is planting corn seeds in a field. It needs to plant 10,000 seeds in a 1-acre area. If the recommended planting density is 30,000 seeds per acre, how many seeds should the robot plant per square foot?

Solution: Planting density = 30,000 seeds/acre. Seeds per square foot = (30,000 seeds/acre)/(43,560 sq ft/acre) ≈ 0.689 seeds/sq ft.

Robotic Weeding

Problem 17. A weeding robot can cover a 2-acre field in 5 hours. If the field size is 10 acres, how many hours will it take the robot to complete the entire weeding operation?

Solution: Time to complete 10 acres = (10 acres)/(2 acres/5 hours) = 25 hours.

Autonomous Irrigation

Problem 18. A crop field has an area of 50,000 square meters. A robotic irrigation system is programmed to deliver 5 liters of water per square meter when the soil moisture drops below a certain threshold. How much water will the system deliver when the entire field needs irrigation?

Solution: Total water delivered = (5 liters/sq m) * (50,000 sq m) = 250,000 liters

Pest Detection Drone

Problem 19. A pest-detection drone covers an area of 2 square kilometers in 30 minutes. If a 10-square-kilometer field needs monitoring, how many hours will it take the drone to cover the entire field?

Solution: Time to cover 10 sq km = (10 sq km)/(2 sq km/30 min) = 150 minutes = 2.5 hours.

Robotic Harvesting Efficiency

Problem 20. A robotic harvesting system harvested 800 kilograms of apples in 4 hours. What is the harvesting rate in kilograms per hour?

Solution: Harvesting rate = (800 kg)/(4 hours) = 200 kg/hour.

Soil Nutrient Management

Problem 21. A robot analyzed soil data and recommended applying 100 kg of nitrogen fertilizer to a 2-hectare field. What is the recommended application rate in kg/hectare?

Solution: Application rate = (100 kg)/(2 hectares) = 50 kg/hectare.

Robotic Greenhouse Management

Problem 22. In a greenhouse, a robot maintains the temperature at 25°C by turning on the heating system when the temperature drops to 23°C and turning it off at 27°C. If the temperature is currently 22°C, how long will the heating system stay on to reach the target temperature?

Solution: Time heating system stays on = (25°C – 22°C)/(27°C – 23°C) = 0.75 (or 45 minutes).

Aquatic Weed Removal

Problem 23. An aquatic weed removal drone can clear 100 square meters of water surface in 20 minutes. If a pond is 500 square meters, how many hours will the drone take to clear the entire pond?

Solution: Time to clear 500 sq m = (500 sq m)/(100 sq m/20 min) = 2 hours.

Robotic Pollination

Problem 24. A robotic pollinator takes 10 seconds to move pollen from one flower to another. If there are 300 flowers in a greenhouse, how long will it take the robot to pollinate all the flowers?

Solution: Time to pollinate 300 flowers = (300 flowers) * (10 seconds/flower) = 3000 seconds = 50 minutes.

Precision Agriculture ROI

Problem 25. A farmer invested $10,000 in a robotic precision planting system. The system increased crop yield by 15%, resulting in an additional profit of $5,000. Calculate the return on investment (ROI).

Solution: ROI = (Additional Profit – Investment)/Investment ROI = ($5,000 – $10,000)/$10,000 = –0.5 or –50% (negative ROI).

Short Questions and Answers

Q.1. What is precision agriculture?

Ans. Precision agriculture is the use of technology, including robotics, to optimize farming practices and resource utilization on a field-specific basis.

Q.2. How can robotics help in planting crops?

Ans. Robotic planting systems can ensure precise seed placement, spacing, and depth, leading to uniform and efficient crop growth.

Q.3. What is the purpose of robotic weeding?

Ans. Robotic weeding helps control weeds in fields by using sensors and algorithms to selectively target and remove unwanted plants.

Q.4. How do robots assist in pest control?

Ans. Robotic pest control systems use sensors and cameras to identify pests and apply targeted interventions, reducing the need for chemical treatments.

Q.5. What is the benefit of using robots for harvesting crops?

Ans. Robotic harvesting systems can improve efficiency and reduce labor costs while ensuring produce is harvested at optimal ripeness.

Q.6. How do robots contribute to data collection in agriculture?

Ans. Robotic systems equipped with sensors and cameras collect data on soil conditions, crop health, and environmental factors.

Q.7. What is the role of robotics in livestock management?

Ans. Robotics can automate tasks like feeding, health monitoring, and milking, improving efficiency and animal welfare.

Q.8. How do robots optimize water usage in agriculture?

Ans. Robotic irrigation systems can monitor soil moisture levels and apply water only where needed, conserving water resources.

Q.9. What is GIS in precision farming?

Ans. Geographic Information Systems (GIS) are used to create maps that show the spatial variability of factors like soil types and yields in precision agriculture.

Q.10. How can robots reduce chemical usage on farms?

Ans. Robotic technologies enable targeted application of chemicals, reducing the amount of pesticides and fertilizers used.

Q.11. What is the main advantage of using robotics for farming tasks?

Ans. The main advantage is increased efficiency, as robots can work consistently and precisely without tiring.

Q.12. What role do sensors play in agricultural robotics?

Ans. Sensors provide data about soil moisture, temperature, crop health, and other parameters, helping robots make informed decisions.

Q.13. How do robots assist in greenhouse management?

Ans. Robotic systems can control temperature, humidity, lighting, and irrigation in greenhouses, optimizing plant growth conditions.

Q.14. What is the potential ROI of adopting agricultural robotics?

Ans. Return on Investment (ROI) can include savings from reduced labor costs, increased yield, and optimized resource usage.

Q.15. How do robotic systems navigate in fields?

Ans. Robotic navigation uses GPS technology and path planning algorithms to move autonomously through fields.

Q.16. What is variable rate technology in agriculture?

Ans. Variable rate technology involves applying inputs like water and fertilizers at different rates across the field based on data insights.

Q.17. How can robotics improve soil health?

Ans. Robotic systems can analyze soil data and apply nutrients where needed, preventing soil degradation and nutrient imbalances.

Q.18. What is the concept of precision planting?

Ans. Precision planting involves accurately placing seeds at the optimal depth and spacing for uniform crop growth.

Q.19. How can robotic systems contribute to waste reduction?

Ans. Robotic harvesting ensures only ripe produce is collected, reducing waste due to overharvesting or damage.

Q.20. What is the benefit of robotic milking in dairy farming?

Ans. Robotic milking systems can increase efficiency by automating the milking process and improving cow comfort.

Q.21. How can robotics enhance resource efficiency in irrigation?

Ans. Robotic irrigation systems optimize water usage by delivering water precisely where it's needed, reducing waste.

Q.22. How do robots assist in waste management on farms?

Ans. Robotic waste management systems can clean animal living spaces and manage waste, promoting hygiene.

Q.23. What is the advantage of using drones in precision agriculture?

Ans. Drones can provide aerial imagery and data for field analysis, allowing farmers to monitor crop health and assess field conditions.

Q.24. How can robotics contribute to sustainable farming practices?

Ans. Robotic technologies reduce resource waste, minimize chemical usage, and promote efficient land and water management.

Q.25. How does automated data collection benefit farmers?

Ans. Automated data collection provides real-time insights into crop health, allowing farmers to make informed decisions for timely interventions.

Q.26. How do robots improve labor shortage challenges in agriculture?

Ans. Robotic systems can perform tasks that require manual labor, addressing labor shortages and reducing dependency on human workers.

Q.27. What is the role of AI in agricultural robotics?

Ans. Artificial Intelligence (AI) enables robots to analyze data, make decisions, and adapt to changing conditions, improving their efficiency and effectiveness.

Q.28. How can robotics help with sustainable pest management?

Ans. Robotic pest control systems can identify pest outbreaks early, enabling targeted interventions that minimize chemical usage.

Q.29. What is the potential impact of using robots in reducing greenhouse gas emissions?

Ans. Robotic operations powered by renewable energy sources can reduce greenhouse gas emissions by decreasing reliance on fossil fuels.

Q.30. How do robots contribute to biodiversity preservation on farms?

Ans. Reduced chemical usage and sustainable practices promoted by robots contribute to the preservation of local ecosystems and wildlife.

Q.31. What is the main advantage of using robotics for data collection?

Ans. Robotic data collection is more accurate and consistent compared to manual methods, leading to better decision-making.

Q.32. How do robots optimize planting density?

Ans. Robotic planting systems calculate and adjust planting density based on the desired spacing and crop requirements.

Q.33. How can robots help with crop monitoring during growth stages?

Ans. Robotic systems equipped with cameras and sensors can monitor crop growth, detect abnormalities, and assess nutrient needs.

Q.34. What is the benefit of using autonomous tractors for plowing fields?

Ans. Autonomous tractors can operate without human intervention, improving efficiency and reducing labor requirements.

Q.35. How do robots contribute to water quality improvement?

Ans. Robotic systems can minimize nutrient runoff by precisely applying fertilizers, reducing water pollution.

Q.36. What is the concept of controlled environment agriculture?

Ans. Controlled environment agriculture involves using robotics and technology to optimize indoor farming conditions for year-round production.

Q.37. How do robots enhance data-driven decision making in agriculture?

Ans. Robots collect data on soil conditions, weather, and crop health, which can be analyzed to make informed decisions about planting and resource allocation.

Q.38. What is the potential economic impact of adopting robotics in agriculture?

Ans. Adopting robotics can lead to increased yield, reduced operational costs, and improved profitability for farmers.

Q.39. How can robotics improve food safety in agriculture?

Ans. Robotic systems ensure that produce is harvested, sorted, and packaged under controlled conditions, reducing the risk of contamination.

Q.40. What role do robotic sensors play in irrigation management?

Ans. Sensors monitor soil moisture levels and transmit data to robotic irrigation systems, enabling precise water application.

Q.41. How do robotic systems contribute to efficient fertilizer application?

Ans. Robots can apply fertilizers based on soil nutrient levels, optimizing distribution and reducing overuse.

Q.42. What is the benefit of using robotics for labor-intensive tasks?

Ans. Robots can perform tasks consistently and without breaks, reducing the need for human labor and addressing labor shortage challenges.

Q.43. How can robotic systems improve traceability in agriculture?

Ans. Robotic data collection and labeling systems can track the origin and journey of produce, enhancing traceability and quality control.

Q.44. What is the role of robotics in reducing soil compaction?

Ans. Robotic vehicles can distribute weight evenly, reducing soil compaction caused by heavy machinery and preserving soil structure.

Q.45. How do robots contribute to reducing waterborne diseases in aquaculture?

Ans. Robotic systems can monitor water quality and oxygen levels in aquaculture ponds, preventing conditions that promote disease outbreaks.

Q.46. What is the benefit of using robotics for post-harvest sorting and packaging?

Ans. Robotic systems ensure accurate sorting and packaging, reducing waste and improving the quality of packaged produce.

Q.47. How do robotic systems optimize crop rotation strategies?

Ans. Robotic data collection and analysis help farmers make informed decisions about crop rotation based on soil health and historical yield data.

Q.48. What is the role of robotics in reducing food waste?

Ans. Robotic systems can detect and sort damaged or spoiled produce, ensuring that only high-quality items reach consumers.

Q.49. How can robotic systems help with invasive species management?

Ans. Robotic technologies can identify and remove invasive plant species, preventing them from spreading and affecting native ecosystems.

Q.50. What is the potential impact of robotics on soil erosion prevention?

Ans. Robotic planting and cultivation techniques can be optimized to minimize soil disturbance and erosion, preserving soil health and structure.

Multiple Choice Questions and Answers

Q.1. What is precision agriculture?

(*a*) Large-scale farming
(*b*) Random farming practices
(*c*) Data-driven and precise farming
(*d*) Traditional farming techniques

Q.2. Which technology is commonly used for navigation in agricultural robotics?

(*a*) Bluetooth
(*b*) WiFi
(*c*) GPS
(*d*) NFC

Q.3. How do robots contribute to precision planting?

(*a*) By planting seeds randomly
(*b*) By planting seeds at precise depths and spacing
(*c*) By using excessive amounts of seeds
(*d*) By avoiding planting altogether

Q.4. Robotic systems can be used for:

(*a*) Inefficient resource usage
(*b*) Random chemical spraying
(*c*) Precision application of inputs
(*d*) Overuse of fertilizers

Q.5. What is the purpose of robotic weeding?

(*a*) To promote weed growth
(*b*) To selectively target and remove weeds
(*c*) To increase chemical usage
(*d*) To automate soil tilling

Q.6. How can robots help in pest control?

(*a*) By attracting pests
(*b*) By promoting pest growth
(*c*) By using targeted interventions
(*d*) By ignoring pest issues

Q.7. Robotic systems are primarily designed to:

(*a*) Replace human workers entirely
(*b*) Increase labor costs
(*c*) Reduce efficiency
(*d*) Improve efficiency and precision

Q.8. What is the main advantage of using robots for harvesting crops?

(*a*) Decreased efficiency
(*b*) Increased labor costs
(*c*) Improved efficiency and optimal harvesting time
(*d*) Increased waste

Q.9. How do robotic systems assist in data collection in agriculture?

(*a*) By performing manual data entry (*b*) By analyzing satellite images

(*c*) By using sensors and cameras (*d*) By relying on human observations

Q.10. What is the primary goal of using robotics in livestock management?

(*a*) To reduce efficiency (*b*) To increase labor demand

(*c*) To improve animal welfare and efficiency (*d*) To decrease animal care

Q.11. What role do sensors play in agricultural robotics?

(*a*) They make robots visually appealing (*b*) They help robots fly

(*c*) They provide data for decision-making (*d*) They are used for decoration

Q.12. Robotic technologies can be used to optimize:

(*a*) Resource waste (*b*) Chemical overuse (*c*) Planting density (*d*) Human errors

Q.13. What is GIS in precision farming?

(*a*) Global Inventory System (*b*) Geographical Imaging System

(*c*) Geographic Information Systems (*d*) Geological Identification System

Q.14. How do robotic systems navigate in fields?

(*a*) By following human guidance (*b*) By using radar

(*c*) By GPS technology and algorithms (*d*) By using magnetic fields

Q.15. What is variable rate technology in agriculture?

(*a*) Using different types of vehicles

(*b*) Applying inputs at varying rates across the field

(*c*) Planting different crops in the same field

(*d*) Changing the field location frequently

Q.16. What is the benefit of robotic milking in dairy farming?

(*a*) Decreased efficiency (*b*) Reduced cow comfort

(*c*) Improved efficiency and cow comfort (*d*) Increased milk wastage

Q.17. What is the role of AI in agricultural robotics?

(*a*) Artificial Intelligence helps robots fly

(*b*) AI prevents robots from performing tasks

(*c*) AI analyzes data and makes decisions

(*d*) AI increases robotic inefficiency

Q.18. How do robotic systems optimize water usage in agriculture?

(*a*) By over-irrigating fields (*b*) By using excessive water

(*c*) By applying water precisely where needed (*d*) By avoiding water usage

Q.19. What is the concept of controlled environment agriculture?

(*a*) Using random environmental conditions

(*b*) Optimizing indoor farming conditions using robotics

(*c*) Ignoring environmental factors in agriculture
(*d*) Using outdoor conditions for farming

Q.20. What is the potential impact of using robots on soil erosion prevention?

(*a*) Increase soil erosion
(*b*) No impact on soil erosion
(*c*) Decrease soil erosion
(*d*) Promote soil compaction

Key to Multiple Choice Questions

1. (*c*) **2.** (*c*) **3.** (*b*) **4.** (*c*) **5.** (*b*) **6.** (*c*) **7.** (*d*) **8.** (*c*) **9.** (*c*)
10. (*c*) **11.** (*c*) **12.** (*c*) **13.** (*c*) **14.** (*c*) **15.** (*b*) **16.** (*c*) **17.** (*c*) **18.** (*c*)
19. (*b*) **20.** (*c*)

Project Works

1. **Precision Planting Robot:** Develop a robot that can precisely plant seeds in a field based on soil data and planting density requirements.

   Hints: The robot can be equipped with sensors to measure soil properties and use algorithms to determine the optimal seed placement, depth, and spacing.

2. **Robotic Weed Removal System:** Create a robotic system that can identify and remove weeds in a crop field without affecting the crops.

   Hints: The robot can use image recognition techniques to identify weeds and robotic arms to remove them while avoiding damage to the main crops.

3. **Autonomous Irrigation System:** Build an autonomous robot that monitors soil moisture levels and irrigates crops as needed to ensure efficient water usage.

   Hints: The robot can use soil moisture sensors to collect data and algorithms to control irrigation based on real-time conditions.

4. **Pest Detection and Monitoring Drone:** Develop a drone that can fly over fields, detect pests, and collect data on pest populations for targeted intervention.

   Hints: The drone can be equipped with cameras and image analysis software to identify pests and calculate their density in different parts of the field.

5. **Robotic Harvesting Machine:** Design a robot capable of autonomously harvesting crops such as fruits or vegetables to improve efficiency and reduce labor costs.

   Hints: The robot can use sensors and cameras to detect ripe produce, robotic arms for gentle harvesting, and algorithms to optimize the harvesting process.

6. **Soil Health Monitoring Rover:** Create a rover that can move through fields to collect data on soil health parameters, such as nutrient levels and compaction.

   Hints: The rover can carry sensors to measure soil properties, and the data can be used to make recommendations for soil management practices.

7. **Robotic Greenhouse Management System:** Build a system of robots that collectively manage various aspects of a greenhouse, including temperature, humidity, and irrigation.

   Hints: The robots can communicate with each other and use sensors to monitor environmental conditions, adjusting them as needed for optimal plant growth.

8. **Automated Fertilization and Nutrient Management:** Develop a robot that analyzes soil nutrient data and applies fertilizers precisely where and when they are needed.

   Hints: The robot can use soil testing results to calculate the right amount and type of fertilizers, minimizing waste and improving nutrient utilization.

9. **Aquatic Weed Removal Drone:** Create a drone that can navigate over water bodies and remove aquatic weeds to maintain water quality and aquatic ecosystems.

   Hints: The drone can be equipped with tools to cut or pull out weeds while avoiding damage to desirable plants and aquatic life.

10. **Robotic Pollination System:** Design a robot that mimics pollination by bees, aiding in pollinating crops in controlled environments.

    Hints: The robot can be programmed to move pollen from one flower to another, ensuring effective pollination and promoting crop yield.

6 Robots in Education

6.1 INTRODUCTION

At recent years, robots have moved beyond their traditional industrial uses and taken centre stage at educational institutions. The interactive and programmable nature of educational robots enables students to direct their actions and behavior using coding and programming languages. These robots come in a variety of levels of complexity, from easy-to-use tactile bots for young children to sophisticated platforms that require students to solve challenging challenges. Robotics in education are revolutionizing learning approaches by encouraging participation, creativity, and problem-solving abilities. Educational robots are forming future generations of learners who are prepared for the challenges of the digital age by promoting a thorough understanding of STEM ideas and encouraging teamwork. The use of robots in teaching is poised to grow as technology develops, enriching the learning experiences of students around the world.

6.2 INTEGRATION OF ROBOTS IN EDUCATION

The integration of robots in education offers a multitude of benefits that foster holistic learning:

1. **Active Learning:** Students actively engage with concepts through constructing, programming, and experimentation when using robots to learn.
2. **STEM Engagement:** Robotics piques interest in STEM fields by offering practical experiences and real-world applications that simplify complicated theory.
3. **Problem Solving:** As students work to fix the behaviors of their robots, they develop resilience and adaptability by learning to solve difficulties via trial and error.
4. **Creativity:** As they design and program their robots to carry out certain tasks, students can use robotics as a platform to think creatively and innovatively.
5. **Collaboration:** Robotics frequently requires cooperation, which fosters collaborative abilities as students cooperate to accomplish shared objectives.
6. **Critical Thinking:** When programming robots, students must exercise critical thought, foresee difficulties, and plan ahead to obtain desired results.

7. **Career Readiness:** Students that are exposed to robotics technology have the abilities necessary to meet the demands of the modern workforce, particularly in vocations involving technology.
8. **Inclusivity:** Robots level the playing field for learning by taking into account a range of learning styles and aptitudes.

6.3 EDUCATIONAL ROBOTS

Some very popular educational robots are described below.

6.3.1 Educational Robots: LEGO Mindstorms

With the help of LEGO bricks and a visual programming interface, these kits let students construct and program robots. With the help of robotics, coding, and LEGO building blocks, the well-known educational program LEGO Mindstorms teaches children about engineering, coding, and problem-solving. It includes bricks, sensors, and motors as well as software tools for controlling and programming robots. The platform is a great resource for hands-on STEM teaching since it allows students to design, construct, and program their robots to carry out different jobs.

Let's explore the technology, theory, and related mathematics behind LEGO Mindstorms:

Technology:

1. **Brick:** The programmable brick serves as the primary controller in LEGO Mindstorms. Students may build responsive robots using it since it has a microprocessor, memory, inputs (sensors), and outputs (motors).
2. **Sensors:** There are several different types of sensors available with LEGO Mindstorms, including touch sensors (buttons), light sensors, ultrasonic sensors (used for measuring distance), and gyro sensors (used for measuring rotation). The robot can interact with its environment thanks to the input from these sensors.
3. **Motors:** The robot can move thanks to its motors. The robot can move, turn, or carry out certain tasks thanks to the students' ability to program the motors to rotate at various rates and directions.
4. **Building Components:** Students can design and construct their robots using a variety of LEGO building components, such as gears, axles, beams, and connectors.
5. **Programming Software:** Students can build sequences of actions and behaviours for their robots using the visual programming interface provided by the LEGO Mindstorms software.

Theory:

1. **Robotics Concepts:** Students are introduced to robotics concepts like sensors, actuators, control loops, and feedback mechanisms through the use of LEGO Mindstorms. They discover how robots perceive their surroundings and modify their behaviour in response to sensor data.
2. Visual programming blocks are used to teach students programming principles. They comprehend coding fundamentals like sequences, loops, conditions, and events.
3. **Sensor Integration:** By incorporating sensor data into their programmes, students learn how to create robots that can react to their immediate environment. For instance, a robot can be made to follow a dark line on the ground thanks to a light sensor.
4. **Mechanical Engineering:** Students learn about mechanical concepts like leverage, gear ratios, and stability by designing robots with LEGO pieces.

Related Mathematics:

1. **Geometry:** Students plan and create their robots using geometry. To guarantee that their designs function as planned, they calculate distances, angles, and measurements.
2. **Measuring:** Students work with measurements of distance, rotation, and time while utilizing sensors like gyro or ultrasonic ones.
3. **Mathematical Operations:** In order to control motor speed, determine distances, and process sensor data, students use mathematical operations including addition, subtraction, multiplication, and division in programming.
4. **Proportional Control:** Students learn about proportional control algorithms in advanced programming. These algorithms use mathematical ideas to modify motor speed in response to sensor feedback.
5. **Patterns and Sequences:** Creating patterns and sequences of operations is a common task in programming logic, which helps students hone their pattern recognition abilities.

Teaching children with LEGO Mindstorms is a fun method to introduce them to a variety of mathematical and engineering ideas in addition to robotics and programming. Students build and program robots to solve problems in the real world, gaining practical experience that combines theoretical knowledge with real-world application.

6.3.2 Educational Robots: Ozobot

Small, programmable robots that respond to color codes and follow lines drawn on paper are used to introduce students to the fundamentals of programming. Ozobot is a diminutive, programmable robot created to help kids learn coding and fundamental programming principles through engaging tactile activities. It is a handy tool for instructing young students about robotics and coding because it obeys lines written on paper and responds to color codes.

Let's explore the technology, theory, and related mathematics behind Ozobot:

Technology:

1. **Sensors:** Ozobot has sensors that can recognize lines and colors. It is able to distinguish between several color codes and detect differences in color intensity.
2. **LED Lights:** Ozobot incorporates color-changing LEDs that enable it to show visual feedback in response to user inputs.
3. **Microprocessor:** The Ozobot's microprocessor interprets the data from its sensors and LEDs and carries out the preprogrammed behaviors as necessary.
4. **Motor:** Ozobot is equipped with teeny motors that allow it to follow lines and react to colour codes.
5. **Color Codes:** Ozobot can identify certain color codes written on paper that direct it on how to move and behave.

Theory:

1. **Fundamental Programming Ideas:** Through the use of Ozobot, students are introduced to programming ideas like commands, sequences, loops, and conditional expressions. They design colour code sequences to program Ozobot's behavior.
2. **Algorithmic Thinking:** Students learn how to organize tasks into logical sequences and divide larger jobs into smaller phases (color codes) in order to accomplish desired results.

3. **Cause and Effect:** Students learn about cause-and-effect linkages in programming via Ozobot's responses to color codes. They observe how particular inputs result in particular outputs.
4. **Patterns:** To control Ozobot's movements, students employ patterns of colour codes. This explains the idea of identifying patterns and developing them.

Related Mathematics:

1. Geometry and spatial thinking are necessary for designing paths and layouts for Ozobot. Lines and color codes are arranged to create the routes by the students.
2. **Colour Mixing:** Ozobot can decipher colour codes by recognising combinations of fundamental colors. This explains fundamental colour mixing ideas.
3. **Numerical Patterns:** By using colour codes, students can make patterns that reinforce the ideas of counting and numerical sequencing.
4. **Measuring:** To ensure that Ozobot moves smoothly, students measure distances, angles, and sizes when building the robot's tracks.
5. **Problem-Solving:** Solving problems is a necessary step in debugging Ozobot's actions. Students analyze their code for problems and conduct experiments to find fixes.

Students learn through entertaining, intuitive interactions with Ozobot that encourage computational thinking and mathematical comprehension. Young learners can study the fundamental ideas of robotics, coding, and mathematics using an approachable pathway made possible by the combination of hands-on robot interactions and the visual language of color codes.

6.3.3 Educational Robots: Sphero

A spherical robot for teaching coding and fundamental robotics that can be operated with a smart phone app. A fun educational tool to teach coding and robotics principles, Sphero is a spherical robot that can be controlled via a smartphone app. Sphero gives pupils an engaging and hands-on introduction to programming logic and computational thinking because to its many capabilities and interactive interface.

Let's delve into the technology, theory, and related mathematics behind Sphero:

Technology:

1. **Spherical Design:** Sphero is incredibly manoeuvrable due to its distinctive spherical shape, which enables it to move in any direction.
2. Sphero contains motors that enable it to roll in different directions, giving children feedback on their physical movement.
3. **LED Lights:** Sphero includes color-changing LEDs that provide visual feedback and customization.
4. **Bluetooth Connectivity:** Sphero may be remotely controlled by students by pairing with a smartphone or tablet over Bluetooth.
5. Sensors Sphero has sensors that can measure its direction and speed, providing information for movement programming and control.

Theory:

1. **Introduction to Coding:** By enabling students to write programs that direct Sphero's movements and behaviors, Sphero teaches them the fundamentals of coding.
2. **Problem-Solving:** Students experiment with various instructions and steps to use Sphero to overcome obstacles and accomplish particular objectives.

3. **Algorithmic Thinking:** Sphero programming requires segmenting tasks into logical sequences in order to accomplish desired results.
4. **Conditional Statements:** With the aid of conditional statements, students can program Sphero to react in different ways in response to certain circumstances, such as collisions or sensor readings.

Related Mathematics:

1. **Geometry and Angles:** In order to precisely control Sphero's movements, students use angles, which helps them comprehend ideas like degrees and rotations.
2. **Measurement:** To ensure that Sphero follows their desired path, students measure distances, boosting their understanding of length and units.
3. **Patterns:** By programming Sphero with patterns of movement and commands, students are introduced to the idea of identifying and imitating patterns.
4. **Numerical Relationships:** Students use calculations and numerical relationships when programming Sphero's speed or duration.
5. **Data Interpretation:** Students have the chance to understand data from Sphero's sensor data, such as speed readings, and modify their programming accordingly.

The educational philosophy behind Sphero enables children to experiment with mathematics, robotics, and coding in a fun and engaging way. Students get a deeper knowledge of fundamental ideas while improving their problem-solving and computational thinking skills by engaging with a physical robot via an intuitive interface.

6.3.4 Educational Robots: VEX Robotics

Students in middle and high schools can construct and program robots to compete in challenges using kits. Middle and high school students can design, construct, and program robots using the comprehensive platform provided by VEX Robotics in order to compete in a variety of tasks. In addition to teaching students about engineering and programming, this instructional tool also gives them practical experience with problem-solving, teamwork, and critical thinking. Let's examine the science, philosophy, and associated mathematics that underlie VEX Robotics:

Technology:

1. **Building System:** Students can construct complex robotic structures using VEX Robotics' adaptable building system, which includes metal parts, gears, motors, and sensors.
2. **Motors and Sensors:** VEX provides a variety of motors and sensors, including motors for movement and sensors for sensing touch, distance, and other things.
3. **Microcontroller:** VEX robots are run by microcontrollers, which interpret sensor input and carry out instructions to operate motors and other parts.
4. **Programming Software:** Students can program their robots using graphical or text-based programming languages with the help of VEX's programming software.

Theory:

1. **Mechanical Engineering:** As they design and construct their robots, students gain knowledge of mechanical ideas such as gear ratios, leverage, torque, and structural integrity.
2. **Electronics and Circuits:** Students learn the fundamentals of electronics and circuits by connecting motors, sensors, and microcontrollers.

3. **Engineering Design Process:** The engineering design process, which includes brainstorming, prototyping, testing, and refining, is used in the design and construction of a VEX robot.

Related Mathematics:

1. **Geometry and Angles:** In order to precisely plan and carry out robot motions, students require geometry. To navigate accurately, one must have a solid understanding of angles and distances.
2. **Measuring:** To ensure accurate construction, students measure lengths, dimensions, and angles when designing and manufacturing robots.
3. **Algebra and Equations:** When programming motors, mathematical correlations are reinforced by setting speeds and distances using algebraic equations.
4. **Trigonometry:** In more difficult situations, students can use trigonometry to determine the angles and distances needed for intricate robot motions.
5. **Data Analysis:** During robot testing, students gather and analyse data on performance measures, such as the amount of time it takes to accomplish a task or the distance travelled.
6. **Mathematical Logic:** Students must use their logical thinking skills and mathematical knowledge when programming logic elements such conditional statements and loops.

Engineering, programming, and mathematics are all integrated into the multidisciplinary learning experience provided by the VEX Robotics platform. Students not only gain technical proficiency by designing, creating, and programming robots to take on specified tasks, but they also acquire problem-solving techniques, teamwork, and project management. Students gain practical knowledge through this intensive and hands-on approach that goes beyond the classroom, preparing them for future employment in STEM industries.

6.3.5 Educational Robots: Robolink CoDrone

A drone that uses drone flight and customisation to teach programming. An educational drone called the Robolink CoDrone was created to help students learn coding and programming through the fascinating world of aerial robotics. CoDrone is a fun tool for teaching a variety of STEM ideas because of its interactive features and hands-on learning style.

Let's examine the science, philosophy, and associated mathematics that underlie the Robolink CoDrone:

Technology:

1. **Quadcopter Design:** The quadcopter design of the CoDrone allows for controlled flight, hovering, and vertical launch.
2. **Microcontroller:** A microcontroller is a feature of the CoDrone that interprets commands and manages the drone's sensors and motors.
3. **Infrared Sensors:** The CoDrone has infrared sensors that allow it to maintain altitude and detect obstacles.
4. **LED Indicators:** The CoDrone's LED lights offer visual input and serve to distinguish between various flight modes or statuses.
5. CoDrone has Bluetooth connectivity, enabling remote programming and control from a computer or mobile device.

Theory:

1. **Aerial Robotics Concepts:** Through hands-on experience flying a drone, students get a practical understanding of the concepts of lift, thrust, drag, and gravity.

2. **Programming Logic:** As they develop flying routines and behaviours for CoDrone, students are introduced to programming ideas including sequences, loops, and conditionals.
3. **Practical Uses:** Students investigate practical uses for drones, including aerial photography, surveying, and environmental monitoring.

Related Mathematics:

1. **Geometry and Angles:** Students utilize geometry to determine rotational angles and regulate the orientation of the drone while it is in flight.
2. **Measuring:** To better understand units and conversions, students measure distances, altitudes, and other flight-related parameters.
3. **Trigonometry:** Recognizing trigonometric principles and functions enables pupils to plan aircraft routes, compute travel times, and regulate movement.
4. **Data Interpretation:** Analyzing sensor data, flight records, and telemetry data requires understanding numerical values and adjusting as necessary.
5. **Graphing and Plotting:** Students may use graphing and plotting to visualize patterns and trends when choosing flight paths or analyzing data.

With the pleasure of operating a drone, Robolink CoDrone provides an immersive learning environment that integrates coding, robotics, and mathematics. Students receive practical experience with applied mathematics and engineering principles by creating flight courses, managing manoeuvres, and experimenting with various sensors. The interactive and practical aspects of CoDrone encourage a thorough comprehension of STEM concepts while inspiring a passion for aerial robotics and technology.

6.4 HOW ROBOTICS INTEGRATED INTO EDUCATIONAL CURRICULA

From elementary schools to universities, robotics is being incorporated into educational curricula on a wide range of educational levels. With the help of this integration, teachers hope to improve students' comprehension of STEM courses (Science, Technology, Engineering, and Mathematics), encourage practical learning, and foster the growth of vital abilities like problem-solving, creativity, and teamwork. Here is an example of how robotics is being included in school curricula.

6.4.1 STEM-focused Courses

Numerous schools provide clubs or specialized STEM courses with an emphasis on coding and robotics. Through practical projects and activities, these courses teach students to the fundamentals of programming, engineering, and robotics. Designing educational experiences that include students in practical activities, coding challenges, and robotics-related problem-solving is necessary to incorporate robots into STEM-focused courses.

Especially in STEM-focused courses, robotics is being incorporated into educational curricula in the following ways:

1. **Introduction to Robotics:** Many STEM courses begin with a discussion of the fundamental ideas of robotics. Students gain knowledge of robot parts, fundamental programming, and how robots interact with their surroundings.
2. **Interactive Projects:** Students complete assignments that require them to construct robots out of LEGO Mindstorms or VEX Robotics kits. Students are introduced to mechanical design, electronics, and assembly through these projects.

3. **Coding Challenges:** A key part of robotics instruction is coding. Students learn how to program robot behavior and responds to stimuli using languages like Python, Scratch, or Blockly.
4. **Algorithmic Thinking:** Students learn algorithmic thinking techniques through robotics tasks. They divide up jobs into detailed directions that their robots can follow.
5. **Sensors and Actuators:** Students learn about the many sensors that robots employ to monitor their surroundings and interact with them, including ultrasonic, touch, light, and servos.
6. **Scenarios for Problem-Solving:** Students are given real-world issues to address with their robots in robotics challenges. This promotes critical thinking and inventive problem-solving.
7. **Design Iteration:** Based on performance comments and test results, students refine their robot designs. They discover the value of honing and enhancing their creations.
8. **Collaborative Projects:** Working in teams is essential to many robotics projects. Students work together to design, construct, program, and test their robots, developing their communication and collaboration abilities.
9. **Mathematical Integration:** Students integrate mathematics into their projects by calculating angles and distances in geometry, measuring the dimensions of robot parts, and even writing equations in programming.
10. **Physics Fundamentals:** Torque, force, and motion are all physics ideas that are used in robotics. When creating and operating their robots, students learn how to use these ideas.
11. **Engineering Design Process:** Students develop, test, and refine their robots as part of the engineering design process, which entails brainstorming, planning, and building.
12. **Project Documentation:** Students write up the code they use to program their robots and the methods they use to solve issues. This paperwork develops a portfolio of their work and strengthens communication abilities.
13. **Ethical Considerations:** The curriculum includes discussions of the ethical implications of robotics, including how automation will affect jobs and the ethics of AI.
14. Robotics contests are a common STEM course capstone in which students demonstrate the capabilities of their robots, compete against one another, and apply what they have learned in a competitive setting.
15. **Integration with Other Subjects:** Robotics projects can be combined with other STEM disciplines including physics (mechanics), math (geometry), and even biology (biomechanics and bio-inspired robotics).

Teachers may give students a comprehensive learning experience that integrates theory and practice by incorporating robotics into STEM-focused curricula. These opportunities not only help kids better comprehend STEM subjects but also give them the critical thinking, problem-solving, cooperation, and technology literacy skills needed in today's workforce.

6.4.2 Coding and Programming

Python, Scratch, and Blockly are just a few of the computer languages that are taught to kids using robotics kits and platforms. Students write programs to control robots, which promotes analytical reasoning and algorithmic problem-solving. Teaching students how to program robots to carry out various activities and behaviors is a key component of integrating robotics into curricula through coding and programming. With this method, you can develop your computational thinking, your ability to solve problems, and your comprehension of algorithms. Here is an example of how robotics is included into curricula at schools that emphasize coding and programming:

1. **Introduction to Coding:** Using straightforward, visual programming environments like Blockly or Scratch, students are exposed to coding languages and ideas. For beginners, these platforms' usage of block-based coding makes it simple and fun.
2. **Programming Languages:** Students may go from object-oriented programming languages like C++ to text-based programming languages like Python or JavaScript as they advance, earning more flexibility and exposure to real-world coding languages.
3. **Robot Simulation:** Students can utilise software simulations to programme virtual robots before working with actual robots. This allows for experimentation and learning without the usage of actual hardware.
4. **Controlled Movements:** Students are taught how to code fundamental motions including moving forward, moving backward, turning, and halting. They gain knowledge on how to manage actuators and motors.
5. **Conditional Statements:** Students program robots to make decisions based on specific situations or sensor readings using if-else statements. A robot might, for instance, alter its course when it runs into a barrier.
6. **Loops:** By introducing loops (such for and while loops), students can repeat action sequences, which enables robots to carry out jobs repeatedly.
7. **Functions and Procedures:** In order to organize and streamline programming, students write reusable code segments (functions). They learn about modular programming from this.
8. **Event-Driven Programming:** Students design robots to react to particular occasions or triggers, like a touch, light, or sound sensor.
9. **Problem-Solving Tasks:** Robotics projects and challenges demand on students to use critical thinking and coding expertise to address actual issues.
10. **Algorithmic Thinking:** By breaking complex tasks down into smaller, more manageable parts, students develop algorithms that direct the actions of robots.
11. **Debugging:** In coding, debugging is a crucial ability. To ensure that robots operate as intended, students learn how to spot and remedy problems in their code.
12. **Project-Based Learning:** To further their comprehension of programming ideas, students work on projects that require them to program robots to carry out particular tasks.
13. **Creative Expression:** Coding enables students to create original robot interactions and behaviours, stimulating creativity.
14. **Sensor Integration:** Students incorporate sensor information into their code to enable robots to interact with and react to their surroundings in a variety of ways.
15. **Cross-Curricular Integration:** Coding and robotics can be incorporated into other courses, including math (using coding to model mathematical concepts) and art (producing robot-generated art).
16. **Real-World Applications:** Educators emphasise the connections between coding and programming and real-world uses, including automation, data analysis, and the Internet of Things (IoT).

Coding and programming are used to include robotics into the curriculum, giving students the tools they need to create technology on their own. Students build fundamental digital literacy skills, problem-solving attitudes, and logical and systematic thinking abilities—a set of abilities that are more important in our technology-driven society.

6.4.3 Electronics and Engineering

Electronics and engineering ideas are frequently covered in robotics curriculum. Students create and customize own robots, learning about sensors, actuators, motors, circuitry, and mechanics in the process.

Teaching students about the mechanical, electrical, and electronic components of robots is a necessary step in integrating robotics into curricula through electronics and engineering education. With this method, students may build, assemble, and troubleshoot robotic devices practically. Here is an example of how robotics is included into curricula in schools that emphasize electronics and engineering:

1. **Introduction to Robotics Components:** Students learn about the different parts that go into building a robot, such as the motors, sensors, actuators, microcontrollers, and structural components.
2. **Mechanical Design:** As they plan and construct their robots, students gain knowledge of mechanical engineering concepts including leverage, torque, stability, and structural integrity.
3. **Building Systems:** Educational robotics kits like VEX Robotics and LEGO Mindstorms give students a platform to build robots using standardised parts and building blocks.
4. **Electronics Fundamentals:** Students gain an understanding of the fundamental ideas in electronics, including circuits, current, voltage, and resistance. They are aware of how parts are interconnected to form working systems.
5. **Soldering Capabilities:** Complex robotics projects may call for the assembly of custom circuits by soldering parts to printed circuit boards (PCBs).
6. **Motor Control:** Using microcontrollers and programming, students investigate how to control the speed, direction, and behavior of motors.
7. **Sensor Integration:** Students gain knowledge of the operation of various sensors, including gyro, ultrasonic, light, and touch sensors, as well as how to incorporate them into robotic systems.
8. **Power Management:** Power sources, battery life, and power distribution must all be understood in order to construct effective and useful robots.
9. **Wiring and Connections:** In order to ensure correct communication and functionality, students learn how to connect various components using wires, connectors, and soldering procedures.
10. **Prototyping and Iteration:** Students participate in the engineering design process by developing prototypes, testing them, and then improving them in response to feedback.
11. **Kinematics and Mechanics:** Students learn about concepts including rotational and translational motion, wheel configurations, and kinematic limitations that are relevant to robot movement.
12. **Robot Construction Methods:** Students learn about several robot construction methods, including the use of gears, pulleys, belts, and linkages.
13. **Design Constraints:** As they construct working robots, students gain experience working under design restrictions including size, weight, and material availability.

14. **Troubleshooting and Debugging:** As they face obstacles throughout construction, students hone their problem-solving abilities by locating and resolving problems.
15. **Collaborative Engineering:** Since many robotics projects need teamwork, they encourage students to work together while creating intricate robotic systems.
16. **Integration with Programming:** As students build robots that can move physically and respond to commands and sensor inputs, programming and electronics are merged.

Teachers may give students a thorough grasp of the design, construction, and operation of robots by including robotics into their electronics and engineering courses. This hands-on training develops practical abilities, critical thinking, and an understanding of the multidisciplinary character of robotics as a discipline.

6.4.4 Problem-Solving Challenges

Robot races, hackathons, and other challenges that involve applying knowledge and creativity to real-world problems motivate students to do so.

Students are involved in practical and dynamic learning experiences when robotics is incorporated into educational curricula through problem-solving exercises. These tasks allow students to use their understanding of robotics, coding, and engineering to address issues in the real world, encouraging critical thinking, cooperation, and creativity. Following are some examples of how robotics is incorporated into curricula for schools with an emphasis on problem-solving tasks:

1. **Real-World Context:** Problem-solving exercises put students in circumstances modeled after actual ones, which makes learning more applicable and meaningful.
2. **Project-Based Learning:** Students work on assignments that call for them to create, assemble, and program robots to take on particular jobs or problems.
3. **Scenario-based Challenges:** Students can create creative solutions by simulating scenarios like crisis response, environmental monitoring, or space exploration in these challenges.
4. **Open-Ended Tasks:** Open-ended tasks give pupils a variety of options on how to approach and resolve a problem.
5. **Collaborative Problem Solving:** Many tasks are meant to be completed in teams. Students work together to generate ideas, pool knowledge, and develop solutions as a class.
6. **Critical Thinking:** Students examine problems, pinpoint potential solutions, and assess the most effective methods for achieving their objectives.
7. **Adaptation and Iteration:** As students face obstacles or failures, they learn to modify their plans, make adjustments, and improve their answers.
8. **Restrictions:** Challenges frequently have restrictions attached to them, challenging pupils to think imaginatively within the confines of their available resources, time, or rules.
9. **Algorithmic Design:** Students develop programming routines and algorithms that let robots carry out specific jobs or negotiate challenging surroundings.
10. **Testing and Optimization:** Students evaluate the performance of their robots, collect data, and improve their designs and programs to get better outcomes.
11. **Failure as Learning:** Fostering a growth mentality, resourcefulness, and resilience in kids by allowing them to learn from their mistakes and failures.

12. **Presentation and Communication:** Students frequently show their ideas to peers or judges, improving their ability to explain their design decisions and communication abilities.
13. **Cross-Disciplinary Integration:** Tasks may include ideas from several disciplines, such as science (understanding the environment) or math (calculations for robot movement).
14. **Creativity and Innovation:** Students are given the flexibility to develop original, creative solutions, which promotes creativity and outside-the-box thinking.
15. **Robotics Contests:** Events like FIRST Robotics, VEX Robotics, and RoboCup give students a stage on which to demonstrate their problem-solving abilities in a competitive setting.
16. **Reflection and Analysis:** Students evaluate their methods, achievements, and areas for development after completing problems.

Students are given the opportunity to think critically, use theoretical knowledge in real-world situations, and develop abilities that go beyond robotics, such teamwork, adaptability, and effective communication, by incorporating robotics into educational curricula through problem-solving tasks. These questions help develop a thorough comprehension of subjects as well as a sense of pride and assurance in solving difficult issues.

6.4.5 Interdisciplinary Integration

Math, science, art, and even literature projects can all use robotics components. For instance, students might construct robots to investigate physics principles or create robotic art. Robotics concepts and activities are connected with various areas outside of STEM through multidisciplinary robotics integration into educational curriculum. This method highlights how information is interconnected and encourages students to use robots in a variety of settings.

Here's how robotics is integrated into educational curricula with a focus on interdisciplinary integration:

1. **Mathematical Ideas:** Mathematical ideas are used in robotics projects to calculate angles and distances, write programs to solve equations, and analyze data gathered from sensors.
2. **Science:** Students use scientific theories to comprehend how robots use motors and actuators to interact with forces, perceive their environment using light sensors, and navigate using inertial sensors.
3. **Language Arts:** To improve their writing and communication abilities, students create reports, documents, or narratives outlining their robotic projects.
4. **Art:** Students could design aesthetics for robots, produce artwork produced by robots, or investigate the connection between technology and artistic expression.
5. **History:** Students research the evolution of robotics throughout history, learning about important individuals, significant events, and the effects of robotics breakthroughs on society.
6. **Societal Studies:** The curriculum includes discussions about the ethical and societal ramifications of robotics, including how automation will affect jobs and AI ethics.
7. In geography, students might research the potential use of robots in the mapping and environmental monitoring of geographic data.
8. **Physical Education:** Combining robotics with physical activity inspires students to create robots that imitate human motion or engage with their surroundings.

9. **Environmental Studies:** Robotics can be used to collect data and monitor ecosystems as part of ecological research, conservation efforts, and ecosystem studies.
10. **Music:** Students investigate how robotics and music can be combined by building robots that can play instruments or even write music.
11. **Foreign Languages:** Students could write computer instructions in a different language, strengthening their language abilities in a technical context.
12. **Cultural Studies:** Students study how various communities interpret and interact with technology as they investigate cultural perspectives on robotics.
13. **Economics:** Topics on the economics of robotics are incorporated, including manufacturing costs, labor markets, and sectors affected by automation.
14. **Health and Medicine:** Students investigate the use of robots in healthcare, including telemedicine, robotic prostheses, and robotic surgery.
15. **Ethics and Philosophy:** Conversations regarding the ethical implications of artificial intelligence, robotics, and automation encourage philosophical debates.
16. **Business Potential:** Students examine business and entrepreneurship potential in relation to robots and technology startups.

By using multidisciplinary techniques to include robots into educational courses, educators extend students' views and demonstrate how robotics connects with numerous facets of their lives. This method encourages a comprehensive grasp of the world and the influence that technology has on various academic disciplines. Additionally, it promotes innovative thinking and problem-solving in a variety of settings.

6.4.6 Project-Based Learning

In project-based learning activities involving robotics, students identify a problem, create a solution, construct a robot, program it, and test its functionality. This all-encompassing strategy promotes collaboration and critical thinking. By using project-based learning, it is possible to include robotics into school curricula by organizing the learning process around practical tasks where students create, construct, program, and test robots to solve problems in the real world. This strategy encourages participation, critical thinking, teamwork, and the use of STEM principles in real-world situations.

Here's how robotics is integrated into educational curricula with a focus on project-based learning:

1. **Projects that mirror Real-World Problems or Scenarios:** Projects that mirror real-world problems or scenarios make learning more relevant and usable.
2. **Student-centered Approach:** As they plan, create, and carry out their robotics projects, students take responsibility of their learning.
3. **Active Engagement:** Students actively participate in the learning process, which fosters a better comprehension of ideas and tenets.
4. **Multidisciplinary Integration:** In order to complete projects, students frequently need to use knowledge from a variety of disciplines, including science, math, engineering, and coding.
5. **Problem Identification:** To promote critical thinking and problem-solving, students choose a problem or challenge that their robot can handle.
6. **Design and Prototyping:** Students create robot prototypes, testing and improving their concepts as they go.

7. **Building and Construction:** Students use kits or materials to physically put together robot parts while learning about mechanics and engineering concepts.
8. **Programming and Coding:** Using coding principles they have learnt throughout the process, students program robots to carry out particular actions or behaviors.
9. **Trial and Error:** As kids test their robots, they run into difficulties and learn by making mistakes, which helps them become resilient and adaptable.
10. **Collaborative Learning:** Projects frequently require teamwork, which motivates students to interact, exchange ideas, and work efficiently as a team.
11. **Data Collection and Analysis:** Students collect information about their robot's performance, examine the findings, and come to well-informed conclusions on how to improve.
12. Fostering creativity and inventive thinking, students are given the flexibility to generate original solutions.
13. **Documentation and Reflection:** Students record the details of their project's design, coding, and results. Self-evaluation is encouraged by reflection.
14. **Presentation Skills:** A lot of projects end with presentations where students display their robots and discuss design decisions and functioning.
15. **Critical Thinking:** Students examine problems, assess solutions, and draw conclusions based on facts and information.
16. **Technology Literacy:** Students gain technical expertise, digital literacy, and a working knowledge of the platforms and technologies used in robotics.
17. **Career Preparation:** By giving students hands-on experience, project-based learning in robotics prepares students for future professions in STEM industries.
18. **Inquiry-based Learning:** Students engage in inquiry by posing problems, looking for answers, and carefully examining ideas.
19. **Societal Impact:** Projects may examine how automation affects occupations or how AI is used in daily life as it relates to society.

Teachers give students a dynamic and engaging learning experience by incorporating robotics into project-based learning programs. In addition to improving topic knowledge, this strategy also develops crucial abilities like critical thinking, teamwork, problem-solving, and effective communication—abilities that are beneficial in both academic and practical contexts.

6.4.7 Inquiry-Based Learning

Open-ended questions, practical exercises, and research are all used to enable students to explore and study issues in robotics. Encouraging a student-driven approach where students investigate robotics concepts, pose questions, carry out investigations, and look for solutions through practical experimentation and research is a key component of integrating robots into educational curricula through inquiry-based learning. This approach fosters interest, skepticism, problem-solving abilities, and a deeper comprehension of robotics fundamentals.

Here is an example of how robotics is incorporated into lesson plans with an emphasis on inquiry-based learning:

1. **Curiosity and Exploration:** Students are invited to research any aspects of robotics that catch their interest.

2. **Open-Ended Questions:** Open-ended questions like "How can robots assist in disaster relief?" or "What factors affect a robot's speed?" are used to kick off robotics initiatives.
3. **Problem Identification:** Students discover actual issues that robots might be able to assist with, which inspires investigation and testing.
4. **Research and Investigation:** To learn about robotics principles, technology, and applications, students obtain knowledge from a variety of sources, including books, journals, and online resources.
5. **Development of Hypotheses:** Students develop hypotheses in light of their research and preliminary comprehension of robotics principles.
6. **Construct and Experimentation:** Using hands-on experience with robots, code, and sensors, students construct experiments or projects to test their assumptions.
7. **Data Collection and Analysis:** Students gather information for analysis during experiments by employing sensors and measurements.
8. **Interpretation and Conclusion:** Students explain their findings and form conclusions about their hypothesis based on data analysis.
9. **Reflection and Iteration:** Students ponder the results of their trials, take into account any drawbacks, and make suggestions for enhancements or new strategies.
10. **Independent Learning:** Students are free to select their own research questions, plan their own experiments, and investigate any ideas about robotics that catch their attention.
11. **Cross-Disciplinary Integration:** In robotics inquiry-based learning, courses like math (measuring and data analysis) and science (physics and mechanics) are naturally integrated.
12. **Critical Thinking:** Students conduct logical experimentation, critically evaluate information sources, and analyze data.
13. **Problem-Solving Techniques:** As they tackle challenging robotics topics, students who participate in inquiry-based learning gain efficient problem-solving techniques.
14. **Communication Skills:** Through presentations, reports, and discussions, students hone their communication skills as they learn to explain their findings.
15. **Technology Literacy:** Students become accustomed to using robotics tools, platforms, and software for research and experimentation.
16. **Real-World Relevance:** Inquiry-based learning makes learning more meaningful by connecting robotics ideas to real-world applications.
17. Students think about the ethical ramifications of robots, including AI prejudice, privacy issues, and social effects.
18. **Scaffolding:** Teachers help students' enquiries by offering advice and materials, ensuring they have access to reliable information.

Through the use of inquiry-based learning, educators may empower students to become active learners, researchers, and problem solvers by incorporating robots into their curricula. This method promotes a deeper comprehension of robotics ideas, promotes lifelong learning, and gives students the tools they need to deal with difficult problems in a technology environment that is continually changing.

6.4.8 Higher Education Programs

In the domains of robotics engineering, automation, AI, and other relevant areas, universities offer specialized degrees and programs. Students in these programs are prepared for careers in robotics research and business. Offering specialized courses, programs, and research opportunities that explore cutting-edge robotics concepts and applications is necessary to integrate robotics into higher education programs. Students enrolled in these programs are prepared for employment in academia, industry, and robotics research.

Here's how robotics is integrated into higher education curricula:

1. Universities provide bachelor's, master's, and doctoral programs in robotics engineering where students can pursue advanced studies in robotics design, control, artificial intelligence (AI), and automation.
2. **Robotics Specializations:** Within more general engineering fields like mechanical, electrical, or computer engineering, robotics can be a specialization.
3. **Curriculum Organization:** Robotics programs integrate electives in sensors, AI, machine learning, and automation with core engineering courses and advanced robotics courses.
4. **Advanced Coding and Algorithms:** Advanced programming for robotic systems, path planning, and complicated behavior control are the main topics of higher-level coding and algorithm courses.
5. **Advanced Robotics Technologies:** Students investigate cutting-edge robotics innovations such swarm robotics, bio-inspired robotics, human-robot interaction, and autonomous systems.
6. **Machine Learning and Artificial Intelligence:** Robotics programs encompass machine learning strategies and AI algorithms utilized in robotic perception, judgment, and control.
7. Universities offer specialized robotics labs that are outfitted with cutting-edge technology and software for research, experimentation, and project creation.
8. **Capstone Projects:** In their capstone projects, students design, create, and test sophisticated robotic systems.
9. **Research Possibilities:** Under the guidance of faculty, students work on research initiatives that enhance robotics theory and technology.
10. **Robotics Clubs and Competitions:** Some universities have robotics teams that compete in events like the DARPA Robotics Challenge and the RoboCup.
11. **Guest Lectures and Industry Collaboration:** Industry and academic experts may present guest lectures, workshops, and seminars to students, exposing them to real-world applications.
12. **Internships and Co-op Programs:** Through internships and co-op programs with robotics firms and research organizations, students receive real-world experience.
13. **Ethics and Societal Implications:** The ethical issues, political ramifications, and societal effects of robotics and AI are covered in the courses.
14. **Interdisciplinary Collaboration:** To explore interdisciplinary applications, robotics programs work with other departments like computer science, physics, biology, and healthcare.
15. **Advanced Robotics Research Centers:** A number of universities create research hubs for cutting-edge robotics, bringing together academics, scientists, and students.
16. **Curriculum Driven by Industry:** Programs follow industry trends and requirements, ensuring that graduates are well-prepared for professions in domains related to robotics.
17. **Innovation and Entrepreneurship:** A number of programs provide courses on entrepreneurship to help students launch their own robotics businesses.
18. **Graduate Research Opportunities:** Graduate students frequently publish papers and create new technologies as part of cutting-edge research.

19. **Postgraduate Opportunities:** Robotics program graduates work in academia, robotics research, engineering, and technology businesses.

By integrating robotics into higher education curricula, universities provide students with advanced knowledge and skills needed to excel in the field of robotics. These programs contribute to technological advancements, innovation, and the development of solutions that address real-world challenges across industries.

6.4.9 Online Resources and Platforms

Students can remotely access robotics classes, tutorials, and virtual simulations on educational platforms and websites. Students can access robotics education and learning materials remotely by integrating robotics into curricula through online resources and platforms. These tools offer adaptability, accessibility, and chances for self-guided learning.

Here's how robotics is integrated into educational curricula using online resources and platforms:

1. **Online Courses:** Numerous educational websites provide online courses that cover a range of robotics-related topics, from the fundamentals to more complex ones. These courses frequently have homework, interactive quizzes, and video lectures.
2. **Massive Open Online Courses (MOOCs):** MOOCs give students access to robotics courses from universities and organizations all over the world and let them study at their own speed.
3. **Educational Websites:** Websites devoted to robotics provide tutorials, how-tos, and other materials for learning about robotics principles, projects, and programming.
4. **Robotics Simulators:** Without using actual robots, online simulators allow students to realistically create, program, and test robots in a digital environment.
5. **Video Lessons:** There are a tone of robotics lessons, demos, and project walkthroughs available on YouTube and other video-sharing websites.
6. **Interactive Platforms:** A few platforms provide coding environments where students can code and program virtual robots in real time.
7. **Coding Challenges:** Students can use their coding expertise to tackle robotics-related problems by participating in online platforms' coding challenges and competitions.
8. **Virtual Laboratories:** Without the need for actual tools, virtual labs let students explore robotics principles and carry out virtual experiments.
9. **Discussion Boards:** Online discussion boards and communities give students a place to exchange knowledge, ask questions, and work together on robotics projects.
10. **Open-Source Robotics Software:** Students can use open-source robotics frameworks, libraries, and software for coding and experimentation in online repositories.
11. **Robotics Kits and Simulations:** A number of sites sell robotics kits that may be bought and utilized with virtual reality training programs.
12. **Project Repositories:** Students can explore a variety of real-world applications by accessing and contributing to repositories of robotics projects.
13. **Specialized Platforms:** Tools for studying and using cutting-edge robotics technology are available on platforms like ROS (Robot Operating System).
14. **Coding Challenges:** Coding challenges with an emphasis on robotics and algorithms can be found on websites like Codecademy and LeetCode.

15. **Virtual Tournaments:** Students can compete with peers from across the world in online robotics tournaments, showing their robotics expertise and creativity.
16. Online forums for robotics experts, educators, and students come together to share information, analyse trends, and work together.
17. Online workshops and webinars offer interactive learning opportunities that let students interact with professionals and their peers.
18. **Self-Paced Learning:** Students can use online tools and platforms whenever they want to, allowing them to suit their education around their schedules.

Robotics education is made more widely available to students outside of typical classroom settings by instructors who incorporate it into educational curriculum via online resources and platforms. These tools enable lifelong learning in the subject of robotics, accommodate various learning preferences, and offer chances for independent investigation.

6.4.10 Hands-On Labs and Workshops

Under the supervision of professors, students can work on robotics projects at workshops, camps, and labs hosted by educational institutions. Students get the opportunity to learn practically and experientially by participating in hands-on labs and workshops that incorporate robotics into educational courses. Students can work with others to create projects, interact directly with robotic devices, and more through these activities.

Here's how robotics is integrated into educational curricula using hands-on labs and workshops:

1. **Hands-on Learning:** Students have the opportunity to work with actual robots, sensors, actuators, and other hardware components in labs and workshops, which help them grasp robotics topics better.
2. **Project-Based Approach:** Using theoretical knowledge to design, construct, program, and test robotic systems, students work on practical projects.
3. **Experiential Education:** Through hands-on experiences, students can apply abstract ideas to actual situations, honing their problem-solving abilities.
4. **Collaborative Learning:** Group work in labs and workshops frequently encourages cooperation, communication, and the sharing of ideas.
5. **Skill Development:** Students learn how to put together parts, wire them together, program, troubleshoot, and test robotic systems.
6. **Sensory Experience:** By exposing students to the physical and sensory aspects of robotics, they are better able to comprehend how robots relate to their surroundings.
7. Quick iteration and development are made possible by the students' quick feedback, which they obtain as they observe the results of their robot's activities.
8. **Project Documentation:** Students keep track of the development of their projects, reflecting on design decisions, difficulties encountered, and workarounds applied.
9. **Expert Direction:** Teachers or facilitators offer direction, respond to inquiries, and help students with the practical exercises.
10. **Problem-Solving Obstacles:** Students face obstacles during the building and programming phases, which promotes original problem-solving.
11. **Hardware Integration:** Students are introduced to a variety of hardware elements in labs and workshops, including motors, sensors, microcontrollers, and actuators.

12. **Circuit Design:** As they connect components, comprehend interconnection, and debug electronic systems, students gain knowledge of circuitry.
13. **Programming Proficiency:** By creating code to direct the actions and behaviour of their robots, students hone their programming abilities.
14. **Prototyping:** Before completing their work, students experiment with several design iterations and investigate various strategies.
15. **Interactive Demonstrations:** Live instructor demonstrations of advanced methods and skills are frequently included in workshops.
16. **Interdisciplinary:** Labs and workshops can bring together students from different academic fields, encouraging the exchange of knowledge and abilities.
17. **Robotics Competitions:** Some courses close with friendly competitions where students present the talents and tactics of their robots.
18. **Creating Interest:** Especially for pupils that learn more tactilely, hands-on activities can create interest in robotics and engineering jobs.

Teachers can give students immersive learning experiences that combine theory and practice by incorporating robots into educational curricula through hands-on labs and workshops. In addition to honing fundamental abilities, these activities also encourage creativity, confidence, and a love of robotics and technological exploration.

6.4.11 Teacher Training

In order to successfully integrate robots into their teaching, educators get training and professional development in robotics education approaches. In order to successfully teach robotics principles, programming, and hands-on activities, it is important to equip teachers with the knowledge and tools they need. The provision of effective and engaging robotics instruction to pupils is ensured by teacher preparation.

Here is an example of how teacher training might incorporate robots into educational curricula:

1. **Professional Growth Seminars:** To help instructors learn about robotics concepts, tools, and teaching techniques, educational institutions and organizations offer seminars and training sessions.
2. **Curriculum Integration Training:** Teachers receive instruction on how to effortlessly incorporate robotics into current curricula in a variety of areas.
3. **Pedagogical Approaches:** Training includes project-based learning, inquiry-based learning, and practical exercises for teaching robotics.
4. **Robotic Kits and Tools:** Teachers acquire the skills necessary to effectively use and maintain robotics kits, components, sensors, and programming platforms.
5. **Development of Coding Skills:** Teachers who receive training are better equipped to help students through programming problems.
6. **Practical Experience:** Teachers engage in practical exercises and projects to get first-hand knowledge of designing and automating robots.
7. **Lesson Planning:** Teachers gain knowledge of how to create lesson plans and learning activities that combine robotics principles and are in line with educational objectives.
8. **Classroom Management:** Training includes techniques for overseeing robotics-related classroom activities, such as group projects and troubleshooting.

9. **Techniques for Assessment:** Teachers have received training in how to evaluate students' robotics projects and determine how well they are picking up robotics principles.
10. **Ethical Considerations:** The training covers ethical issues associated to robots and AI, assisting teachers in addressing related subjects in the classroom.
11. **Support Resources:** To improve their instruction, teachers have access to teaching materials, lesson plans, coding tutorials, and instructional videos.
12. **Ongoing Support:** Online communities and programs for continuous professional development offer ongoing support and networking possibilities.
13. **Collaborative Learning:** Training activities may include opportunities for teachers to work together to solve problems and share insights.
14. **Integration with Core Courses:** Teacher training teaches teachers how to include robots into core courses including maths, physics, coding, and engineering.
15. **Interdisciplinary Approaches:** Using robots to teach wider topics, teachers learn how to investigate interdisciplinary connections.
16. **Hands-On Labs:** To practice effective teaching techniques, teacher training programs may have participants take part in hands-on robotics workshops and labs.
17. **Diversity Sensitivity:** Training places a strong emphasis on building welcoming environments that motivate all students, especially those from varied backgrounds, to participate in robotics education.
18. **Innovation and Adaptation:** Teachers are urged to modify their methods as robotics technology advances in order to foster an innovative culture.

Institutions equip teachers to instruct students in robotics concepts and abilities by offering teacher training for incorporating robotics into curricula. This strategy guarantees that kids receive a top-notch education in robotics, encouraging them to investigate STEM subjects and fostering the next generation of creative thinkers and problem solvers.

6.4.12 Maker Spaces and Fab Labs

For students to design, create, and experiment with robotics projects, schools frequently offer maker rooms that are furnished with the necessary equipment. By incorporating maker spaces and fab labs into educational curricula, students will have access to interactive, collaborative settings where they may design, prototype, and construct robotic projects. These settings offer the equipment, materials, and assistance required to promote innovation, creativity, and the growth of useful skills.

To use maker spaces and fab labs to incorporate robots into educational curricula, consider the following:

1. **Hands-On Investigation:** Maker spaces and fab labs give students the chance to participate in hands-on robotics learning activities, encouraging active learning.
2. **Prototyping and Iteration:** Using the various materials, parts, and equipment in the area, students can prototype and iterate their robotic creations.
3. **Design Thinking:** Students are encouraged in maker spaces to approach challenges with a design thinking perspective, taking user demands into account and iteratively solving problems.
4. **Collaborative Learning:** Students take part in group projects that encourage communication, teamwork, and peer learning.

5. **Access to Tools:** Tools like 3D printers, laser cutters, CNC machines, and electronics equipment are accessible at maker spaces and can be used to build robotic parts.
6. **Coding and Programming:** Students program the behaviors and capabilities of their robots using coding tools and software.
7. **Sensors and Actuators:** Students can incorporate a range of sensors, actuators, motors, and other parts into their robotic creations by using the maker spaces' available actuators and sensors.
8. **Expert Guidance:** Trained personnel or mentors offer direction, help with problem-solving, and knowledge on how to use the tools and equipment.
9. **Project-Based Learning:** Students do robotics projects that entail designing, constructing, programming, testing, and improving their inventions.
10. **Multidisciplinary Integration:** Student exploration of the connections between robotics, art, engineering, and other fields is frequently encouraged in maker spaces.
11. **Entrepreneurship:** By assisting students in transforming their robotic ideas into marketable goods, maker spaces can encourage students in the development of entrepreneurial abilities.
12. **Open Source Collaboration:** Students can work together on open-source robotics projects and add to the body of knowledge within the maker community.
13. **Innovation and Creativity:** Maker spaces give pupils a platform to experiment with novel robotics concepts and imaginative fixes.
14. **Practical Applications:** Students can work on robotics projects that have real-world uses, like automation, environmental monitoring, or assistive devices.
15. **Safety and Ethics:** Maker spaces instruct students on ethical issues related to robotics as well as safety procedures when using equipment and materials.
16. **Experiential Learning:** Maker spaces encourage experiential learning by letting students experience the visible outcomes of their work.
17. **Inclusive Learning:** Maker spaces place a strong emphasis on inclusion by offering a setting where students from all backgrounds and academic abilities can participate in practical learning.
18. **Project Showcases:** Maker spaces frequently arrange occasions where students can present their robotic projects, allowing the community to see their innovations.

Institutions enable students to become active creators, problem-solvers, and innovators by integrating robotics into educational courses through maker spaces and fab laboratories. Students are more equipped for the opportunities and challenges presented by the quickly changing technology landscape thanks to these environments that foster a culture of innovation, cooperation, and lifelong learning.

6.4.13 Educational Robotics Kits

To make learning robotics easier, kits like LEGO Mindstorms, VEX Robotics, and others offer pre-designed parts and software. Students have structured hands-on learning opportunities when robotics is included into curricula through educational robotics kits. Students can design, construct, and program robots using these kits, which come with hardware parts, sensors, actuators, and programming platforms.

Here's how educational robotics kits are integrated into educational curricula:

1. **Practical and Engaging Learning Resources:** Educational robotics kits give students real resources to investigate robotics ideas in a hands-on manner.

2. **Structured Learning:** Kits frequently include curricular materials and step-by-step instructions that direct students through the construction and programming of robots.
3. **Interactive Exploration:** Students build robots out of the kit's parts, developing a deeper comprehension of mechanics and electronics.
4. **Teaching Students Code Skills:** Kits come with code interfaces or platforms that help students learn the coding skills necessary to govern robot behavior.
5. **Sensors and Actuators:** Students can experiment with a range of sensors (like light or distance sensors) and actuators (like motors) included in kits.
6. **Problem-Solving:** Using kits to build and program robots encourages students to use their problem-solving abilities to solve obstacles.
7. **Project-Based Learning:** Kits frequently come with assignments that let students construct particular kinds of robots, including line-following or obstacle-avoidance robots.
8. **Creative Exploration:** In addition to following directions, kids can change and customize their robots to test out original concepts and inventions.
9. **Teamwork:** As students assemble and program their robots in pairs or groups, kits promote teamwork.
10. **Conceptualization:** As they construct and program robots, students conceptualize theoretical ideas from STEM fields to actual situations.
11. **STEM Integration:** Educational robotics kits integrate science, technology, engineering, and math principles and are aligned with STEM curricula.
12. **Digital Literacy:** Kits instruct students on how to use software platforms, digital interfaces, and coding languages.
13. Engineering Overview: Kits introduce students to the fundamental ideas, mechanics, and design procedures of engineering.
14. **Instant Feedback:** Students who construct and program robots get instant feedback on the designs and coding decisions they make.
15. **Progressive Learning:** Kits frequently offer various levels of difficulty, allowing students to develop from simple to complicated projects.
16. **Skill Development:** Through practical exercises, students gain proficiency in electronics, mechanics, programming, and problem-solving.
17. **Inclusivity:** Educational robotics kits can be customized to meet the needs of different age groups and ability levels, fostering inclusivity in education.
18. **Career Preparation:** Exposure to robotics kits helps students get ready for professions in technology, engineering, and programming.
19. **Innovation and Creativity:** As students interact with tangible components, kits inspire them to think creatively and come up with novel solutions.

By incorporating educational robotics kits into academic curriculum, organizations give students a hands-on and engaging approach to learn about robotics, igniting their interest in STEM subjects and developing skills that are essential in the current technological environment.

6.4.14 Collaborative Learning

In designing and building robots, students frequently collaborate in teams, developing their project management, communication, and cooperation abilities. Collaborative learning is a method for incorporating robotics into school curricula that entails creating an environment where students can study robotics principles, work through issues, and construct projects. Teamwork, communication, critical thinking, and a deeper comprehension of robotics principles are all encouraged via collaborative learning.

Here's how robotics is integrated into educational curricula using collaborative learning:

1. **Group Projects:** Students work in groups to design, construct, and programme robots to accomplish predetermined objectives.
2. **Peer Teaching:** Through collaborative learning, students are encouraged to share their knowledge and skills with peers, which improves their comprehension of ideas.
3. **Diverse Skill Sets:** Students contribute to the project in a variety of ways, including design, coding, electronics, and mechanics.
4. **Interdisciplinary Teams:** Collaborative initiatives may include students from many academic fields, encouraging the exchange of talents and ideas.
5. **Discussions about Solving Problems:** Students talk about problems, generate ideas, and come up with solutions to overcome difficulties as a class.
6. **Communication Skills:** As students describe their ideas, pose questions, and get comments, collaborative learning improves communication skills.
7. **Conflict Resolution:** Students gain interpersonal skills by learning how to settle disputes that come up during group activity.
8. **Shared Responsibility:** By each team member making a contribution to the project's success, ownership and accountability are fostered.
9. **Peer Support:** Students assist one another in problem-solving, code debugging, and performance enhancement of the robot.
10. **Active Participation:** As students actively engage in conversations, debates, and decision-making, collaborative learning keeps them interested.
11. **Co-Creation:** Students are encouraged to co-create solutions as part of collaborative projects by utilizing their varied perspectives and experiences.
12. **Learning from Peers:** By encouraging a culture of continuous learning, students can benefit from the techniques, strategies, and insights of their peers.
13. **Larger-Scale Projects:** Collaborative projects may entail more complex, larger robots that call for the contributions of a larger crew.
14. **Critical Thinking:** Students assess each other's ideas critically, which helps them make wise decisions and produce superior project results.
15. **Practical Skills:** Collaborative learning helps students develop the skills they'll need in the workplace when they go on to pursue professions in it.
16. **Project Management:** Students gain knowledge about how to divide activities, define objectives, and monitor progress.
17. **Inclusive Learning:** By appreciating various points of view and fostering a positive environment, collaborative learning promotes inclusion.
18. **Leadership Opportunities:** Students assume leadership positions within teams, developing their organizational and leadership abilities.

19. **Reflection and Feedback:** Following the completion of projects, students reflect on their collaborative opportunities and offer helpful criticism.

Through the use of collaborative learning, educators can incorporate robots into their curricula while also fostering the growth of social and interpersonal skills necessary for employment in the twenty-first century. Students benefit from one another's knowledge and help one other grow and develop by working together in collaborative learning environments.

6.4.15 Real-World Applications

Teachers can demonstrate the usefulness of STEM principles by tying robotics projects to real-world situations. Teaching students about the numerous ways robotics technology is applied in different professions and industries is a crucial part of integrating robotics into curricula through real-world applications. This methodology links theoretical ideas with real-world, palpable applications, increasing the relevance and effectiveness of learning.

Here's how robotics is integrated into educational curricula using real-world applications:

1. **Industry Case Studies:** Students learn about actual instances of robotics usage in fields like manufacturing, healthcare, agriculture, and logistics.
2. Inviting experts from fields connected to robotics to speak to students about their positions and projects offers students real-world insights.
3. **Field trips:** Taking students to robotics labs, factories, medical facilities, or research institutions exposes them to robotics applications that are used in the real world.
4. Students take on robotics projects that mimic real-world situations, such as creating robots for disaster response or exploration.
5. **Healthcare Applications:** Students investigate the use of robots in prosthetics, telemedicine, rehabilitation, and surgery.
6. **Manufacturing:** Students gain knowledge of the material handling, quality control, and industrial robots used in assembly lines.
7. **Logistics and Warehousing:** Students learn about the picking, packing, and inventory management processes that robots do in automated warehouses.
8. **Agriculture:** Students learn about the use of robots in precision agriculture, crop monitoring, planting, and harvesting.
9. **Space Exploration:** Students learn about the employment of robots for study, data collection, and exploration in space missions.
10. **Environmental Monitoring:** Students discover how robots are used to explore the ocean, find pollutants, and protect species.
11. **Disaster Response:** Students learn about the use of robots in search and rescue efforts in disaster-affected areas.
12. **Entertainment and Media:** Students look at how robots are used in special effects, virtual reality, and animatronics.
13. **Retail and Customer Service:** Students investigate how robots are employed in retail settings for delivery, customer service, and inventory management.
14. **Energy Sector:** Students research the use of robotics in the energy sector, including pipeline inspection and maintenance in nuclear power facilities.

15. **Ethical Considerations:** This section includes discussions of the ethical implications of robotics, such as those related to AI ethics, privacy issues, and employment displacement.
16. **Sustainability:** Students look at how robots may help with sustainable practises including energy conservation and trash reduction.
17. **Global Challenges:** Students engage in robotics projects that tackle issues including pollution, climate change, and access to healthcare.
18. **Policy and Regulation:** It is possible to incorporate discussions on the legal and regulatory aspects of robotics, such as liability and safety requirements.
19. **Entrepreneurship:** Students investigate how technological advancements in robotics might result in business opportunities and technology enterprises.

The incorporation of robotics into educational curricula through practical applications motivates students by demonstrating how robotics technology is actively changing numerous industries and tackling pressing issues. This method encourages students to see themselves as potential contributors to these fascinating subjects in the future and creates a greater awareness of the practical consequences of robotics.

Institutions are educating students for the changing technology landscape and developing skills necessary for success in the twenty-first century by incorporating robots into educational courses. This interactive and engaging method of teaching not only improves topic knowledge but also helps students acquire vital life skills that they may use in both their academic and everyday lives.

6.5 ROBOTICS IMPACT ON TEACHING AND LEARNING

Robotics has a big impact on teaching and learning because it changes how teachers present material and how students interact with different disciplines. The following are some significant ways that robotics affects instruction and learning:

1. **Engagement and Motivation:** Robotics increases the motivation and engagement of learning. Robot building, programming, and experimentation are engaging hands-on activities that spark kids' curiosity and encourage a love of learning.
2. **Experiential Learning:** Students gain understanding of complex ideas in science, technology, engineering, and math (STEM) topics through hands-on experience with robots.
3. **Critical Thinking:** Designing and programming robots involves the ability to solve problems, thus teachers should encourage their pupils to think critically and come up with logical solutions to problems.
4. **Innovation and Creativity:** As students design and perfect robots, robotics projects push them to think creatively, come up with original solutions, and innovate.
5. **Collaborative Learning:** Robotics frequently requires teamwork, encouraging the development of interpersonal, communication, and collaboration skills necessary for future jobs.
6. **Cross-Disciplinary Learning:** Robotics integration enables students to investigate connections between several disciplines, fostering a comprehensive knowledge of ideas.
7. **Practical Application:** Robotics gives theoretical ideas real-world perspective, assisting students in understanding how STEM knowledge is applied to everyday situations.
8. **Personalized Learning:** Students can advance at their own rate, tailoring the designs and programming of their robots to fit their interests and ability levels.
9. **Multisensory Learning:** Robotics stimulates the visual, aural, and kinesthetic senses, improving memory and accommodating various learning preferences.

10. **Computer and Technology Literacy:** Students who deal with the hardware, software, sensors, and coding platforms frequently utilized in robotics can increase their technological literacy.
11. **Career Preparation:** Exposure to robotics prepares kids for careers in STEM-related sectors and is in line with new job market trends.
12. Robotics allows students to take the initiative in autonomously exploring issues, experimenting, and troubleshooting.
13. **Inclusivity:** Robotics encourages diverse student participation and success in inclusive learning environments.
14. **Adapting to Challenges:** Students develop resilience and problem-solving skills by learning to adapt when robots do not perform as expected.
15. **Concept Demonstration:** Robots can be used by teachers to make abstract ideas more concrete and understandable.
16. **Digital Citizenship:** Robotics education promotes ethical considerations, responsible technology use, and awareness of the societal consequences of robotics.
17. **Teacher Professional Development:** Including robotics in the classroom necessitates instructors to improve their knowledge and abilities, encouraging continuous professional development.
18. **Parental Involvement:** Robotics programs frequently involve parents or legal guardians in assisting their children's learning, promoting family involvement.
19. **Lifelong Learning:** Robotics encourages pupils to learn outside of the classroom and piques their interest in lifelong learning.

Robotics-based teaching and learning increase the educational experience by providing dynamic, engaging, and useful methods that meet the needs of a technologically advanced society. It gives students the knowledge and skills they need to succeed in a world that is rapidly changing.

Numerical Problems with Solutions

Robotics Competition Scores

Problem 1. In a robotics competition, Team A scores 150 points in the obstacle course, 200 points in the maze challenge, and 100 points in the coding challenge. What is the total score for Team A?

Solution: Total score = 150 + 200 + 100 = 450 points.

Robot Building Costs

Problem 2. A school purchases robotics kits for a class. Each kit costs $150, and they buy 15 kits. What is the total cost of the kits?

Solution: Total cost = $150 * 15 = $2250.

Robot Speed Calculation

Problem 3. A robot travels a distance of 50 meters in 10 seconds. Calculate its average speed in meters per second.

Solution: Average speed = Distance/Time = 50 m/10 s = 5 m/s.

Robot Battery Life

Problem 4. A robot's battery has a capacity of 5000 mAh (milliampere-hours). If the robot consumes 200 mA of current, how many hours will the battery last?

Solution: Battery life = Battery capacity/Current consumption = 5000 mAh/200 mA = 25 hours.

Programming Time Calculation

Problem 5. A student spends 3 hours each day for 5 days programming a robot. How many hours in total did the student spend programming?

Solution: Total programming time = 3 hours/day * 5 days = 15 hours.

Robot Arm Angle

Problem 6. A robot arm is positioned at an angle of 45 degrees to the horizontal. Calculate the sine and cosine of the angle.

Solution: Sine (45 degrees) = 0.7071, Cosine (45 degrees) = 0.7071.

Robot Sensor Range

Problem 7. A robot's ultrasonic sensor has a range of 2 meters. If the robot detects an object at a distance of 1.5 meters, how far is the object from the sensor?

Solution: Object's distance from sensor = Sensor range – Detected distance = 2 m – 1.5 m = 0.5 meters.

Robot Wheel Circumference

Problem 8. A robot's wheel has a diameter of 8 centimeters. Calculate the circumference of the wheel.

Solution: Circumference = π * Diameter = 3.1416 * 8 cm = 25.1328 cm.

Coding Challenge Completion Time

Problem 9. A student completes a coding challenge in 45 minutes. How many seconds did it take to complete the challenge?

Solution: Completion time in seconds = 45 minutes * 60 seconds/minute = 2700 seconds.

Robot Velocity Calculation

Problem 10. A robot travels a distance of 100 meters in 20 seconds. Calculate its velocity in meters per second.

Solution: Velocity = Distance/Time = 100 m/20 s = 5 m/s.

Short Questions and Answers

Q.1. What are educational robots?

Ans. Educational robots are robots designed for teaching and learning purposes, used to engage students in STEM subjects and develop various skills.

Q.2. How do educational robots benefit students?

Ans. Educational robots enhance critical thinking, problem-solving, coding skills, teamwork, and creativity among students.

Q.3. What subjects can educational robots be used for?

Ans. Educational robots can be used to teach STEM subjects, including science, technology, engineering, and math.

Q.4. What is the purpose of integrating robots into education?

Ans. Integrating robots into education enhances engagement, makes learning fun, and prepares students for future careers in technology.

Q.5. How can robots promote hands-on learning?

Ans. Robots require physical assembly, programming, and interaction, offering practical, hands-on learning experiences.

Q.6. What is the role of coding in educational robots?

Ans. Coding allows students to program robots to perform specific tasks, helping them understand logic and algorithms.

Q.7. How do robots foster creativity in students?

Ans. Building and customizing robots encourage students to think creatively and find innovative solutions.

Q.8. What are the benefits of collaborative learning with robots?

Ans. Collaborative learning with robots promotes teamwork, communication, and the exchange of ideas among students.

Q.9. What skills can students develop through educational robots?

Ans. Students develop skills in coding, problem-solving, critical thinking, communication, and project management.

Q.10. What are the different types of educational robots?

Ans. Educational robots include programmable robots, robotics kits, humanoid robots, and interactive robot platforms.

Q.11. Can educational robots be used in different age groups?

Ans. Yes, educational robots are designed for various age groups, from elementary to higher education.

Q.12. How do educational robots support inclusive learning?

Ans. Educational robots can be adapted to accommodate students with diverse abilities and learning styles.

Q.13. What is the role of robotics competitions in education?

Ans. Robotics competitions promote creativity, teamwork, and practical application of STEM concepts among students.

Q.14. What is the significance of ethics discussions in robotics education?

Ans. Ethics discussions help students consider the societal and moral implications of robotics and AI.

Q.15. What are some real-world applications of robotics in education?

Ans. Robotics is used in teaching subjects like physics (mechanics), geometry (shapes), and programming (coding skills).

Q.16. How can robots help teach abstract concepts?

Ans. Robots can provide tangible, visual representations of abstract concepts, making learning more concrete.

Q.17. What is the role of robotics in preparing students for future careers?

Ans. Robotics education equips students with skills relevant to fields such as engineering, programming, and technology.

Q.18. How do educational robots contribute to technological literacy?

Ans. By working with robots and programming, students gain familiarity with technology, coding, and engineering concepts.

Q.19. What are some popular educational robot platforms?

Ans. LEGO Mindstorms, VEX Robotics, and Ozobot are popular educational robot platforms.

Q.20. What is the impact of educational robots on student engagement?

Ans. Educational robots enhance student engagement through interactive, hands-on learning experiences.

Q.21. What do robotics labs and maker spaces offer to students?

Ans. Robotics labs and maker spaces provide tools, resources, and an environment for students to design and build robots.

Q.22. How do robots support project-based learning?

Ans. Robots serve as subjects for project-based learning, where students design, build, and program robots to solve specific challenges.

Q.23. How can teachers integrate robotics into various subjects?

Ans. Teachers can incorporate robotics projects and challenges into subjects like math, science, art, and even history.

Q.24. What is the significance of teacher training in robotics education?

Ans. Teacher training ensures educators are equipped to effectively teach robotics concepts and guide students in hands-on activities.

Q.25. What are some examples of real-world challenges that robotics can address?

Ans. Robotics can address challenges such as environmental monitoring, disaster response, and healthcare assistance.

Q.26. How can educational robots prepare students for a technology-driven world?

Ans. By familiarizing students with robotics and automation, educational robots prepare them for a technology-centered future.

Q.27. What is the connection between robotics and critical thinking skills?

Ans. Designing, programming, and troubleshooting robots encourage students to think critically and develop logical solutions.

Q.28. What is the role of robotics in fostering gender diversity in STEM fields?

Ans. Robotics education encourages inclusivity and can help bridge the gender gap in STEM subjects and careers.

Q.29. How do educational robots cater to different learning styles?

Ans. Educational robots engage visual, auditory, and kinesthetic learners through hands-on activities and interactive experiences.

Q.30. Can educational robots be integrated with online learning platforms?

Ans. Yes, educational robots can be used in conjunction with online platforms to provide remote learning opportunities.

Q.31. What is the significance of ethics discussions in robotics education?

Ans. Ethics discussions help students consider the societal and moral implications of robotics and AI.

Q.32. How can robots enhance communication skills among students?

Ans. Collaborative robot projects require students to communicate ideas, delegate tasks, and work as a team.

Q.33. What role does educational robotics play in introducing students to programming?

Ans. Educational robots provide a practical context for students to learn programming concepts and apply them in real-world scenarios.

Q.34. How do robotics competitions impact student motivation?

Ans. Robotics competitions challenge students to apply their skills in a competitive context, boosting motivation and engagement.

Q.35. Can educational robots be customized for specific learning goals?

Ans. Yes, educational robots can be adapted and programmed to align with specific learning objectives and curriculum goals.

Q.36. What is the future potential of robotics in education?

Ans. Robotics is expected to play an increasingly significant role in shaping innovative, technology-focused education.

Q.37. How do educational robots contribute to STEM-focused learning?

Ans. Educational robots integrate science, technology, engineering, and math concepts into hands-on, experiential learning.

Q.38. What impact do robotics labs have on student creativity?

Ans. Robotics labs encourage students to experiment, explore, and come up with imaginative solutions to robotics challenges.

Q.39. How do educational robots encourage problem-solving skills?

Ans. Students encounter challenges while building and programming robots, fostering problem-solving skills to overcome obstacles.

Q.40. What is the significance of robotics in project-based assessments?

Ans. Robotics projects offer a dynamic way for students to showcase their understanding of concepts and practical skills.

Q.41. How do robotics competitions promote student innovation?

Ans. Robotics competitions encourage students to push boundaries, experiment with new ideas, and innovate in robot design.

Q.42. How can educational robots foster a lifelong love for learning?

Ans. By making learning engaging and enjoyable, educational robots instill a passion for continuous exploration and growth.

Q.43. What is the role of educational robots in teaching computational thinking?

Ans. Educational robots help students develop computational thinking skills by designing algorithms and logical sequences.

Q.44. What are some global initiatives promoting robotics in education?

Ans. Initiatives like FIRST Robotics and RoboCup aim to inspire students to engage with robotics and technology.

Q.45. How do robots contribute to fostering a growth mindset among students?

Ans. Robotics challenges encourage students to embrace failures as learning opportunities and persevere in problem-solving.

Q.46. What role do robots play in introducing advanced concepts to students?

Ans. Educational robots can introduce advanced concepts like artificial intelligence, machine learning, and automation in a tangible way.

Q.47. Can robotics be integrated into arts and creative subjects?

Ans. Yes, robotics can be used in art projects, installations, and performances to blend technology with creative expression.

Q.48. How do robotics kits support self-directed learning?

Ans. Robotics kits offer students the autonomy to explore concepts at their own pace, promoting self-directed learning.

Q.49. What are some considerations for evaluating the effectiveness of robotics in education?

Ans. Assessments should consider students' engagement, learning outcomes, problem-solving abilities, and real-world application.

Q.50. What is the long-term impact of robotics education on society?

Ans. Robotics education cultivates a generation of technologically literate individuals who contribute to technological advancements and innovation.

Multiple Choice Questions and Answers

Q.1. What are educational robots designed for?

(*a*) Entertainment (*b*) Industrial tasks
(*c*) Teaching and learning (*d*) Medical procedures

Q.2. Which skills can students develop through educational robots?

(*a*) Baking (*b*) Painting (*c*) Coding (*d*) Gardening

Q.3. Educational robots enhance which of the following skills?

(*a*) Lack of engagement (*b*) Passive learning
(*c*) Critical thinking (*d*) Isolation

Q.4. What subjects are educational robots commonly used to teach?

(*a*) Music and literature (*b*) Art and fashion
(*c*) History and geography (*d*) STEM subjects

Q.5. How can educational robots make learning more engaging?

(*a*) By reducing interactions (*b*) By replacing teachers
(*c*) By promoting rote memorization (*d*) By offering hands-on activities

Q.6. What is a primary benefit of using educational robots in classrooms?

(*a*) Decreased student participation (*b*) Limited creativity
(*c*) Increased teacher workload (*d*) Enhanced student motivation

Q.7. Which of the following skills can students develop by building and programming robots?

(*a*) Writing poetry (*b*) Riding bicycles
(*c*) Playing musical instruments (*d*) Problem-solving

Q.8. Educational robots can be integrated into teaching which type of subjects?

(*a*) Non-technical subjects (*b*) Humanities only
(*c*) Science, technology, engineering, and math (STEM)
(*d*) Social sciences only

Q.9. What role does coding play in using educational robots?

(*a*) It is irrelevant to robotics education (*b*) It helps students learn foreign languages
(*c*) It allows students to program robot behaviors
(*d*) It promotes outdoor activities

Q.10. What is a common outcome of integrating robots into education?

(*a*) Decreased student engagement (*b*) Limited student interaction
(*c*) Reduced teacher involvement (*d*) Improved student learning outcomes

Q.11. How do educational robots encourage creativity?

(*a*) By providing fixed solutions (*b*) By restricting student ideas
(*c*) By promoting rote memorization
(*d*) By allowing customization and innovation

Q.12. Collaborative learning with robots promotes which of the following skills?

(*a*) Isolation (*b*) Competition
(*c*) Communication and teamwork (*d*) Rote memorization

Q.13. What type of learners can benefit from hands-on learning with robots?

(*a*) Visual and kinesthetic learners (*b*) Auditory learners only
(*c*) Tactile learners only (*d*) Linguistic learners

Q.14. What is a potential impact of integrating robots into education on future careers?

(*a*) No impact on career choices (*b*) Reduced demand for technology skills
(*c*) Preparation for technology-driven careers (*d*) Preparation for non-technical careers

Q.15. What do robotics competitions encourage among students?

(*a*) Passive learning (*b*) Isolation

(*c*) Problem-solving, teamwork, and innovation (*d*) Rote memorization

Q.16. What is the significance of ethics discussions in robotics education?

(*a*) Irrelevant to student learning

(*b*) Encourages students to ignore ethical considerations

(*c*) Helps students consider the societal impact of robotics

(*d*) Focuses solely on technical skills

Q.17. How do educational robots support personalized learning?

(*a*) By providing a one-size-fits-all approach (*b*) By offering a fixed curriculum

(*c*) By allowing students to progress at their own pace

(*d*) By eliminating the need for teachers

Q.18. What impact can educational robots have on gender diversity in STEM fields?

(*a*) None (*b*) Exclusively negative impact

(*c*) Encourage inclusivity and bridge the gender gap

(*d*) Limit opportunities for both genders

Q.19. Which types of learners can educational robots cater to?

(*a*) Visual learners only (*b*) Auditory learners only

(*c*) Kinesthetic learners only (*d*) A variety of learning styles

Q.20. How can educational robots be integrated with online learning?

(*a*) They cannot be integrated with online learning

(*b*) Through virtual reality simulations only

(*c*) By providing remote learning opportunities

(*d*) By replacing online platforms

Q.21. What skills can students develop through robotics competitions?

(*a*) Isolation and rote memorization (*b*) Problem-solving and teamwork

(*c*) Passive learning and competition (*d*) Technical skills only

Q.22. What is the role of educational robotics in nurturing a growth mindset?

(*a*) Reinforces fixed mindset (*b*) Discourages creative thinking

(*c*) Encourages learning from failure and resilience

(*d*) Promotes isolation

Q.23. How can educational robots contribute to the introduction of advanced concepts?

(*a*) By avoiding advanced concepts altogether

(*b*) By making advanced concepts difficult to understand

(*c*) By providing hands-on experience with advanced concepts

(*d*) By focusing exclusively on basic concepts

Q.24. Which subjects can robots be integrated into aside from STEM?

(*a*) Non-technical subjects only
(*b*) Humanities only
(*c*) Arts and creative subjects as well
(*d*) Social sciences only

Q.25. How can educational robots be customized for specific learning objectives?

(*a*) They cannot be customized
(*b*) By ignoring curriculum goals
(*c*) By adapting their behaviors and programming
(*d*) By providing a fixed set of functionalities

Q.26. What is the long-term impact of robotics education on society?

(*a*) No impact on society
(*b*) Restricts technological advancements
(*c*) Fosters a generation of technologically literate individuals
(*d*) Has negative implications for the job market

Q.27. What impact do robotics competitions have on student motivation?

(*a*) Decreases motivation
(*b*) Has no impact on motivation
(*c*) Increases motivation and engagement
(*d*) Promotes individual competition only

Q.28. How do educational robots prepare students for a technology-driven world?

(*a*) By avoiding technology altogether
(*b*) By isolating students from technology
(*c*) By providing exposure to robotics and automation
(*d*) By focusing solely on traditional subjects

Q.29. How do educational robots contribute to computational thinking?

(*a*) By discouraging logical thinking
(*b*) By teaching irrelevant skills
(*c*) By promoting creativity
(*d*) By requiring students to design algorithms and logical sequences

Q.30. What are some global initiatives promoting robotics in education?

(*a*) Robotics has no global initiatives
(*b*) These initiatives only focus on competitive robotics
(*c*) Initiatives like FIRST Robotics and RoboCup inspire students to engage with robotics
(*d*) Initiatives are exclusively limited to technology companies

Key to Multiple Choice Questions

1. (*c*)	**2.** (*c*)	**3.** (*c*)	**4.** (*d*)	**5.** (*d*)	**6.** (*d*)	**7.** (*d*)	**8.** (*c*)	**9.** (*c*)
10. (*d*)	**11.** (*d*)	**12.** (*c*)	**13.** (*a*)	**14.** (*c*)	**15.** (*c*)	**16.** (*c*)	**17.** (*c*)	**18.** (*c*)
19. (*d*)	**20.** (*c*)	**21.** (*b*)	**22.** (*c*)	**23.** (*c*)	**24.** (*c*)	**25.** (*c*)	**26.** (*c*)	**27.** (*c*)
28. (*c*)	**29.** (*d*)	**30.** (*)						

Project Works

1. **Robotics Science Fair Project: Line Following Robot Project Description:** Build a robot that can follow a black line on a white surface using sensors.

Hints: This project teaches students about sensors, programming, and basic robotics concepts. The robot follows the line by detecting the contrast between the black line and white background.

2. **Robot-Assisted Storytelling Project Description:** Create a robot that narrates stories to young children when prompted.

   Hints: This project encourages creativity and interdisciplinary learning. Students can design a robot, program it to tell stories, and incorporate elements of language arts and communication.

3. **Environmental Monitoring Robot Project Description:** Design a robot that collects environmental data such as temperature, humidity, and pollution levels.

   Hints: This project combines robotics with environmental science. Students can program sensors to collect data, analyze trends, and explore the impact of the environment on various factors.

4. **Robotic Art Installation Project Description:** Build a robot that creates artworks using various mediums and techniques.

   Hints: This project blends art and technology, fostering creativity and innovation. Students can design a robot capable of painting, drawing, or creating sculptures.

5. **Educational Quizbot Project Description:** Develop a robot that quizzes students on different subjects and provides instant feedback.

   Hints: This project enhances students' knowledge retention and programming skills. The robot can ask questions and engage students in interactive learning.

6. **Interactive History Tour Guide Robot Project Description:** Construct a robot that guides users through historical sites and provides information using voice commands.

   Hints: This project combines history, programming, and human-robot interaction. Students can program the robot to narrate historical facts based on users' questions.

7. **Robot-Assisted Language Learning Project Description:** Create a robot that helps students practice foreign language pronunciation and vocabulary.

   Hints: This project supports language learning and human-robot interaction skills. The robot can prompt students to repeat words and phrases, providing feedback on pronunciation.

8. **Robotic Math Tutor Project Description:** Develop a robot that assists students in solving math problems and explains concepts visually.

   Hints: This project reinforces math skills and communication abilities. The robot can interact with students, display mathematical operations, and guide them through problem-solving steps.

9. **Robotic Coding Challenge Project Description:** Organize a coding challenge where students program robots to navigate through a maze or complete tasks.

   Hints: This project promotes teamwork, coding skills, and problem-solving. Students work together to develop algorithms that guide robots through challenges.

10. **Robotics for Social Change Project Description:** Design a robot that addresses a social issue, such as helping people with disabilities or assisting the elderly.

    Hints: This project highlights the ethical and social impact of robotics. Students can create robots that enhance accessibility, independence, or quality of life for individuals with specific needs.

7

Robotics in Space Exploration

7.1 INTRODUCTION

Robotics has revolutionized our ability to explore and comprehend the cosmos beyond our planet by making significant contributions to space exploration. Robotic technologies have been essential to scientific investigation, data gathering, spaceship operations, and even the prospect for human colonization of other celestial worlds since the early days of space exploration. Robotics has changed how we view the universe, from the early success of the Mars rovers Spirit and Opportunity to the current exploration of Mars by the Curiosity and Perseverance rovers, as well as missions like the Juno spacecraft studying Jupiter and the New Horizons mission to Pluto and beyond. Robotics will continue to be a crucial instrument for astronomical exploration and the advancement of human knowledge thanks to planned missions like the Europa Clipper, the Artemis program to send people back to the Moon, and plans for manned missions to Mars.

7.2 SIGNIFICANCE OF ROBOTICS IN SPACE EXPLORATION

Here's an outline of the significance of robotics in space exploration:

7.2.1 Remote Exploration

Robotic spacecraft have made it possible for mankind to explore dangerous and remote areas of space, such as planets, moons, asteroids, and comets. These robots have the capacity to collect information, take pictures, and carry out tests that provide details on the make-up, atmosphere, geology, and possible habitability of these celestial worlds. In the context of space exploration, remote exploration refers to the utilization of robotic spacecraft and tools for exploring and gathering data from far-off celestial bodies without the physical presence of humans. Researchers, engineers, and scientists can gather information, conduct experiments, and investigate space habitats that could be difficult or dangerous for people to explore in person. The foundation of space exploration activities has been remote exploration, which has produced a wealth of new knowledge and technological breakthroughs.

Here are some key points about remote exploration using robotics in space exploration:

1. **Unmanned Missions:** Unmanned robotic missions are sent to a variety of celestial bodies, such as planets, moons, asteroids, and comets, as part of remote exploration. For the purpose of

gathering information about the target environment, these missions are outfitted with a variety of scientific instruments and sensors.

2. **Robotic Rovers and Landers:** These machines are made to land on the surface of a celestial planet and carry out scientific research. Rovers can travel, take pictures, examine samples of rock and soil, and send information back to Earth.
3. **Sample Gathering:** Gathering samples from the target celestial body is a common component of remote exploration missions. These samples can reveal important details on the make-up, architecture, and past of the body. The goal of sample return missions is to return these samples to Earth for further study.
4. **Teleoperation and Autonomy:** Teleoperation and autonomous operations are both used in remote exploration. Robotic devices on faraway bodies can be remotely controlled by engineers and scientists on Earth. Robots can work autonomously under the guidance of pre-programmed commands and feedback from sensors.
5. **Interplanetary Communication:** A network of communication relays, including orbiters and ground stations, is used to establish contact with distant robotic missions. These relays make sure that instructions are given to the spacecraft and that information is sent back to Earth.
6. **Space Telescopes:** Using space telescopes positioned in Earth's orbit is another aspect of remote exploration. Images and data from far-off galaxies, stars, and other celestial objects are captured by telescopes like the Hubble Space Telescope and the future James Webb Space Telescope.
7. **Environmental Monitoring:** Robotic systems with a variety of sensors can keep an eye on things like radiation levels, temperature, the makeup of the atmosphere, and surface characteristics.
8. **Dangerous Places:** Remote exploration is very useful for investigating dangerous places where human presence is dangerously inappropriate because of harsh conditions including strong radiation, extremely low temperatures, and an absence of breathable atmosphere.
9. **Continuous Monitoring:** Robotic missions can offer long-term observations and continuous monitoring of celestial bodies, allowing scientists to track changes and occurrences throughout time.
10. **Data Gathering and Analysis:** Remote exploration produces enormous volumes of data, which are then examined by scientists to learn more about the formation and evolution of celestial bodies as well as to provide answers to important puzzles regarding the nature of the universe.

The Mars rovers Spirit, Opportunity, Curiosity, and Perseverance, the Voyager spacecraft, the Juno mission to Jupiter, the New Horizons mission to Pluto, and the forthcoming Europa Clipper mission to examine Jupiter's moon Europa are a few examples of remote exploration projects. These missions have offered important insights on the past, present, and possible habitability of many celestial worlds, as well as their geology and atmosphere. Our grasp of the universe continues to grow thanks to remote exploration, which also influences our plans for further investigation.

7.2.2 Reducing Risk

A crucial part of space research is minimizing danger, especially when using robotics to explore and collect data from far-off celestial bodies. Robotic missions let us gather information, test technology, and make decisions before sending people into potentially dangerous situations, which is crucial for reducing the dangers involved with human spaceflight. Here is how robots contributes to lowering space exploration risks:

1. Robotic missions are dispatched to investigate hazardous settings, such as those with high radiation levels, sweltering temperatures, and thin atmospheres, that are too harsh or dangerous for human presence. This protects against exposing people to unneeded hazards.
2. **Pre-Exploration:** Before human expeditions, robotic missions act as scouts, gathering crucial information and insights about a celestial body. This information aids in locating prospective perils, landing locations, and practical difficulties.
3. **Technology Validation:** Robotic missions test and validate technologies that are essential for a successful human exploration before deploying humans. This comprises landing methods, propulsion technology, and life support and communication systems.
4. **Learning from Failures:** Robotic mission failures and difficulties provide important lessons that can be applied to future missions, including those manned by people, to make them safer and more successful.
5. **Data Gathering:** Robotic missions collect information on the target environment, assisting scientists and engineers in identifying potential threats and modifying mission plans accordingly.
6. **Experimentation:** Robotic missions can conduct research experiments to examine how the environment in space affects equipment, living creatures, and materials. Strategies for human protection are informed by this information.
7. **Remote Operations:** Robots can carry out activities in real time while posing the fewest risks to human operators by permitting remote operations from Earth.
8. **Sample Return:** Robotic missions are able to gather samples from celestial bodies and bring them back to Earth for further investigation, hence removing the necessity for human presence at the site.
9. **Improved Mission Planning:** The information gathered by robotic missions enables mission planners to foresee difficulties and hazards, enabling improved human mission preparation and decision-making.
10. **Lengthy-term Monitoring:** Robots can operate continuously for lengthy periods of time, gathering information over time to comprehend long-term environmental conditions and potential risks.
11. **Redundancy and Reliability:** Redundancy and fail-safes are built into robotic systems, which lower the chance of system failure and guarantee mission completion.

Robotic missions can be more economical than human missions, enabling more frequent exploration and experimentation without endangering human lives.

Robotic missions, including the Mars rovers, Voyager spacecraft, and the impending Europa Clipper mission, have greatly increased our understanding of the universe while lowering the hazards connected with sending humans into space. These missions open the door for safe and fruitful human exploration by disseminating vital data and ideas on the difficulties of space habitats.

Before sending humans into a target region, we can obtain vital information by using robotic missions. By identifying probable risks and difficulties, this promotes the security of manned missions in the future.

7.2.3 Versatility

Robotic systems are adaptable tools for various exploratory missions since they may be built for specific purposes. For instance, rovers can travel across the surface of Mars, landers can examine the makeup of the atmospheres of other planets, and orbiters may observe far-off celestial bodies from space. Our ability to study and comprehend the cosmos has been fundamentally changed by the adaptability of robotics in space research. Robotic systems are flexible, scalable, and capable of a wide range of tasks, making them essential tools for astronomical exploration and conducting in-space research.

Here's how robotics showcases its versatility in space exploration:

1. **Multi-Tasking Capability:** Robots can be created and programmed to carry out numerous activities simultaneously, such as gathering samples, carrying out tests, taking pictures, and analyzing data.
2. Robotics enables exploration of a variety of habitats, including the depths of asteroids and the outer reaches of the solar system in addition to the surfaces of planets and moons.
3. **Remote Control:** Robots can be controlled remotely from Earth, allowing for real-time control and modifications in response to shifting circumstances and goals.
4. **Accurate Movements:** Robotics allows for accurate movements and actions, which are necessary for delicate activities like collecting samples or precisely positioning instruments.
5. **Autonomous Operations:** A lot of robots have the ability to act independently and adjust to unforeseen circumstances without the need for human interaction.
6. **Adaptation to Harsh Environments:** Robots are built to resist harsh environments like microgravity, strong radiation, vacuum, and temperature changes.
7. **Long-duration Missions:** Robotic systems are able to run continuously, enabling the collection of long-term data and observations.
8. **Customizability:** By giving robots varied sensors, equipment, and tools based on the scientific goals, robots can be adapted for certain missions.
9. **Risk Reduction:** Robotic systems can be used to evaluate dangers and difficulties before sending people into uncharted territory, protecting their safety.
10. **Exploration of Inaccessible Areas:** Robots can explore hazardous or difficult-to-reach places, like deep craters, steep cliffs, or places with extreme weather.
11. **Complex Operations:** Robots are capable of carrying out complicated tasks including drilling, building structures, and analyzing complicated geological formations.
12. **Sample Collection and Return:** Robotic missions can take samples from other heavenly bodies and bring them back to Earth, revealing important details about their makeup and past.
13. **Telepresence and Teleoperation:** Robots with cameras and sensors enable humans to virtually explore far-off places, giving them a sense of presence and assisting with decision-making.
14. Robotic spacecraft can act as relays for communication between Earth and other missions, ensuring continuous data delivery.
15. **Space Observatories:** To accurately position their equipment and take pictures of far-off astronomical objects, space observatories like Hubble and the James Webb Space Telescope use robots.

Outstanding accomplishments have been made possible by the versatility of robots in space exploration, including the successful exploration of Mars, the analysis of Jupiter and its moons, the study of asteroids, and the deep-space observations of galaxies and stars. Robotics will become more and more crucial to our continuous cosmological investigation as technology develops.

7.2.4 Unmanned Spacecraft

Robotic unmanned spacecraft can function over extended periods of time without the requirement for life support systems, allowing them to concentrate only on exploration and scientific research goals. Unmanned spacecraft are robotic ships built to conduct space exploration and collect data for scientific purposes without having human astronauts on board. These spacecraft, also known as robotic probes or robotic missions, are essential for space travel because they allow us to investigate far-off celestial bodies, carry out experiments, and gather important data about the cosmos.

Here are some notable examples of unmanned spacecraft used in space exploration:

1. **Voyager Spacecraft:** The Voyager spacecraft, launched in 1977, are among the most recognizable unmanned missions. By performing flybys of Jupiter and Saturn, they were able to collect hitherto unheard-of photographs and data on these planets and their moons. Voyager 1 is now the farthest object produced by humans from Earth after travelling into interstellar space.
2. **Mars Rovers:** NASA's Mars rovers (Spirit, Opportunity, Curiosity, and Perseverance) have been crucial in helping to explore the Martian surface. These rovers have carried out geological studies, examined samples of rock and soil, and looked for indicators of past or present life. The 2020 Perseverance rover's mission is to investigate Jezero Crater on Mars and gather materials for potential return to Earth.
3. **Cassini-Huygens:** The Cassini spacecraft studied Saturn and its moons, including the famous moon Titan, in great detail. The Huygens probe, carried by Cassini, made a successful landing on Titan's surface and revealed information about the atmosphere and makeup of the planet.
4. **New Horizons:** The 2006-launched New Horizons probe made a historic flyby of Pluto in 2015, providing information on the planet's surface and makeup. It is currently travelling to the Kuiper Belt, which is an area beyond Neptune, for investigation.
5. **Hubble Space Telescope:** The Hubble Space Telescope is a robotic observatory that orbits the Earth but is not a conventional spacecraft. Beautiful pictures and insightful information on far-off galaxies, nebulae, and other cosmic objects have been made available by it.
6. **Juno:** NASA's Juno mission is investigating the makeup, magnetic field, gravitational field, and polar magnetosphere of Jupiter. It seeks to learn more about the structure, history, and origins of the planet.
7. **Rosetta and Philae:** The European Space Agency's (ESA) Rosetta mission examined Comet 67P/Churyumov-Gerasimenko. The first-ever soft landing on a comet's surface was accomplished by the Philae lander, which was carried by Rosetta.
8. **OSIRIS-REx:** NASA's OSIRIS-REx mission is investigating the asteroid Bennu that is close to Earth. A sample from Bennu's surface was taken by the spacecraft, and it will be brought back to Earth for examination.
9. **Chang'e Missions (China):** The Chang'e missions from China have orbited and landed on the Moon, where they have conducted in-depth geological and scientific investigations.
10. **Mars Orbiter:** The NASA Mars Reconnaissance Orbiter has been orbiting Mars since 2006 and carries out atmospheric observations, high-resolution imagery, and communication relay for other Mars missions.

Unmanned spacecraft offer a flexible and affordable way to explore the cosmos. They have offered insightful information on the creation of the solar system, the features of distant planets and moons, the make-up of asteroids and comets, and the nature of space itself. These missions are advancing our knowledge of the cosmos and laying the foundation for further exploration projects, perhaps even involving sending people to other planets and celestial bodies.

7.2.5 Space Telescopes

Robotics is used by space telescopes like the Hubble Space Telescope and the soon-to-be-built James Webb Space Telescope to precisely position their instruments, take pictures, and send data back to Earth. Robotic observatories called space telescopes are placed in Earth's orbit or other parts of space to watch and study celestial objects and phenomena that occur outside of our planet's atmosphere. By taking high-resolution pictures, gathering information at different wavelengths, and conducting scientific research free from the disturbance of Earth's atmosphere, these telescopes have substantially improved our understanding of the cosmos.

Here are some notable examples of space telescopes used in space exploration:

1. **Hubble Space Telescope:** One of the most well-known and significant space telescopes is the Hubble Space Telescope (HST), which was launched in 1990. Beautiful photographs taken by Hubble of galaxies, nebulae, stars, and planets have revealed intricacies and phenomena that were previously invisible to the human eye. Hubble has been instrumental in the identification of exoplanet atmospheres and the measurement of the pace of the universe's expansion, among other findings.
2. **Chandra X-ray Observatory:** Designed to study X-rays from high-energy objects in the universe, such as black holes, neutron stars, and supernova remnants, Chandra was launched in 1999. It has given us important new understandings about how these energy events work.
3. **Spitzer Space Telescope:** Spitzer is an infrared space telescope that was launched in 2003 with the goal of examining the universe at infrared wavelengths. It has provided information on exoplanets, distant galaxies, and star formation.
4. **James Webb Space Telescope (JWST):** As the Hubble's replacement, JWST is slated to launch in December 2021. In order to research the early cosmos, the development of galaxies, and the characteristics of exoplanet atmospheres, it will largely work in the infrared region.
5. **Fermi Gamma-ray Space Telescope:** Fermi is a gamma-ray observatory that was launched in 2008 and is used to find high-energy gamma rays coming from pulsars, supernova remnants, and gamma-ray bursts, among other sources. It aids in the study of cosmic rays and extreme phenomena.
6. **XMM-Newton:** The European Space Agency (ESA) launched the XMM-Newton X-ray observatory in 1999. It has significantly advanced our understanding of the X-ray radiation from different astronomical bodies.
7. **NuSTAR (Nuclear Spectroscopic Telescope Array):** Founded in 2012, NuSTAR is a high-energy X-ray observatory that focuses on researching supernova remnants, neutron stars, and black holes. It has shed light on the processes taking place around such extreme objects.
8. **AstroSat, India's Multi-Wavelength Space Observatory:** AstroSat, India's first Multi-Wavelength Space Observatory, was launched in 2015. It uses ultraviolet, visible, and X-ray wavelengths to detect astronomical objects.
9. **Gaia:** The European satellite Agency (ESA) launched Gaia in 2013 as a satellite observatory that precisely examines the locations, distances, and motions of over a billion stars in the Milky Way galaxy.

The benefit of using a space telescope is that it can observe celestial processes and objects without being distorted by the Earth's atmosphere. They give astronomers an unmatched view of the universe at many wavelengths, enabling them to make ground-breaking discoveries, investigate fundamental issues regarding the universe's origins, and investigate far-off galaxies, stars, and other cosmic structures.

7.2.6 Sample Collection

Robotic spacecraft are capable of retrieving samples from asteroids and bringing them back to Earth. For instance, the asteroid Bennu was sampled by the OSIRIS-REx mission, and the future Mars Perseverance rover will gather soil and rock samples. A key component of robotic space exploration missions is sample collecting. Robotic spacecraft are outfitted with specialized equipment for collecting surface samples from a variety of celestial bodies, including planets, moons, asteroids, and comets. The composition, history, and circumstances of these objects are revealed by the samples that have been obtained, offering important knowledge on the formation and development of our solar system and other systems.

Here's how sample collection works in robotic space exploration:

1. **Surface Contact Mechanisms:** Robotic spacecraft are outfitted with tools like robotic arms, drills, scoops, and collectors that are intended to physically interface with the target celestial body's surface.
2. **Sampling Instruments:** Depending on the surface's characteristics, various sampling instruments are utilized. For instance, scoops are used to collect loose materials like regolith or soil, whereas drills are used to retrieve samples from solid materials.
3. **Handling and Containment:** Samples that are collected are handled carefully to prevent contamination and stored in sealed containers to preserve their integrity for the return trip to Earth.
4. **Analysis On-Site:** To obtain results right away, several missions use built-in instruments to analyze samples right away. This is crucial in ensuring that the appropriate kinds of samples are obtained.
5. **Sample Return Missions:** Several missions entail returning the samples that were obtained to Earth for in-depth examination in facilities outfitted with cutting-edge equipment. Complex planning is needed for sample return missions, which also include spacecraft rendezvous and re-entry into Earth's atmosphere.
6. **Minimizing Contamination:** It is crucial to ensure that the samples were not contaminated by substances from the Earth. To keep any undesired elements from interfering with the analysis, special precautions are taken.
7. **Planetary Protection:** Space organizations follow planetary protection guidelines to avoid the possibility of introducing Earth-based species to other celestial bodies. Sterilization standards for sample collection missions must be strictly adhered to.
8. **Analyses and Discoveries:** Examining the samples that have been collected can reveal information about the celestial body's past, the forces that shaped it, its makeup, and more.

Notable missions that have involved sample collection include:

- **Stardust:** This company brought back to Earth dust from the comet Wild 2's coma.
- The Japanese missions Hayabusa and Hayabusa2, which gathered samples from the asteroids Ryugu and Itokawa, respectively.

- **OSIRIS-REx:** Designed to bring samples back to Earth after collecting them from the asteroid Bennu.
- **Mars Sample Return (coming soon):** NASA and ESA intend to work together on missions to gather samples from Mars and return them to Earth for examination.

The gathering of samples is an essential step in understanding the nature of celestial bodies and solving the puzzles surrounding the genesis and evolution of the solar system. It necessitates interdisciplinary cooperation between scientists, engineers, and mission planners as well as precise engineering, careful planning, and teamwork.

7.2.7 Teleoperation

Robots in orbit can be remotely operated from Earth by engineers and scientists, allowing for data collecting and real-time decision-making. This is essential when exploring areas where communication delays make quick control impossible. Through the method of teleoperation, operators on Earth may manage robots and carry out activities in far-off or dangerous locations, such as space exploration. Teleoperation is the process of directing robotic systems from a remote location. Human interaction with robotic spacecraft, rovers, and instruments placed in orbit depends heavily on teleoperation.

Here's how teleoperation works in the context of space exploration:

1. **Remote Control:** Teleoperation refers to the real-time control of robotic systems' motions, actions, and functions from a control centre on Earth. This enables operators to carry out duties as if they were actually there where the robot is.
2. **Communication Link:** The robotic system in space is connected to the control centre by a communication link, which is commonly established using radio waves or other communication protocols. Using this link, operators and the robot can communicate commands, data, and telemetry.
3. **Human-Machine Interface:** To see the robot's surroundings and direct its movements, operators employ specialised interfaces that frequently include screens, joysticks, haptic devices, and cameras.
4. **Management of Latency:** Since signal transmission speeds are limited, managing latency during teleoperation is a difficulty. When managing the robot, especially when performing dynamic or delicate tasks, operators must take these delays into consideration.
5. **Operations Requiring Accuracy:** Teleoperation is very helpful for operations requiring accuracy, such as handling sensitive instruments, gathering samples, and conducting scientific research.
6. **Real-time Decision-Making:** Teleoperation enables operators to make decisions in real-time based on observations made by the robot and data from its sensors. This is essential for coping with unforeseen circumstances.
7. **Flexibility and Adaptability:** The robot's actions can be dynamically changed in response to new information learned throughout the operation, allowing operators to adapt to changing circumstances.
8. **Dangerous Situations:** Teleoperation permits exploration of dangerous situations that are dangerous for people, including hot or cold places.
9. **Training and Expertise:** Teleoperation requires a specific set of abilities, including spatial awareness, scenario analysis, and hand-eye coordination, which operators gain through specialized training.

10. **On-board Autonomy:** While teleoperation requires human supervision, certain robotic systems feature some degree of on-board autonomy to manage routine tasks and avoid collisions in the event of prolonged connection delays.

Examples of teleoperation in space exploration include:

(*a*) Managing the Mars rovers' (Spirit, Opportunity, Curiosity, and Perseverance) motions as they travel across the Martian surface, take pictures, and carry out scientific experiments.

(*b*) Using robotic arms on spacecraft like the International Space Station (ISS) to deploy instruments, carry out maintenance procedures, and capture cargo capsules.

(*c*) Using space telescopes like the Hubble Space Telescope to manoeuvre and align them in order to take exact pictures of faraway astronomical objects.

Humans can explore and interact with distant locations, collect data, and carry out difficult activities over great distances in space thanks to the essential tool known as teleoperation. It improves our capacity for cosmic exploration and comprehension while lowering the risks to manned explorers.

7.2.8 Future Human Exploration

Robotics is a prerequisite for human spaceflight. Robotic systems assist in the development of habitats, test technologies, and scout potential landing sites to prepare the way for human missions. The continuing development and integration of robotic systems will be crucial to the success of human space travel. Robotics will be a key component of preparing for and assisting human trips to a variety of celestial worlds, allowing for safer, more productive, and scientifically beneficial exploration of the solar system and beyond.

Here's how robotics will contribute to the future of human space exploration:

1. **Pre-Mission Survey and Preparation:** Robotic missions will keep scouting and surveying target locations to gather vital data on potential hazards, landing places, and the local ecosystem. This information will help mission designers plan more safely for manned missions.
2. **Habitat Construction and Maintenance:** Prior to the arrival of humans, robots can help build habitats, shelters, and infrastructure on distant planets or moons. They can also carry out maintenance procedures to keep habitats functional.
3. **Resource Utilization:** To assist human activities and lessen the need to ship supplies from Earth, robots can be built to extract and process nearby resources, including water or minerals.
4. **Teleoperation and Remote Assistance:** Robots will enable distant operations and aid in jobs that may be hazardous or tiresome for astronauts, even when human personnel are present. Exploring potentially dangerous regions, using tools, and running tests all fall under this category.
5. **Scientific Research:** Robots will continue to carry out experiments and scientific research, both independently and with human supervision, deepening our understanding of remote settings and phenomena.
6. **Sample Collection and Return:** Robotic missions have the ability to gather samples before human missions, giving astronauts already-gathered materials for examination and research. Robots can help bring these samples back to Earth as well.
7. **Human-Robot Collaboration:** As robots support astronauts in challenging activities including carrying out repairs, putting up studies, and assisting in medical procedures, human-robot collaboration will become more and more crucial.

8. Robots will act as communication relays, enabling human crews to stay in touch with mission control on Earth even when they are on the opposite side of celestial bodies.
9. **Exploratory of Dangerous Environments:** Robots can investigate and evaluate dangerous places on behalf of people, protecting their safety and enabling astronauts to concentrate on important scientific and exploratory goals.
10. **Enhanced AI and Autonomy:** It's conceivable that future robotic systems will have enhanced artificial intelligence and autonomy, allowing them to make choices and respond to unforeseen circumstances without continual human supervision.
11. **Lunar and Martian Bases:** Robots will be critical for establishing necessary infrastructure, carrying out first experiments, and preparing the environment for human habitation as ideas for lunar and Martian bases develop.

Deep space missions, such as those to asteroids or outer planets, will require the use of robotic missions for data gathering, route planning, and reconnaissance.

The success of upcoming space exploration projects will ultimately depend on the cooperation between humans and robots. Robotic technologies will help us explore more difficult settings, collect information before humans do, improve mission safety, and spread human presence throughout the solar system. The limits of human space exploration will be pushed as technology develops, creating new vistas for research and understanding.

7.2.9 Interplanetary Communication

Robotic spacecraft act as relays for communication, allowing information to be sent between Earth and far-off probes and facilitating ongoing contact with missions around the solar system. The ability of robotic spacecraft to transfer data, commands, and information over great distances between Earth and other celestial bodies is a key component of interplanetary communication. Controlling robotic missions, obtaining scientific data, and staying in touch with spacecraft exploring distant planets, moons, asteroids, and comets all depend on reliable communication.

Here's how interplanetary communication works in the context of robotics in space exploration:

1. **Communication Relays:** Satellites or orbiters placed orbiting a celestial body are common examples of communication relays used in interplanetary communication. By receiving messages from robotic spacecraft and delivering them to Earth, these relays serve as middlemen.
2. **Deep Space Network (DSN):** The Deep Space Network is a network of ground-based antenna complexes that are strategically positioned all over the world and is run by space agencies like NASA. These antennas traverse the solar system, receiving and sending signals to and from spacecraft.
3. **Signal Delays:** Communication between Earth and robotic spacecraft is delayed because of the limited speed of light. As the distance between the spaceship and Earth widens, this delay becomes more noticeable. For instance, a signal travels in between 4 and 24 minutes.
4. **Earth-based Control:** Earth-based ground control centers use the DSN to transmit commands to robotic spacecraft. These commands give the spacecraft instructions for a variety of tasks, such as trajectories adjustments, scientific observations, and sensor operations.
5. **Data Transmission:** During their missions, robotic spacecraft gather enormous volumes of scientific data. Scientists and mission planners can analyze this data by sending it back to Earth via the DSN.

6. **Data Compression:** Spacecraft frequently utilize data compression techniques to minimize the quantity of information provided without sacrificing crucial scientific content in order to maximize the effectiveness of data transmission.
7. **Autonomy and Decision-Making:** Onboard autonomy enables robotic systems to make decisions in the present without relying on directives from Earth during missions with considerable communication delays.
8. **Space Weather and Noise:** Space weather and cosmic noise, which can introduce interference and decrease signal quality, have an impact on interplanetary communication.
9. **Communication Protocols:** To ensure dependable and error-free data transfer, spacecraft and ground stations employ specialized communication protocols and encoding techniques.
10. **Emergencies and Abnormalities:** Communication with the spacecraft is essential for troubleshooting and remotely addressing difficulties in the event of emergencies or unanticipated abnormalities.

The Mars rovers, the New Horizons trip to Pluto, the Voyager spacecraft, and prospective missions like the Mars Sample Return campaign are notable robotic missions that have relied on interplanetary communication. The success of interplanetary exploration depends on keeping contact with these missions, which is made possible by the Deep Space Network and other networks of a similar design run by other space agencies.

7.2.10 Advancing Technology

Advancements in a number of disciplines, such as artificial intelligence, autonomy, remote sensing, communication, and materials science, are fueled by the development of robots for space exploration. The development of robotics in space exploration has been fueled in part by advances in technology. Robotic systems are becoming more sophisticated, adaptable, and efficient as technology advances, allowing us to explore and understand the cosmos in ways that were previously considered to be unthinkable. Here are some significant areas of robotics technology development for space exploration:

1. **Miniaturization:** The creation of smaller, lighter robotic systems that may be used on missions with constrained payload capacities is made possible by advances in miniaturization.
2. **AI and Autonomy:** Enhanced AI algorithms and autonomy allow robotic systems to make complicated judgments, adapt to changing situations, and carry out tasks without continual human supervision.
3. **Sensing and Perception:** Advanced sensor and perception technologies enable robots to precisely perceive their environment, traverse treacherous terrains, and collect high-quality data for scholarly research.
4. **Mobility and Locomotion:** Cutting-edge locomotion technologies, such as wheeled rovers, walking robots, and flying drones, allow robots to explore a variety of settings on celestial worlds with various topographies.
5. **Manipulation and Interaction:** Technological advancements in robotic manipulation enable robots to wield tools, gather samples, carry out sensitive jobs, and interact with the environment more skillfully.
6. Real-time control and data transfer are supported by advances in communication technology, which enable dependable and high-bandwidth data transmission between robotic spacecraft and Earth.

7. **Power Generating and Efficiency:** New energy storage and generating technologies allow robotic systems to operate for longer durations and in more challenging environments, such as those with scarce resources or sunlight.
8. **Sample Analysis and Instrumentation:** State-of-the-art analytical tools on robotic missions can conduct in-depth analyses of samples, revealing details about the composition and past of celestial bodies.
9. **3D Printing and In-Situ Manufacturing:** Technologies for in-situ 3D printing and manufacturing can create tools, structures, and replacement parts on demand, decreasing the need to ship anything from Earth.
10. **Modular Design:** By allowing robotic systems to be tailored for certain missions, modular designs make them more adaptable to different goals and conditions.
11. **Virtual Reality and Teleoperation:** By giving users immersive experiences, technologies like virtual reality improve teleoperation and let users manage robots as if they were physically present.
12. **Radiation and Thermal Protection:** Technological developments in the field of materials science enable the development of robotic systems that can withstand the harsh radiation and temperatures found in space.
13. **Swarming and Collaboration:** By allowing numerous robots to cooperate, swarm robotics and collaborative techniques increase efficiency and durability in difficult conditions.
14. **Data Analytics and Machine Learning:** The large amounts of data that robotic missions collect can be effectively analyzed thanks to advanced data analytics and machine learning techniques.

Together, these developments make robotic missions more effective and successful, improving our capacity to investigate far-off planets, moons, asteroids, and other celestial entities. As technology develops, it will open up new avenues for research, deepen our understanding of the cosmos, and open the door for future robotic and human exploration missions.

7.3 ROBOTIC ROVERS ON OTHER PLANETS

Unmanned robots called robotic rovers are built to go to other planets, moons, and celestial bodies, and examine their surfaces. These rovers are outfitted with cutting-edge engineering, analytical tools, and communication systems that let them manoeuvre, process data, and send it back to Earth. Engineering principles, scientific theories, and mathematics are all used to successfully operate rovers. An overview of the science, philosophy, and mathematics underlying robotic rovers on other worlds is given below:

1. **Mobility and Locomotion:**

   Technology: Robotic rovers can move on wheels or tracks. The rover is able to handle challenging terrain since each wheel is frequently separately powered and steerable.

   Theory: In order to compute movements, wheel rotations, and steering angles for the best navigation, rovers use the concepts of mechanics, kinematics, and dynamics.

   Mathematical Considerations: Kinematic equations for motion, vector calculus for calculating velocities and accelerations, and trigonometry for calculating angles all come into play.

2. **Sensors and Perception:**

   Technology: To perceive their surroundings, gather photos, and analyse topography and geological features, rovers are outfitted with a variety of sensors, including cameras, LIDAR (Light Detection and Ranging), and spectrometers.

3D Maps: Theoretically, sensor data is processed to produce 3D maps, identify barriers, and recognize landmarks using computer vision algorithms and sensor fusion techniques.

Mathematics: Algorithms for feature detection, camera calibration, and depth estimation involve linear algebra, geometry, and image processing techniques.

3. **Communication:**

   Technology: To transmit data and receive instructions from Earth, rovers require communication systems. Relays and antennas can be used to provide dependable communication.

   Theory: To ensure effective data transmission and error correction in the presence of noise and interference, signal processing and communication theory are used.

   Mathematics: For signal encoding, data compression, and error detection/correction, Fourier analysis, modulation methods, and information theory are used.

4. **Power Generation and Management:**

   Technology: Nuclear or solar energy is used to power rovers. In times of limited sunshine, energy storage systems are employed to maintain continuous operation.

   Theory: Power generation, distribution, and management use electrical circuit theory and concepts of energy conversion.

   Optimize Energy: Ohm's law, Kirchhoff's rules, and calculus are used in mathematics to calculate and optimize energy.

5. **Remote Operation:**

   Technology: Teleoperated rovers allow for real-time control and decision-making by operators from Earth.

   Theoretically, teleoperation theory incorporates ideas from ergonomics, human-computer interface, and control systems.

   Teleoperation involves the use of control theory, feedback systems, and operator interface design principles.

6. **Scientific Experiments:**

   Technology: Scientific equipment on rovers is used to examine the soil, rocks, and atmosphere. Spectrometers, microscopes, drills, and environmental sensors are some of these tools.

   Theory: Geological, chemical, and physics-related scientific theories direct the choice of instruments and the interpretation of evidence that is gathered.

   Mathematics: To understand scientific data, modeling, statistical analysis, and spectroscopy methods are used.

Integration of engineering, physics, mathematics, and computer science is required for robotic rovers to operate successfully on distant worlds. These rovers are designed, constructed, and operated by engineers and scientists in order to collect useful data, advance our understanding of the cosmos, and prepare the path for upcoming human exploration.

7.4 ROBOTIC SYSTEMS SATELLITE MAINTENANCE AND ASSEMBLY

Advanced technologies, engineering principles, scientific ideas, and mathematical concepts are all used in robotic expeditions to other planets and in the upkeep and construction of satellites in orbit. An overview of the technology, theory, and related mathematics that underlie these efforts is given below. Precision movements and robotic systems are needed for satellite assembly and maintenance in order to assure the longevity and proper operation of these spacecraft.

1. **Orbital Mechanics:**
   - Theoretically, satellite orbital manoeuvres are guided by Kepler's equations of planetary motion and Newtonian mechanics.
   - Mathematical computations using orbital equations, velocity, and acceleration.
2. **Robotics and Manipulation:**
   - Technology: Satellite assembly and maintenance are done with robotic arms and tools.
   - Theory: Planning and carrying out accurate robotic motions requires knowledge of kinematics, dynamics, and control theory.
   - Mathematics: Control systems theory, differential equations, and linear algebra.
3. **Teleoperation and Control:**
   - Technology: For satellite repair, remote control of robotic devices is essential.
   - Theory: Effective robot control is made possible by control theory and concepts of human-computer interaction.
   - Mathematical topics include stability analysis, operator interface design, and feedback control systems.
4. **Energy Systems:**
   - Technology: Solar energy is used to power satellites.
   - Theory: Energy production and distribution are governed by electrical circuit theory and photovoltaic principles.
   - Mathematics: Electrical circuit analysis and calculus for energy calculations.
5. **Communication and Data Transmission:**
   - Satellites and ground stations can communicate thanks to technology.
   - Theory: Reliable data transmission is made possible by communication theory, modulation, and encoding techniques.
   - Signal processing and information theory in mathematics.

The maintenance and assembly of satellites as well as robotic expeditions to other planets need interdisciplinary work that combines engineering, physics, math, and mechanical concepts. These initiatives stretch the limits of human understanding and technology, allowing us to explore the cosmos and use space for scientific study and exploration.

Numerical Problems with Solutions

Lunar Rover Navigation

Problem 1. A lunar rover is moving across the surface of the Moon at a velocity of 2 m/s. If it travels for 1800 seconds, how far will it have traveled?

Solution: Distance = Velocity × Time = 2 m/s × 1800 s = 3600 meters.

Robot Power Consumption

Problem 2. A robotic spacecraft exploring Mars consumes an average power of 50 watts. If the mission lasts for 30 days, how much energy does the robot consume?

Solution: Energy = Power × Time = 50 W × (30 days × 24 hours/day) = 36,000 watt-hours.

Teleoperation Latency

Problem 3. A robot on Mars is controlled from Earth. Due to communication delays, the round-trip signal travel time is 20 minutes. Calculate the total latency in teleoperation.

Solution: Total latency = 2 × Signal travel time = 2 × 20 minutes = 40 minutes.

Solar Panel Efficiency

Problem 4. A Mars rover's solar panels have a total area of 3 square meters and receive an average solar radiation of 400 W/m^2. Calculate the maximum power the solar panels can generate.

Solution: Power = Area × Solar radiation = 3 m^2 × 400 W/m^2 = 1200 watts.

Robotic Arm Load Capacity

Problem 5. A robotic arm on a spacecraft can lift a maximum load of 500 kg. If it needs to lift a satellite component weighing 300 kg, can the robotic arm handle it?

Solution: Yes, the robotic arm can handle it since the load (300 kg) is less than the maximum load capacity (500 kg).

Planetary Exploration Time

Problem 6. A Mars rover travels at an average speed of 0.1 m/s. If it needs to cover a distance of 500 meters, how long will it take to complete the journey?

Solution: Time = Distance/Speed = 500 m/0.1 m/s = 5000 seconds.

Power Generation on Mars

Problem 7. A Mars rover's solar panels generate an average power of 150 watts. If the rover operates for 12 hours during daylight on Mars (which has a day length of about 24.6 hours), how much energy is generated?

Solution: Energy = Power × Time = 150 W × 12 hours = 1800 watt-hours.

These numerical problems provide a practical understanding of various concepts related to robotics in space exploration, including velocity, power, acceleration, distance, latency, and efficiency.

Mars Rover's Travel Distance

Problem 8. The Mars rover Opportunity traveled 45 kilometers during its mission. If each wheel of the rover has a diameter of 0.5 meters, how many wheel rotations did Opportunity complete during its mission?

Solution: Given the diameter of each wheel = 0.5 meters. Circumference of a wheel = π × diameter = π × 0.5 meters ≈ 1.57 meters.

Total distance traveled = 45 kilometers = 45,000 meters. Number of wheel rotations = Total distance traveled/Circumference of a wheel = 45,000 meters/1.57 meters ≈ 28,662 wheel rotations.

Teleoperation Time Delay

Problem 9. A robot on Mars is teleoperated from Earth. Due to the signal delay, the round-trip communication time is 20 minutes. If the robot performs an action immediately upon receiving a command, how long does it take for the robot to respond to a command from Earth?

Solution: Round-trip communication time = 20 minutes One-way communication time (to Mars) = Round-trip communication time/2 = 20 minutes/2 = 10 minutes

Time taken for the robot to respond = One-way communication time + Action time = 10 minutes + 0 (immediate action) = 10 minutes

Satellite Assembly Task

Problem 10. A robot is tasked with assembling a satellite in space. It takes the robot 2 hours to attach a solar panel and 1.5 hours to install communication equipment. If the robot works continuously without breaks, how long does it take to complete the entire assembly?

Solution: Time to attach a solar panel = 2 hours. Time to install communication equipment = 1.5 hours.

Total assembly time = Time to attach a solar panel + Time to install communication equipment = 2 hours + 1.5 hours = 3.5 hours

Space Debris Collection Speed

Problem 11. A space debris collection robot is moving at a constant speed of 0.1 meters per second. If it needs to cover a distance of 500 meters to reach a piece of debris, how long will it take for the robot to reach the debris?

Solution: Robot's speed = 0.1 meters per second Distance to the debris = 500 meters.

Time taken to reach the debris = Distance/Speed = 500 meters/0.1 meters per second = 5000 seconds.

Lunar Cave Exploration Time

Problem 12. A lunar cave exploration robot moves at a speed of 0.02 meters per second. If it needs to navigate through a cave with a length of 150 meters, how long will it take for the robot to traverse the entire cave?

Solution: Robot's speed = 0.02 meters per second Length of the cave = 150 meters.

Time taken to traverse the cave = Length/Speed = 150 meters/0.02 meters per second = 7500 seconds

Sample Collection Depth

Problem 13. A robotic arm drills into the surface of Mars to collect a soil sample. If the arm's drill can penetrate to a depth of 2 meters and it drills at a speed of 0.01 meters per second, how long will it take to collect a sample at the maximum depth?

Solution: Drill's speed = 0.01 meters per second. Maximum sample collection depth = 2 meters.

Time taken to collect a sample at the maximum depth = Depth/Speed = 2 meters/0.01 meters per second = 200 seconds.

Short Questions and Answers

Q.1. What is robotics in space exploration?

Ans. Robotics in space exploration involves the use of robotic systems and technologies to explore, analyze, and perform tasks in outer space.

Q.2. Why use robots in space exploration?

Ans. Robots are used in space exploration to conduct tasks in hazardous environments, gather data, and assist human astronauts.

Q.3. What are some examples of robotic space missions?

Ans. Examples include Mars rovers (Curiosity, Perseverance), lunar rovers (Yutu-1, Yutu-2), and space telescopes (Hubble, James Webb).

Q.4. How do rovers move on other planets?

Ans. Rovers use wheels or tracks to move, driven by electric motors controlled from Earth.

Q.5. What role do robots play in satellite maintenance?

Ans. Robots can perform repairs, upgrades, and maintenance tasks on satellites in orbit.

Q.6. How do robots communicate in space?

Ans. Robots communicate with Earth using antennas and communication relays.

Q.7. What is the Deep Space Network (DSN)?

Ans. DSN is a network of ground stations used for communication with spacecraft in deep space.

Q.8. What are the benefits of using robots for space exploration?

Ans. Robots can work in extreme conditions, reduce risks to humans, and extend the reach of exploration.

Q.9. How do robots gather scientific data in space?

Ans. Robots carry instruments like cameras, spectrometers, and drills to collect data from their surroundings.

Q.10. What is teleoperation?

Ans. Teleoperation is the remote control of robotic systems from a distance.

Q.11. What challenges do robots face in space?

Ans. Challenges include extreme temperatures, radiation, communication delays, and navigating difficult terrains.

Q.12. How do robots handle sample collection on other planets?

Ans. Robots use drills or scoops to collect samples, which can then be analyzed on-site or returned to Earth.

Q.13. What is the role of AI in space exploration?

Ans. AI helps robots make autonomous decisions, analyze data, and adapt to changing conditions.

Q.14. What's the significance of swarming robots in space exploration?

Ans. Swarming robots can collaborate to perform tasks more efficiently and cover larger areas.

Q.15. How do robots contribute to the assembly of satellites in space?

Ans. Robots can assemble satellite components, perform maintenance, and extend the lifespan of satellites.

Q.16. What is the James Webb Space Telescope?

Ans. The James Webb Space Telescope (JWST) is a next-generation space telescope set to launch to observe the universe in infrared wavelengths.

Q.17. How do robots assist in space telescopes' maintenance?

Ans. Robots can perform repairs, replace components, and ensure the proper functioning of space telescopes.

Q.18. Why is autonomy important for robotic space missions?

Ans. Autonomy allows robots to make decisions in real-time, overcoming communication delays and unexpected situations.

Q.19. How do robots navigate on other planets without GPS?

Ans. Rovers use onboard sensors, landmarks, and pre-planned paths to navigate accurately.

Q.20. What is sample return in space exploration?

Ans. Sample return involves collecting samples from other celestial bodies and bringing them back to Earth for analysis.

Q.21. How do robots handle hazardous materials in space exploration?

Ans. Robots can handle hazardous materials using specialized tools and equipment, reducing risks to humans.

Q.22. What's the role of robotics in interplanetary communication?

Ans. Robots act as communication relays, ensuring reliable data transmission between spacecraft and Earth.

Q.23. How do robots contribute to future human exploration of space?

Ans. Robots can prepare habitats, perform reconnaissance, and assist astronauts in tasks on distant planets.

Q.24. What is the significance of miniaturization in space robotics?

Ans. Miniaturization allows for smaller, more efficient robotic systems with reduced weight and power consumption.

Q.25. How do robots assist in the exploration of lunar caves and tunnels?

Ans. Robots can navigate and explore challenging environments like lunar caves and lava tubes.

Q.26. What is the primary purpose of a lander in space exploration?

Ans. Landers are robotic spacecraft designed to safely deliver payloads to the surface of celestial bodies.

Q.27. How does AI enhance the analysis of data from space missions?

Ans. AI can identify patterns, anomalies, and valuable insights in vast amounts of data collected from space.

Q.28. What is a rover's main source of power on other planets?

Ans. Rovers often use solar panels to generate electricity from sunlight.

Q.29. How do robots handle repairs in the vacuum of space?

Ans. Robots are equipped with specialized tools and technologies designed to function in the harsh space environment.

Q.30. What is the role of robotics in satellite deployment?

Ans. Robots can deploy satellites from launch vehicles, ensuring precise positioning and orbits.

Q.31. How do robots handle assembly tasks in microgravity environments?

Ans. Robots use mechanical grippers, arms, and precision movements to assemble components in microgravity.

Q.32. What is the importance of redundancy in robotic space missions?

Ans. Redundancy ensures that if one component fails, there are backup systems to maintain the mission's success.

Q.33. How can robots help with space debris cleanup?

Ans. Robots can capture and manipulate space debris, helping to reduce the risk of collisions with operational spacecraft.

Q.34. What's the significance of lightweight materials in robotic space missions?

Ans. Lightweight materials enable more efficient propulsion, reduced launch costs, and improved mission capabilities.

Q.35. What role do robotic arms play in space exploration?

Ans. Robotic arms can manipulate objects, collect samples, and perform delicate tasks in space.

Q.36. How do robots adapt to different levels of gravity on other planets?

Ans. Robots adjust their movements and control algorithms to account for variations in gravity.

Q.37. What are some challenges in controlling robots with teleoperation from Earth?

Ans. Challenges include communication delays, limited real-time feedback, and the need for precise control.

Q.38. How can robots assist in the study of asteroids and comets?

Ans. Robots can land on, analyze, and even redirect the trajectories of asteroids and comets for scientific study.

Q.39. What is the role of robots in planetary defense against asteroid impacts?

Ans. Robots could be used to intercept and deflect potentially hazardous asteroids to protect Earth.

Q.40. How do robots assist in the search for signs of past or present life on other planets?

Ans. Robots analyze surface materials, study geology, and detect chemicals that could indicate the presence of life.

Q.41. What are some potential applications of swarm robotics in space exploration?

Ans. Swarm robots can collectively map terrain, search for resources, and perform distributed tasks.

Q.42. How do robots handle extreme temperatures in space?

Ans. Robots are designed with thermal insulation, heating, and cooling systems to withstand extreme temperature variations.

Q.43. What is the concept of in-situ resource utilization (ISRU) in space exploration?

Ans. ISRU involves using local resources to produce fuel, oxygen, and materials for sustained exploration.

Q.44. How can robots assist in the construction of space habitats?

Ans. Robots can 3D-print structures, assemble components, and prepare habitats for human habitation.

Q.45. What is the significance of precision engineering in robotic space missions?

Ans. Precision engineering ensures that robots can perform delicate tasks and maneuvers accurately.

Q.46. What are some ethical considerations in using robots for space exploration?

Ans. Ethical considerations include avoiding contamination of other planets and protecting the environment of celestial bodies.

Q.47. How do robots handle dust and regolith on other planets?

Ans. Robots have brushes, drills, and vacuum systems to handle and analyze surface materials.

Q.48. What role do robots play in the study of planetary atmospheres?

Ans. Robots equipped with instruments can study atmospheric composition, pressure, and weather patterns.

Q.49. How do robots contribute to the search for exoplanets?

Ans. Robots assist in the operation and maintenance of space telescopes used to observe and analyze distant exoplanets.

Q.50. What is the future of robotics in space exploration?

Ans. The future involves more advanced AI, autonomous systems, human-robot collaboration, and missions to distant destinations.

Multiple Choice Questions and Answers

Q.1. What is the primary purpose of using robotics in space exploration?

(*a*) To replace human astronauts
(*b*) To reduce mission costs
(*c*) To enhance scientific discoveries
(*d*) To test new technologies

Q.2. Which of the following is NOT an example of a robotic space mission?

(*a*) Mars Rover
(*b*) International Space Station
(*c*) Hubble Space Telescope
(*d*) Apollo Moon Landing

Q.3. What is the main advantage of using robotic systems in hazardous environments in space?

(*a*) They require less energy
(*b*) They can withstand radiation
(*c*) They reduce risks to human life
(*d*) They are more cost-effective

Q.4. Which technology is commonly used for remote control of robotic systems in space?

(*a*) AI algorithms (*b*) Teleoperation

(*c*) Neural networks (*d*) Autonomous navigation

Q.5. What is the Deep Space Network (DSN) used for in space exploration?

(*a*) Studying deep space phenomena (*b*) Communicating with robotic spacecraft

(*c*) Launching space missions (*d*) Observing black holes

Q.6. Which of the following is a challenge for robotic missions on other planets?

(*a*) Lack of gravity (*b*) Presence of liquid water

(*c*) Communication delays (*d*) Predictable terrain

Q.7. What role does AI play in robotic space missions?

(*a*) AI has no role in space missions

(*b*) AI helps robots make autonomous decisions

(*c*) AI prevents robots from malfunctioning

(*d*) AI improves communication with Earth

Q.8. How do rovers typically move on the surfaces of other planets?

(*a*) By flying (*b*) By swimming

(*c*) By crawling (*d*) By using wheels or tracks

Q.9. What is the main purpose of communication relays in space exploration?

(*a*) To provide power to robots (*b*) To serve as backup navigation systems

(*c*) To relay signals between Earth and spacecraft

(*d*) To capture images of celestial bodies

Q.10. What is the significance of swarming robots in space exploration?

(*a*) They create a swarm of robotic bees

(*b*) They can communicate with extraterrestrial beings

(*c*) They can collaborate to perform tasks more efficiently

(*d*) They eliminate the need for human intervention

Q.11. What is the role of robotic arms in space exploration?

(*a*) To simulate human arms for aesthetic purposes

(*b*) To assist astronauts in spacewalks (*c*) To collect samples from asteroids

(*d*) To serve as communication relays

Q.12. How do robots adapt to different levels of gravity on other planets?

(*a*) They change their size (*b*) They use rockets for propulsion

(*c*) They adjust their movements and control algorithms

(*d*) They can only operate on Earth due to gravity constraints

Q.13. What is the main source of power for rovers on other planets?

(*a*) Nuclear reactors (*b*) Solar panels (*c*) Fossil fuels (*d*) Wind turbines

Q.14. How can robots assist in satellite maintenance?

(*a*) By replacing human astronauts (*b*) By repairing damaged solar panels

(*c*) By sending signals to satellites from Earth

(*d*) By providing entertainment for satellite operators

Q.15. How do robots handle sample collection on other planets?

(*a*) They use buckets to scoop up samples

(*b*) They use robotic arms to drill into surfaces

(*c*) They use nets to catch samples in the atmosphere

(*d*) They rely on astronauts to collect samples

Q.16. What is the purpose of in-situ resource utilization (ISRU) in space exploration?

(*a*) To search for intelligent life (*b*) To explore the depths of the ocean

(*c*) To use local resources for sustained exploration

(*d*) To develop new communication technologies

Q.17. What is the primary goal of using AI in robotic space missions?

(*a*) To replace human astronauts

(*b*) To eliminate the need for communication with Earth

(*c*) To make robots more expensive

(*d*) To enable robots to make autonomous decisions

Q.18. How do robots contribute to space debris cleanup?

(*a*) By creating more space debris

(*b*) By capturing and manipulating space debris

(*c*) By avoiding space debris altogether

(*d*) By sending space debris to other planets

Q.19. What is the significance of redundancy in robotic space missions?

(*a*) It increases the risk of system failure (*b*) It reduces mission costs

(*c*) It ensures backup systems in case of failure

(*d*) It decreases the complexity of missions

Q.20. How can robots contribute to the study of exoplanets?

(*a*) By building habitats on exoplanets (*b*) By collecting samples from exoplanets

(*c*) By observing and analyzing data from space telescopes

(*d*) By communicating with potential extraterrestrial beings

Q.21. What role do robots play in interplanetary communication?

(*a*) Robots communicate with each other

(*b*) Robots communicate with humans on Earth

(*c*) Robots relay signals between Earth and spacecraft

(*d*) Robots send signals to distant galaxies

Q.22. How can robots assist in the search for signs of past or present life on other planets?

(*a*) By sending messages to potential extraterrestrial beings
(*b*) By conducting DNA analysis on other planets
(*c*) By analyzing surface materials and chemicals
(*d*) By collecting samples of alien organisms

Q.23. What is the significance of lightweight materials in robotic space missions?

(*a*) They make robots heavier (*b*) They increase mission costs
(*c*) They improve mission capabilities and efficiency
(*d*) They reduce the need for communication relays

Q.24. What is the primary goal of using robotic arms in space exploration?

(*a*) To perform dance routines (*b*) To manipulate objects and perform tasks
(*c*) To take artistic photographs of celestial bodies
(*d*) To replace human astronauts in spacewalks

Q.25. How do robots handle repairs in the vacuum of space?

(*a*) They wear spacesuits (*b*) They are immune to vacuum conditions
(*c*) They use specialized tools and technologies (*d*) They avoid repairs altogether

Q.26. How do robots contribute to the exploration of lunar caves and tunnels?

(*a*) They build shelters inside caves
(*b*) They can navigate and explore these challenging environments
(*c*) They create music inside caves
(*d*) They use lunar caves as communication relays

Q.27. What is the primary goal of satellite maintenance using robots?

(*a*) To clean the surface of satellites (*b*) To simulate human activities in space
(*c*) To ensure the proper functioning of satellites
(*d*) To create decorative patterns on satellites

Q.28. How do robots handle extreme temperatures in space?

(*a*) They wear insulated jackets
(*b*) They hibernate during extreme temperatures
(*c*) They use thermal insulation and cooling systems
(*d*) They generate their own heat

Q.29. What role do robots play in planetary defense against asteroid impacts?

(*a*) They build shields around asteroids
(*b*) They deflect and intercept potentially hazardous asteroids
(*c*) They send asteroids to other planets
(*d*) They collect samples from asteroids for scientific study

Q.30. How can robots assist in the study of asteroids and comets?

(*a*) By creating new asteroids and comets (*b*) By using robotic arms to grab asteroids

(c) By landing on, analyzing, and studying these celestial bodies
(d) By providing transportation for asteroids and comets

Q.31. What is the primary goal of using AI in robotic space missions?
(a) To replace human astronauts
(b) To eliminate the need for communication with Earth
(c) To make robots more expensive
(d) To enable robots to make autonomous decisions

Q.32. What role do robots play in the assembly of space habitats?
(a) They create habitats from scratch
(b) They assist astronauts in habitat assembly
(c) They act as communication relays for habitats
(d) They are responsible for habitat decoration

Q.33. How do robots handle dust and regolith on other planets?
(a) They use specialized vacuum cleaners
(b) They kick the dust away with their feet
(c) They have brushes, drills, and vacuum systems
(d) They simply ignore the dust

Q.34. What is the significance of AI in space exploration?
(a) AI stands for "Alien Interaction"
(b) AI can predict the future of space missions
(c) AI helps robots adapt to changing conditions and make decisions
(d) AI eliminates the need for human operators

Q.35. How do robots assist in the study of planetary atmospheres?
(a) They use fans to create artificial atmospheres
(b) They observe and analyze data from atmospheric sensors
(c) They create new atmospheres on other planets
(d) They communicate with Earth about atmospheric conditions

Q.36. How do robots handle assembly tasks in microgravity environments?
(a) They use magnetic forces for assembly
(b) They use rockets for propulsion
(c) They rely on human astronauts for assembly
(d) They use mechanical arms and precision movements

Q.37. What is the primary goal of using robotic arms in space exploration?
(a) To perform dance routines
(b) To manipulate objects and perform tasks
(c) To take artistic photographs of celestial bodies
(d) To replace human astronauts in spacewalks

Q.38. How do robots adapt to different levels of gravity on other planets?
(a) They change their size
(b) They use rockets for propulsion
(c) They adjust their movements and control algorithms
(d) They can only operate on Earth due to gravity constraints

Q.39. What is the main source of power for rovers on other planets?

(*a*) Nuclear reactors (*b*) Solar panels (*c*) Fossil fuels (*d*) Wind turbines

Q.40. How can robots contribute to satellite maintenance?

(*a*) By replacing human astronauts
(*b*) By repairing damaged solar panels
(*c*) By sending signals to satellites from Earth
(*d*) By providing entertainment for satellite operators

Q.41. How do robots handle sample collection on other planets?

(*a*) They use buckets to scoop up samples
(*b*) They use robotic arms to drill into surfaces
(*c*) They use nets to catch samples in the atmosphere
(*d*) They rely on astronauts to collect samples

Q.42. What is the purpose of in-situ resource utilization (ISRU) in space exploration?

(*a*) To search for intelligent life
(*b*) To explore the depths of the ocean
(*c*) To use local resources for sustained exploration
(*d*) To develop new communication technologies

Q.43. What is the primary goal of using AI in robotic space missions?

(*a*) To replace human astronauts
(*b*) To eliminate the need for communication with Earth
(*c*) To make robots more expensive
(*d*) To enable robots to make autonomous decisions

Q.44. How do robots contribute to space debris cleanup?

(*a*) By creating more space debris
(*b*) By capturing and manipulating space debris
(*c*) By avoiding space debris altogether
(*d*) By sending space debris to other planets

Q.45. What is the significance of redundancy in robotic space missions?

(*a*) It increases the risk of system failure
(*b*) It reduces mission costs
(*c*) It ensures backup systems in case of failure
(*d*) It decreases the complexity of missions

Q.46. How can robots contribute to the study of exoplanets?

(*a*) By building habitats on exoplanets
(*b*) By collecting samples from exoplanets
(*c*) By observing and analyzing data from space telescopes
(*d*) By communicating with potential extraterrestrial beings

Q.47. What role do robots play in interplanetary communication?

(*a*) Robots communicate with each other
(*b*) Robots communicate with humans on Earth
(*c*) Robots relay signals between Earth and spacecraft
(*d*) Robots send signals to distant galaxies

Q.48. How can robots assist in the search for signs of past or present life on other planets?

(*a*) By sending messages to potential extraterrestrial beings
(*b*) By conducting DNA analysis on other planets
(*c*) By analyzing surface materials and chemicals
(*d*) By collecting samples of alien organisms

Q.49. What is the significance of lightweight materials in robotic space missions?

(*a*) They make robots heavier
(*b*) They increase mission costs
(*c*) They improve mission capabilities and efficiency
(*d*) They reduce the need for communication relays

Q.50. What is the primary goal of using robotic arms in space exploration?

(*a*) To perform dance routines
(*b*) To manipulate objects and perform tasks
(*c*) To take artistic photographs of celestial bodies
(*d*) To replace human astronauts in spacewalks

Key to Multiple Choice Questions

1. (*c*)	2. (*d*)	3. (*c*)	4. (*b*)	5. (*b*)	6. (*c*)	7. (*b*)	8. (*d*)	9. (*c*)
10. (*c*)	11. (*b*)	12. (*c*)	13. (*b*)	14. (*b*)	15. (*b*)	16. (*c*)	17. (*d*)	18. (*b*)
19. (*c*)	20. (*c*)	21. (*c*)	22. (*c*)	23. (*c*)	24. (*b*)	25. (*c*)	26. (*b*)	27. (*c*)
28. (*c*)	29. (*b*)	30. (*c*)	31. (*d*)	32. (*b*)	33. (*c*)	34. (*c*)	35. (*b*)	36. (*d*)
37. (*b*)	38. (*c*)	39. (*b*)	40. (*b*)	41. (*b*)	42. (*c*)	43. (*d*)	44. (*b*)	45. (*c*)
46. (*c*)	47. (*c*)	48. (*c*)	49. (*c*)	50. (*b*)				

Project Works

1. **Mars Rover Simulation:** Design and develop a simulation of a Mars rover navigating a terrain similar to the Martian surface. Implement obstacle detection, autonomous navigation, and sample collection.

   Hints: The simulation could be created using a programming language like Python, and libraries like Pygame for graphics. The rover's navigation algorithms could use pathfinding techniques, and sample collection could be simulated through interactive user input.

2. **Lunar Cave Exploration Robot:** Create a robot capable of exploring a model lunar cave environment. The robot should navigate through dark and narrow passages, autonomously mapping the cave layout and collecting data.

   Hints: The robot could be equipped with ultrasonic sensors for distance measurement and a camera for visual data collection. Path planning algorithms such as A* could be used for navigation and mapping.

3. **Satellite Maintenance Mechanism:** Design a robotic mechanism that can simulate the repair and maintenance of a satellite in a microgravity environment. The mechanism should be able to perform tasks like panel replacement and component repairs.

   Hints: The mechanism could use robotic arms with specialized end-effectors for different tasks. It could be controlled using teleoperation techniques and could incorporate simulated satellite components.

4. **Space Debris Collection Game:** Create an educational game where players control a robotic spacecraft tasked with collecting space debris. The game should simulate the challenges of maneuvering in space and capturing debris.

   Hints: The game could be developed using a game engine like Unity. Players could control the spacecraft's thrusters and arms to capture debris. Physics simulations could accurately depict the challenges of moving in microgravity.

5. **Exoplanet Observer Robot:** Build a robot equipped with sensors to observe and analyze different environmental conditions. The robot should simulate the task of studying an exoplanet's atmosphere and surface from a remote location.

   Hints: The robot could have sensors to measure temperature, pressure, and chemical composition of gases. It could also include a camera for visual observations. Data collected could be used to generate simulated atmospheric conditions.

6. **Swarm Robotics for Asteroid Exploration:** Develop a simulation where a swarm of robotic spacecraft collaboratively explore and study an asteroid. The robots should coordinate to map the asteroid's surface and gather scientific data.

   Hints: The simulation could involve multiple robots with communication capabilities and sensors. Algorithms like flocking and coverage path planning could be implemented to showcase cooperative behavior.

7. **Planetary Defense Simulator:** Create a simulation where players must use robotic spacecraft to prevent an incoming asteroid from impacting Earth. The simulation should incorporate orbital mechanics, propulsion, and deflection strategies.

   Hints: The simulator could provide players with various tools, such as thrusters and kinetic impactors, to alter the asteroid's trajectory. It could calculate the effects of these actions on the asteroid's course.

8. **Mars Sample Return Mission Planning:** Design a software application that simulates the planning and execution of a Mars sample return mission. The application should allow users to select landing sites, design mission profiles, and assess technical challenges.

   Hints: The application could have a user interface where users select from a list of possible landing sites. It could simulate the spacecraft's trajectory, communication with Earth, and resource utilization for sample collection and return.

9. **Robotic Space Habitat Assembly:** Create a virtual environment where users can design and assemble modules for a space habitat using robotic arms. The simulation should showcase the challenges of manipulating objects in microgravity.

   Hints: The simulation could use physics-based simulations to model the behavior of objects in microgravity. Users could manipulate virtual objects using mouse or keyboard controls, simulating robotic arms.

10. **Martian Agriculture Robot:** Design a robot capable of simulating agriculture on Mars. The robot should be able to plant seeds, manage water and nutrients, and monitor plant growth in a controlled environment.

    Hints: The robot could have sensors to measure soil moisture, temperature, and light intensity. It could be equipped with a robotic arm for planting and tending to plants. Users could interact with the robot to monitor its progress.

8

Role of Robotics in Disaster Response

8.1 INTRODUCTION

In recent years, robots has played an increasingly important role in disaster response, revolutionizing the way emergency situations are handled and reducing the difficulties faced by human rescuers. Natural or man-made disasters can cause extensive damage, fatalities, and the interruption of vital infrastructure. Using conventional disaster response techniques frequently requires sending human rescuers into dangerous and unexpected conditions, endangering their lives. Robotics has become a useful tool in disaster response due to its capacity to carry out operations remotely, autonomously, and precisely. These machines can access limited locations, traverse treacherous terrain, and handle dangerous materials—all while posing less of a threat to emergency personnel. The phases of the disaster management cycle, such as preparedness, response, recovery, and mitigation, all exhibit the use of robotics in disaster response.

Robotics are extremely important in the data collection and risk assessment phases of preparedness. Drones, or unmanned aerial vehicles, may collect vital information from disaster-prone locations and capture real-time imagery with the help of sensors and cameras. This data is useful for producing precise maps, identifying vulnerable locations, and formulating plans for efficient disaster management. Robots are used to carry out duties that are too dangerous or taxing for humans during the response phase. Search and rescue efforts in collapsed structures, the handling of dangerous materials, and even the provision of communication networks in places where infrastructure has been destroyed are examples of these jobs.

Ground robots can identify survivors, evaluate the structural integrity of buildings, and provide vital information to human responders when outfitted with sensors, cameras, and specialized equipment. After a calamity, the recovery phase focuses on reestablishing normalcy. Here, robots can help with environmental monitoring, infrastructure restoration, and debris collection. Additionally, they can speed up the recovery process, lessen human exposure to potentially hazardous circumstances, and assist in analysing the damage to structures and utilities. Robots help to lessen the effects of potential disasters during the mitigation phase. They can be used to install early warning systems, build flood barriers, and check for vulnerabilities in vital infrastructure. Robots with sensors can also monitor the environment and foresee potential dangers, assisting authorities in taking precautions.

Robotics integration into disaster response is not without difficulties, despite their many advantages. Technical shortcomings including battery life, movement on difficult terrain, and environment adaptability still need to be addressed. Furthermore, widespread acceptance may be hampered by the high cost of creation, maintenance, and training. Robotics' place in disaster response is quickly developing, providing creative answers to the difficult problems faced both during and after disasters. Robotics is anticipated to play a vital role in boosting the effectiveness, efficiency, and safety of disaster management activities as technology develops as research, engineering, and emergency responders continue to work together.

8.2 ROBOTS UTILIZED IN DISASTER RESPONSE

Let's take a closer look at one particular application of robots in disaster relief: search and rescue efforts in ruined structures. We'll examine the mathematics underlying how robots can more effectively assist in finding survivors.

Search and Rescue in Collapsed Buildings Scenario

Buildings may collapse following an earthquake, encasing people under the debris. While looking for survivors in dangerous environments, human rescuers run substantial hazards. Robots are able to move through the rubble, find survivors, and give emergency personnel up-to-date information.

8.2.1 Mathematical Model: Localization and Mapping

Robot localization and mapping is a key component of search and rescue operations. This entails locating the robot and mapping out the surrounding area. Simultaneous Localization and Mapping (SLAM) is a popular mathematical method for this. In order to determine the robot's position and create a map of its surroundings, SLAM entails integrating input from sensors (such as cameras, LIDAR, and inertial measurement units). Using Bayesian filtering methods like the Extended Kalman Filter (EKF) or Particle Filter, this is frequently represented mathematically.

Probabilistic computations, matrix operations, and sensor fusion techniques are all included in the equations and algorithms used in SLAM. The Kalman filter equations, for instance, assist in updating a robot's position estimate based on the uncertainty of the sensor data and the robot's motion model. This is done given the robot's initial position estimate and the sensor readings it gets.

Robot Path Planning: A Algorithm*

The robot must plan its path once it obtains a preliminary map of the area in order to efficiently seek for survivors. For path finding, the A algorithm* is frequently employed. Heuristics are used to prioritize investigation in regions most likely to lead to the desired outcome (the location of the survivor) while avoiding dangers and impediments.

Calculations are made using the A* algorithm to determine how much it will cost to get to a specific place on the map. The price is determined by adding the distance travelled and a heuristic estimation of the distance left to travel. This aids the robot in efficiently scanning its surroundings while concentrating on promising regions.

8.2.2 Sensor Data Analysis: Image Processing

Robots used for search and rescue frequently use cameras and sensors to collect information from the catastrophe scene. Analysis of these data using image processing techniques is essential for finding survivors and evaluating the surroundings. For instance, possible survivors or impediments are identified using edge detection, object identification, and motion tracking algorithms.

Through a variety of methods, including convolution, Fourier transforms, and statistical techniques, mathematics is used in image processing. For instance, gradient computations are used by edge detection algorithms like the canny edge detector to recognize sharp variations in pixel intensity, which frequently signify object boundaries.

8.2.3 Optimal Sensor Placement: Coverage Algorithms

Robots with sensors must be placed strategically in catastrophe scenarios to maximize coverage and improve the likelihood of finding survivors. Coverage algorithms utilize mathematical optimization methods to choose the ideal locations for the sensors. These algorithms take into account things like sensor range, barriers, and the environment's geometry. They seek to reduce coverage overlap and gaps while maintaining effective resource use. When using mathematical optimization approaches like linear programming or genetic algorithms, the issue is frequently phrased as an optimization problem.

Numerous mathematical techniques and concepts are used when robots are used in disaster response situations. The algorithms and tactics that allow robots to navigate challenging environments and support search and rescue operations are deeply rooted in mathematics, from robot position estimation and map construction using SLAM to path planning with the A* algorithm and sensor data analysis through image processing.

8.3 IN SEARCH AND RESCUE OPERATIONS

In search and rescue efforts, robots are becoming more and more important, particularly in situations involving fallen structures, natural disasters, and dangerous settings. They are essential instruments in such situations because they can reach places that would be too risky or inaccessible for human responders. Let's examine in greater detail the technology, theory, and math underlying robot-assisted search and rescue operations.

1. **Technology:**

   Robotic Platforms: The capabilities and designs of robots employed in search and rescue missions might vary greatly. These can be ground robots, drones, aquatic robots, or even constrained space-specific robots. A variety of sensors, actuators, and communication technologies are present on these platforms.

   Sensors: Sensors are essential for giving robots the knowledge they need to understand their surroundings. These sensors comprise visible light, thermal, multispectral, and LIDAR cameras, as well as ultrasonic, microphone, gas, and other sensors. These sensors aid in finding survivors, determining the stability of the structure, and moving through debris.

   Communication Systems: In order to convey information and receive commands, robots must establish communication links with operators. Wi-Fi, Bluetooth, and even specialized long-range communication systems are utilized as wireless communication methods.

   Interact with Environment: Robots can move and interact with their environment thanks to actuators. Wheels, tracks, legs, manipulator arms, and grippers for picking up objects or removing debris are some examples of these.

2. **Theory:**

   (*a*) **Path Planning and Navigation:** The A* algorithm is a well-known pathfinding method that use heuristics to determine the shortest route between two points while avoiding obstacles. It uses priority queues to investigate the most likely paths first and is based on graph theory.

Dijkstra's Algorithm: Used to determine the shortest path in a graph, Dijkstra's algorithm is comparable to A*. It works well when a robot needs to explore a space without a clear objective.

(*b*) **Simultaneous Localization and Mapping (SLAM):**

Bayesian Filtering: In SLAM, a robot's position is estimated (localised) while it creates a map of its surroundings. Bayesian filtering methods, such as the Extended Kalman Filter (EKF) or Particle Filter, are frequently used for this. These filters refresh the robot's perception of its position and surroundings by fusing sensor measurements and motion models.

(*c*) **Image Processing and Computer Vision:**

Convolution: In image processing, convolution is a mathematical technique that is used to extract information and features from images. Convolution is used in processes like edge detection and image smoothing.

Feature Detection: Algorithms for computer vision identify important features in images, such as edges, corners, or blobs. These characteristics are essential for locating specific items or places.

(*d*) **Communication Relaying:**

Graph Theory: Graph theory is used to model the connectedness between robots and determine the best relaying technique in instances where robots must relay messages to create communication networks.

(*e*) **Sensor Fusion:**

Kalman Filter: To increase precision and reliability, readings from various sensors are combined using the Kalman filter, a recursive mathematical method. It is frequently used in robotics for sensor fusion, in which information from many sensors is combined to produce a more thorough picture of the environment.

3. Mathematics:

(*a*) **Trigonometry and Geometry:**

- Based on sensor measurements, trigonometry is utilized to calculate object distances, angles, and locations.
- Mapping environments, figuring out the distances between the robot and the target, and making travel plans all depend on geometric computations.

(*b*) **Linear Algebra:**

- Sensor measurements, robot positions, and transformations between several coordinate systems are all represented by matrices and vectors.

(*c*) **Statistics and Probability:**

- Statistical approaches are used to quantify uncertainties, such as sensor noise and measurement errors;
- Probability theory is employed in Bayesian filtering methods for state estimation, such as EKF and Particle Filter.

(*d*) **Optimization:**

- Optimization techniques are employed in various aspects, like optimizing robot paths, sensor placement for maximum coverage, and communication relaying.

In conclusion, effective robot design, development, and deployment in search and rescue missions require a sophisticated fusion of technology, theory, and mathematics. Together, these components give robots the ability to navigate challenging settings, find survivors, communicate with operators, and make deft decisions in life-or-death circumstances.

8.3.1 Remote Sensing and Reconnaissance

Drones or unmanned aerial vehicles (UAVs) with cameras, LIDAR, and other sensors can examine disaster zones quickly from the air. They offer real-time aerial imagery and maps that aid first responders in determining the extent of the damage and locating places in need of immediate assistance. Robots on the ground that have cameras and sensors can move around obstacles and dangerous terrain while transmitting photos and data to human controllers.

Let's examine in greater detail the science, philosophy, and mathematics underlying remote sensing and reconnaissance in the context of robot-assisted search and rescue operations.

1. **Technology:**

   Unmanned Aerial Vehicles (UAVs) or Drones: For the purpose of gathering information from the disaster area, drones are fitted with a range of sensors. These sensors include LIDAR (light detection and ranging), cameras (visible light, infrared, multispectral), and GPS receivers for precise positioning.

   Cameras: Visual information about the environment is provided by cameras. Aspects of the scene are captured differently by various camera types, including RGB cameras, thermal cameras, and multispectral cameras.

   LIDAR: LIDAR makes 3D maps of the environment by measuring distances with laser pulses. It aids in locating terrain features and obstructions and offers precise elevation data.

   GPS: Drones equipped with GPS receivers can pinpoint their exact location, which is essential for accurate mapping and navigation.

2. **Theory:**

   (*a*) **Photogrammetry:**

   Triangulation: In photogrammetry, dimensions and distances are calculated using photos obtained at various angles. In order to locate things based on their presence in several photos, the triangulation technique is used.

   (*b*) **Remote Sensing Theory:**

   Reflectance and Emissivity: Various materials have varying reflectivities and emissions of energy. In the theory of remote sensing, interactions between objects and electromagnetic radiation in the visible, infrared, and microwave ranges are taken into account.

   (*c*) **Geometric Concepts:**

   Projection Geometry: This technique is used to convert a 3D environment into a 2D image. For example, perspective and orthographic projections are employed to depict the environment in photographs accurately.

3. **Mathematics:**

   (*a*) **Trigonometry:**

   Parallax and Triangulation: Triangulation uses the parallax effect, or the variation in an object's apparent position when viewed from various angles, to calculate distances and heights.

(*b*) **Geometry:**

Similar Triangles: Similar triangles are used in photogrammetry to accurately measure objects by connecting their dimensions in the actual world to their dimensions in photographs.

(*c*) **Radiometry:**

Radiance and Reflectance: The measurement and quantification of electromagnetic radiation is a part of radiometry. Radiance is the total amount of radiation that an object emits or reflects. Understanding material qualities is made easier by understanding reflection, which is the ratio of reflected brightness to incident radiance.

(*d*) **Image Processing:**

Digital Image Processing: To improve features, reduce noise, and extract useful information from images, techniques including convolution, Fourier transforms, and filters are used.

(*e*) **Spectral Analysis:**

Spectral Signatures: Throughout the electromagnetic spectrum, different objects have distinctive spectral signatures. To identify materials and features, spectral analysis examines these signatures.

(*f*) **Radiometric Calibration:**

Radiometric Correction: Radiometric calibration modifies image pixel values to take into consideration elements like air conditions and sensor sensitivity. This guarantees that pictures capture the brightness of the scene correctly.

(*g*) **Georeferencing:**

Using mathematical transformations like affine transformations or polynomial equations, georeferencing entails translating image coordinates to geographic coordinates.

(*h*) **Digital Elevation Models (DEMs):**

Interpolation: To construct DEMs, which are continuous representations of the topography, elevation values are interpolated between measurement locations using LIDAR data.

In conclusion, a variety of mathematical and theoretical ideas are used in remote sensing and reconnaissance during search and rescue operations. These ideas are utilised to build precise maps, extract useful information from sensor data, and support decision-making during crucial operations. Robots and drones can offer vital information about disaster zones and successfully direct responders because to the confluence of technology, theory, and mathematics.

8.3.2 Survivor Detection and Localization

Robots using sensors, such as thermal cameras, microphones, and motion detectors, can assist in finding indications of life among the debris. Even in low-light situations, thermal cameras can detect heat signatures from human bodies. With the help of acoustic sensors, survivors who may be trapped but can still communicate vocally can be found. Let's examine survivor recognition and localization in robot-assisted search and rescue operations, concentrating on the relevant technology, theory, and math.

1. **Technology:**

   Thermal Cameras: Depending on the object's temperature, thermal cameras may identify infrared radiation that it emits. They are able to recognise heat signatures, making it possible to find survivors even when vision is low.

Microphones: These devices record audio signals, such as screams for aid or tapping noises provided by survivors.

Motion Detectors: Motion detectors can detect minor motions, such as breathing or little movements of people who are confined, and can detect movement.

2. **Theory:**
   (*a*) **Signal Processing:**

   Audio Signal Processing: Signal processing techniques are employed to remove background noise and pinpoint specific sound patterns, such as yells or knocks, that might point to the presence of people.

   Thermal Signal Processing: The processing of thermal photographs identifies areas with temperature fluctuations that might be indicative of a person's presence.

   (*b*) **Signal Analysis:**

   Pattern Recognition: Algorithms for pattern recognition examine auditory and thermal signals to find distinctive patterns connected to human presence.

   (*c*) **Sensor Fusion:**

   Multimodal Fusion: By cross-referencing data from several sources, combining information from many sensors (thermal, auditory, and motion) improves the accuracy of finding and localizing survivors.

3. **Mathematics:**
   (*a*) **Signal Processing:**

   Fourier Transform: Fourier analysis is used to break down complicated signals (like audio signals) into their component frequencies. This can show specific patterns connected to human-generated sounds.

   Wavelet Transform: When analyzing signals at various scales, the wavelet transform is particularly effective for locating rapid shifts or anomalies in audio or thermal data.

   (*b*) **Probability and Statistics:**

   Bayesian Inference: Using Bayesian approaches, data from many sensors may be combined, updating the likelihood that a survivor is there based on fresh sensor readings.

   Statistical Signal Processing: Using statistical approaches to separate out noise from real signals (such human sounds), the accuracy of survival detection is increased.

   (*c*) **Image Processing:**

   Image Enhancement: Methods like histogram equalization and contrast correction can improve thermal images, enhancing their clarity by highlighting minute temperature fluctuations.

   Feature Extraction: Potential survivors can be found using mathematical features retrieved from photos, such as blob analysis for temperature abnormalities.

   (*d*) **Localization:**

   Trilateration and Triangulation: These techniques determine the location of a survivor by using sensor readings (such as audio or thermal signals) from various positions.

   (*e*) **Machine Learning:**

   Pattern Classification: Machine learning systems can learn to distinguish between noises from the environment and sounds from a survivor. Examples of these algorithms are support vector machines and neural networks.

Regression Analysis: Using sensor readings and previous data, regression models can calculate a survivor's distance.

In conclusion, robot-assisted survivor detection and localization uses technology, theory, and mathematics to analyze sensor data, spot trends, and precisely identify the existence and location of survivors in crisis scenarios. The effectiveness of search and rescue operations is increased by the use of various mathematical methodologies, which improves robot capabilities for locating and rescuing trapped people.

8.3.3 Search Operations

Ground robots can use algorithms like the A* algorithm stated previously to autonomously navigate through complicated terrain, including rubble and debris. They are able to cover a sizable portion of the catastrophe scene by methodically exploring different places. Swarm robotics, in which numerous little robots work together to cover a larger area and share information, can improve search capabilities even more. Let's examine in depth the technology, theory, and math underlying robot search operations.

1. **Technology:**

   Ground Robots: These robots can move around on a variety of surfaces, including rocky terrain and places covered with garbage, thanks to their mobility systems (wheels, tracks, legs).

   Autonomous Navigation: Robots can navigate autonomously across challenging surroundings by using algorithms and onboard sensors like cameras, LIDAR, and IMUs (Inertial Measurement Units).

   Swarm Robotics: By dividing the search area among themselves, a swarm of robots may work together to swiftly and effectively cover a greater area.

   Communication Systems: To exchange information and coordinate search efforts, robots may communicate with one another and with human operators.

2. **Theory:**

   (*a*) **Exploration Strategies:**

   Coverage Algorithms: These algorithms decide the best strategy to investigate a region while avoiding unnecessary work. In order to provide thorough coverage, the environment is frequently divided into sectors using Voronoi diagrams and grid-based techniques.

   Graph Theory: Concepts from graph theory are used to model connectedness between various sites in the environment, assisting in the planning of travel and exploration itineraries.

   (*b*) **Path Planning and Mapping:**

   The A* algorithm, which is used for pathfinding and enables robots to design paths to uncharted or possibly dangerous locations while avoiding obstacles, is described earlier.

   Simultaneous Localization and Mapping (SLAM): SLAM lets robots map their surroundings while also determining their own positions. For search operations in uncharted or dynamic environments, this is essential.

   (*c*) **Swarm Robotics:**

   Emergent Behavior: The theory of emergent behavior describes how straightforward individual behaviors can develop into intricate swarm behaviors. The coordination of several robots is improved by having a better understanding of emergent behavior.

3. **Mathematics:**

(*a*) **Path Planning:**

Euclidean Distance: Calculations of distance based on Euclidean geometry aid in identifying the pathways or trajectories that are shortest between two places.

Heuristic Functions: The A* algorithm directs the search towards effective routes by estimating the cost from the current location to the target.

(*b*) **Localization and Mapping:**

Bayesian Filters: Bayesian filters used in simultaneous localization and mapping (SLAM), such as the Extended Kalman Filter (EKF) and Particle Filter, are based on mathematics. Using sensor data and motion models, these filters update robot position estimates.

(*c*) **Swarm Coordination:**

Distributed Algorithms: To ensure that robots in a swarm can make coordinated decisions based on local information without a centralized control, mathematical concepts like consensus algorithms are used.

(*d*) **Graph Theory:**

Shortest Path Algorithms: These algorithms, such as Floyd-Warshall and Dijkstra's algorithms, calculate the shortest path between nodes in a network and direct robots' exploration paths.

(*e*) **Probability and Statistics:**

Probability Density Functions: These functions serve as a foundation for decision-making in uncertain contexts by modeling uncertainty in sensor measurements and estimates.

(*f*) **Optimization:**

Optimization Algorithms: Mathematical optimization techniques like genetic algorithms or simulated annealing can help robots determine optimal paths while considering factors like energy efficiency and terrain traversal.

In conclusion, robot-assisted search operations use a combination of technology, theory, and mathematics to effectively explore and navigate complicated settings. Robots can improve their efficiency in discovering survivors and covering huge regions by utilizing mathematical principles in path planning, swarm coordination, mapping, and other areas, thereby aiding in the success of search and rescue operations.

8.4 COMMUNICATION AND RELAYING INFORMATION

Robots with communication systems can create communication links in places when the infrastructure is disrupted. Between survivors and rescuers, they can serve as communication relays, facilitating communication. These robots assist in establishing crucial links that allow survivors to call for help and first responders to acquire valuable data about the situation on the ground. Let's examine in depth the technology, theory, and mathematics underlying information transmission and communication in robotic search and rescue operations.

1. **Technology:**

Communication Systems: To establish connections with one another and with human operators, robots are outfitted with communication tools including radios, antennas, and transceivers.

Ad Hoc Networks: Robots can build ad hoc networks by relaying signals from one robot to another, expanding the communication range, in crisis scenarios where standard communication infrastructure may be damaged.

Satellite Communication: In remote or isolated locations, robots can link with other devices outside the reach of local communication networks by using satellite communication.

2. **Theory:**

   (*a*) **Network Topology:**

   Mesh networking theory calls for the construction of networks in which each node (robot) passes messages to other nodes, resulting in a strong and adaptable communication structure.

   (*b*) **Routing Algorithms:**

   Distributed Routing: To choose the best path for data packets to take as they are relayed around the network, algorithms like the Ad Hoc On-Demand Distance Vector (AODV) protocol are utilized.

   (*c*) **Communication Relaying:**

   Graph Theory: Concepts from graph theory are used to simulate the connectivity between robots and the probable routes for message relay.

3. **Mathematics:**

   (*a*) **Signal Strength and Attenuation:**

   Signal Propagation Model: Mathematical simulations show how signal amplitudes drop off as they move through the environment. These models aid in determining the dependability and range of communications.

   (*b*) **Graph Theory:**

   Connectivity Analysis: The network's robot connectivity is examined using graph theory to make sure that messages can be transmitted between them effectively.

   (*c*) **Data Routing:**

   Routing Algorithms: Algorithms that find the best routes for relaying messages, minimizing delays, and maximizing network efficiency are implemented with the use of mathematics.

   (*d*) **Error Correction:**

   Error-Correcting Codes: Mathematics is essential in the creation of error-correcting codes that guarantee reliable transmission despite noise and signal interference.

   (*e*) **Information Theory:**

   Channel Capacity: Information theory aids in the optimization of data transfer by determining the maximum data rate that can be reliably transferred over a communication channel.

   (*f*) **Probability and Statistics:**

   Packet Loss and Reliability: Statistical analysis helps determine the reliability of message transmission and estimate the likelihood of packet loss, enabling effective communication tactics.

In conclusion, technology, theory, and math are used to communicate and relay information in search and rescue missions including robots. Robots can create reliable communication links, communicate crucial information, and improve the coordination of search and rescue operations in difficult and dynamic situations by utilizing mathematical ideas in network topology, routing algorithms, and communication reliability.

8.5 HAZMAT AND DANGEROUS ENVIRONMENTS

Robots made for handling hazardous materials, or "Hazmat robots," can go into contaminated areas that are radioactive, chemical, biological, or nuclear. They can handle hazardous materials and execute activities like sample collection and decontamination, which lowers the risk to human responders. Let's explore the science, philosophy, and math behind the use of robots in search and rescue operations to handle hazardous materials (Hazmat) and traverse hazardous situations.

1. **Technology:**

   Hazmat Robots: These specialized robots are furnished with tools and sensors that are intended for the secure handling of hazardous materials. They frequently have manipulator arms or end effectors that are designed specifically for purposes like collecting samples or containment.

   Remote Sensing Equipment: Robots may be fitted with remote sensing equipment, such as spectrometers, to detect and analyze dangerous compounds from a safe distance.

   Hazardous Environment Adaptation: Robots created for hazardous situations are resistant to radiation, severe temperatures, and other unfavorable conditions.

2. **Theory:**

   (*a*) **Material Identification:**

   Spectroscopy: Spectroscopic methods are used to examine how light interacts with various materials. Robots can analyze the produced spectra to identify dangerous compounds.

   (*b*) **Manipulation and Containment:**

   Kinematics and Mechanics: Theoretical principles of kinematics and mechanics are used to develop manipulator arms that can handle hazardous materials precisely and safely.

   Robotic Grasping: According to the theory of robotic grasping, algorithms must be created to let robots grip and manipulate items of various sizes, shapes, and textures.

3. **Mathematics:**

   (*a*) **Spectroscopy:**

   Wavelength Analysis: When doing spectroscopy, mathematical analysis entails examining the wavelengths of light that are absorbed, emitted, or dispersed by various substances. The compounds present can be determined using these patterns.

   (*b*) **Manipulation:**

   Forward and Inverse Kinematics: The forward kinematics equation establishes the joint angles necessary to produce a specified end-effector position, whereas the inverse kinematics equation calculates the location of the end-effector based on those joint angles.

   Inverse Dynamics: Inverse dynamics, the forces and torques needed at each joint to produce a particular trajectory or motion are calculated.

(*c*) **Material Handling:**

Calculations of Forces and Torques: Math is used to determine the forces and torques necessary to handle hazardous materials safely without contaminating or causing harm.

(*d*) **Safety Analysis:**

Risk Assessment: Mathematical modeling aids in determining the dangers of handling hazardous materials and aids in making decisions that will result in safe operations.

(*e*) **Trajectory Planning:**

Path Planning Algorithms: Mathematics is used to construct algorithms that choose the best routes for robots to travel through dangerous areas while avoiding hazards and keeping a safe distance from them.

(*f*) **Control Theory:**

Feedback Control: When designing feedback control systems, control theory concepts ensure that robots maintain stability and stay out of harm's way when interacting with hazardous chemicals.

(*g*) **Probability and Statistics:**

Uncertainty Modeling: To account for uncertainties in sensor data in hazardous situations, mathematical techniques like probabilistic modeling are utilized to improve the accuracy of material identification.

In summary, a solid basis in technology, theory, and mathematics is needed to use robots to handle hazardous materials and navigate risky areas. Robots can help in circumstances where human participation would be too risky, protecting the safety of both rescuers and the environment in search and rescue operations. This is accomplished by using mathematical ideas to material identification, manipulation, safety assessments, and more.

8.6 MAPPING AND NAVIGATION

Robots can build precise maps of their surroundings using mapping algorithms and SLAM techniques. These maps help responders comprehend how collapsed structures are laid out, pinpoint potential rescue routes, and steer clear of hazardous areas. Let's examine in detail the technology, theory, and mathematics underlying robot mapping and navigation.

1. **Technology:**

   Sensors: Robots come with a range of sensors, such as cameras, LIDAR, GPS receivers, IMUs (Inertial Measurement Units), and magnetometers. These sensors offer information about the environment, the robot's location, and its orientation.

   Robots employ algorithms to interpret sensor input and produce maps of their surroundings. A popular method for this is simultaneous localization and mapping, or SLAM.

   Robots employ navigation systems to map out their itineraries and establish their positions. These systems make use of path planning algorithms, maps, and sensor data.

   Localization Beacons: Robots can utilise localization beacons to improve their position prediction in areas where GPS signals may be spotty or unreliable.

2. **Theory:**

   (*a*) **Mapping:**

   Feature Extraction: To recognize landmarks in the environment, robots extract features from sensor data, such as edges and corners in images.

Environment Representation: Maps can include information about barriers, routes, and landmarks by using occupancy grids or feature-based representations.

(*b*) **Navigation:**

Position Estimation: Estimating the robot's position and orientation in relation to a reference frame is a necessary step in navigation. Data from sensors and techniques like SLAM are used to do this.

Path Planning: Taking into account topography and obstructions, algorithms choose the optimum route from the present location to the intended location.

(*c*) **Localization:**

Odometry: Based on wheel rotations, odometry calculates position changes and provides short-term localization data.

- **Sensor Fusion:** To estimate the robot's position while creating a map, SLAM fuses motion models and sensor measurements from devices like LIDAR and IMU.

(*d*) **Landmark-based Navigation:**

Landmark Recognition: Robots can recognize familiar landmarks in the surrounding area and use them to estimate their orientation and position.

Trilateration and Triangulation: Based on distance measurements, robots employ these methods to estimate their position in relation to landmarks.

3. Mathematics:

(*a*) **Geometry:**

Trigonometry: From sensor data, trigonometric functions are employed to compute angles and distances.

Transformation Matrices: Transforms coordinates between various frames of reference by representing translations and rotations in matrices.

(*b*) **Calculus:**

Derivatives: For navigation control, calculus is utilised to calculate velocity and acceleration from location data.

Integration: Calculating displacement through the integration of velocity over time is crucial for path planning and position prediction.

(*c*) **Probability and Statistics:**

Bayesian Filtering: Bayesian filters, such as the Kalman Filter, combine motion models and noisy sensor readings to provide precise location estimation.

Uncertainty Estimation: Using probabilistic methods, uncertainty in sensor readings is quantified and transmitted through navigational algorithms.

(*d*) **Optimization:**

Path Optimization: Algorithms for path optimization determine the quickest or most energy-saving routes between places.

(*e*) **Graph Theory:**

Based on graphs With SLAM, simultaneous mapping and localization are made possible by using graph theory to model the connections between robot poses and landmarks.

(*f*) **Signal Processing:**

Processing of Sensor Data: Mathematical techniques are used to analyze sensor data, remove background noise, and extract pertinent data for mapping and navigation.

In conclusion, using robots for mapping and navigation requires a complex fusion of technology, theory, and mathematics. Robots can precisely map environments, estimate their positions, and navigate over complex terrain by utilizing mathematical ideas from geometry, calculus, probability, and optimization, which helps search and rescue operations.

8.7 STRUCTURAL ASSESSMENT

Robots with sensors can evaluate the strength of buildings and other structures, spotting weak spots and possible collapse threats. Responders can use this information to make educated decisions about where to concentrate their efforts. Let's examine in depth the technology, theory, and mathematics underlying the use of robots in structural assessment.

1. **Technology:**

   Sensors: To collect information about the structural health of buildings and other structures, robots are outfitted with a variety of sensors, including cameras, LIDAR, ultrasound sensors, and vibration sensors.

   Non-Destructive Testing (NDT) Instruments: Specialized instruments, such as ground-penetrating radar (GPR) and ultrasound machines, are used to evaluate the structural integrity of materials without inflicting harm.

   Remote Sensing: Drones with cameras and sensors may collect detailed information from inaccessible locations, assisting in structural evaluation.

2. **Theory:**

   (*a*) **Material Properties:**

   Mechanics of Materials: Knowing a material's elasticity, tensile strength, and density can help determine whether or not beams and columns are structurally sound.

   (*b*) **Vibration Analysis:**

   Vibration theory aids in the analysis of the inherent frequencies, mode forms, and dynamic behavior of structures, assisting in the discovery of flaws or damage.

   (*c*) **Non-Destructive Testing (NDT):**

   Ultrasonic Testing: Ultrasonic waves are used to look for internal flaws, gauge the thickness of materials, and judge the soundness of structural components.

   Ground-Penetrating Radar (GPR): This technology uses electromagnetic pulses to penetrate materials and find cavities, fissures, and hidden items.

   (*d*) **Remote Sensing:**

   Photogrammetric: According to photogrammetric theory, 3D models are made from photographs, enabling in-depth structural study.

   Point Cloud Processing: To produce precise representations of structures, point cloud data from LIDAR or photogrammetric is processed.

3. **Mathematics:**

   (*a*) **Structural Analysis:**

   Stress and Strain Calculations: To determine a structural component's load-bearing capabilities, mathematical formulae that measure stress and strain are used.

   Finite Element Analysis (FEA): FEA simulates structural behavior under various stresses by discrediting complicated structures into smaller elements and solving mathematical equations.

(*b*) **Vibration Analysis:**

Eigenvalue Analysis: This method uncovers possible weak spots in structures by identifying their natural frequencies and matching mode shapes.

(*c*) **Non-Destructive Testing (NDT):**

Wave Propagation: Mathematical models explain how waves move through materials, assisting ultrasonic and GPR techniques in flaw detection.

(*d*) **Remote Sensing:**

Trilateration and Triangulation: These mathematical methods assess the positions and distances of features from photographs, assisting in the creation of precise 3D models.

(*e*) **Data Analysis:**

Statistical Methods: Data gathered from sensors is subjected to statistical analysis in order to spot abnormalities, evaluate the significance of deviations, and make choices concerning the structural integrity.

In conclusion, employing robots to examine structures requires a combination of technology, theory, and maths. Robots can accurately assess the status of buildings and structures by using mathematical principles in mechanics, vibration analysis, NDT procedures, and remote sensing, assisting rescuers in making judgments during search and rescue operations.

8.8 SITUATIONAL AWARENESS

Robots with cameras and sensors can offer human operators real-time video feeds and sensor data. Responders can make better decisions as a result of this real-time knowledge, which improves situational awareness. Let's examine in depth the technology, philosophy, and mathematics underlying robot situational awareness.

1. **Technology:**

Sensors: Robots may collect information about the environment by using a variety of sensors, including as cameras, LIDAR, microphones, gas sensors, and environmental sensors.

Communication Systems: Robots can send real-time video feeds and sensor data to human operators, improving situational awareness.

Generating Picture: Robots use data fusion to generate a thorough picture of the environment by combining information from several sensors.

2. **Theory:**

(*a*) **Data Fusion:**

Sensor Fusion: Data from numerous sensors is combined using mathematical techniques to produce a consistent and precise representation of the environment.

Multi-Sensor Data Association: To accurately identify and monitor objects or occurrences, algorithms help link data from several sensors.

(*b*) **Perception and Interpretation:**

Computer Vision: To analyze visual data, identify objects, spot abnormalities, and evaluate environmental circumstances, computer vision theory is utilized.

Audio Signal Processing: To analyze sounds and pinpoint crucial auditory clues, techniques from audio signal processing are applied.

3. **Mathematics:**

   (*a*) **Sensor Fusion:**

   Kalman Filtering: The Kalman filter, a popular approach for mixing noisy sensor readings to reliably estimate states, is based on mathematics.

   Bayesian Probability: Bayesian techniques are used to determine the likelihood of various states given sensor data, improving situational awareness.

   (*b*) **Computer Vision:**

   Image processing: To process images for object recognition and scene comprehension, mathematical techniques including convolution, edge detection, and feature extraction are used.

   Pattern Recognition: To recognize and categorize items in photos, machine learning and pattern recognition techniques are used.

   (*c*) **Audio Signal Processing:**

   Spectral Analysis: By breaking down audio signals into their frequency components, methods like Fourier analysis can assist identify sounds and anomalies.

   (*d*) **Data Analysis:**

   Statistical Methods: To understand the dynamics of the environment, statistical analysis is utilized to find patterns, anomalies, and correlations in sensor data.

   (*e*) **Communication Analysis:**

   Analyzing Data throughput of Communication Channels: Mathematical analysis of data throughput of communication channels ensures that real-time sensor data can be trustworthily transferred to human operators.

   (*f*) **Uncertainty Management:**

   Probabilistic Models: Probability theory is used to describe uncertainty in sensor readings and forecasts, enabling decision-makers to act with confidence in the face of ambiguous data.

In summary, the integration of technology, theory, and mathematics is required for situational awareness with robots. Robots can offer accurate and thorough information about the environment to rescuers, empowering them to make well-informed judgments during search and rescue operations. This is done by applying mathematical ideas in data fusion, perception, computer vision, and signal processing.

8.9 MEDICAL ASSISTANCE

Some highly developed robots can help survivors with basic medical needs by delivering medical supplies or helping with first aid. Let's take a closer look at the math, theory, and technology that go into using robots to provide medical help.

1. **Technology:**

   Telemedicine Platforms: By carrying cameras and screens, robots can support telemedicine by enabling medical practitioners to assess and communicate with patients from a distance.

   Medical Devices: For remote patient assessment, robots can be fitted with medical devices including stethoscopes, ultrasonography probes, and vital sign monitors.

   Drug Delivery: Robots can help in drug delivery to patients, particularly in risky areas where human access is restricted.

2. **Theory:**

   (*a*) **Telemedicine:**

   Human-Computer Interaction: Robotic interfaces that enable medical practitioners to engage with patients remotely are designed in accordance with the theoretical principles of HCI.

   Communication Theory: To optimize audio and visual streams for successful teleconsultations, theoretical ideas from communication theory are useful.

   (*b*) **Medical Device Operation:**

   Biomechanics: For robots to handle medical devices precisely and securely, it is essential to comprehend human anatomy and biomechanics.

   Medical Procedures: The creation of robots that can replicate actions like taking a blood pressure or giving an injection is guided by theoretical knowledge of medical procedures.

3. **Mathematics:**

   (*a*) **Telemedicine:**

   Audio and Video Compression: To ensure effective real-time transmission of high-quality medical data, mathematics is applied in audio and video compression methods.

   Network Bandwidth Optimization: Using mathematical optimization methods, telemedicine data can be sent efficiently over constrained bandwidth.

   (*b*) **Medical Device Operation:**

   Kinematics and Dynamics: The design of robotic arms and end effectors for precise and secure medical procedures is guided by mathematical concepts of kinematics and dynamics.

   Inverse Kinematics: Algorithms compute the joint angles necessary to precisely position a robotic arm by solving inverse kinematics equations.

   (*c*) **Drug Delivery:**

   Fluid Dynamics: Fluid dynamics mathematical models aid in the design of drug delivery systems, enabling accurate and controlled medicine administration.

   (*d*) **Data Analysis:**

   Vital Sign Monitoring: By mathematically analyzing data on vital signs like blood pressure and heart rate, medical personnel can be made aware of potentially life-threatening conditions and anomalies.

   (*e*) **Decision Support:**

   Machine Learning: To assist medical practitioners in diagnosing and prescribing therapies, algorithms that analyze patient data and medical history are used.

   (*f*) **Probability and Statistics:**

   Diagnostic Accuracy: Probability and statistics are important for determining the validity of diagnostic findings, particularly when using sensor data and imaging in medicine.

In conclusion, utilizing robots for medical help necessitates a fusion of technology, theory, and mathematics. Robots can extend medical care to dangerous or remote locations by using mathematical ideas in telemedicine, medical device operation, drug delivery, and data analysis. This improves patient assessment, treatment, and overall health outcomes during search and rescue operations.

Robotics combines a variety of disciplines, including mechanical engineering, computer science, electronics, and mathematics, in all the aforementioned ways to produce practical solutions for search and rescue missions. Robotic technology integration enhances response operations' speed, accuracy, and safety, eventually improving the likelihood that lives will be saved in urgent circumstances.

8.10 ROBOTS UTILIZED IN DISASTER ASSESS DAMAGE

Robots are being used more frequently in disaster assessment to survey damage, acquire vital data, and help human responders make decisions. Together with their sensor capabilities, their capacity to enter dangerous and challenging locations makes them essential assets in disaster response scenarios. Let's look at how robots are used to assess damage from disasters:

1. **Rapid Assessment:** Robots are capable of moving swiftly through disaster-stricken areas and delivering quick initial inspections. They may infiltrate collapsed structures; look for survivors, and spot potential dangers, allowing first responders to efficiently prioritize rescue operations.
2. **Remote Sensing and Mapping:** Areas damaged by disasters can be mapped in 3D and high-resolution photos can be taken using drones with cameras and LIDAR. These images aid in locating broken buildings, obstacles in the way, and places that are dangerous for people to enter.
3. **Structural Integrity Evaluation:** Robots using sensors like strain gauges and accelerometers can evaluate the structural soundness of infrastructure and structures. To evaluate if a structure is safe to access or whether it needs immediate maintenance, they analyze vibrations, deformation, and stress.
4. **Exploring Hazardous Environments:** Robots can enter dangerous situations like chemical spills or nuclear accidents to determine the amount of damage and weigh the dangers to human responders. Sensors can identify dangerous compounds and keep track of the surroundings.
5. **Subsurface Assessment:** Ultrasound and ground-penetrating radar are utilized to evaluate subsurface damage. Without causing more damage, robots can find holes, cracks, and structural problems below the surface.
6. **Data Gathering and Analysis:** Robots collect a variety of data, such as multispectral, thermal, and visual imaging. Damage assessment reports are produced using this data after analysis, which aids first responders in comprehending the breadth and gravity of the event.
7. **Remote Telepresence:** Remote specialists can evaluate damage and direct on-site responders using teleoperated robots that are outfitted with cameras and microphones. This is particularly useful when human access is limited.
8. **Information Relaying:** Robots can create communication linkages in places where conventional infrastructure is compromised, assisting first responders in more efficiently relaying information and coordinating responses.
9. **Cooperation with Human Teams:** Robots support human responders by clearing debris and getting to difficult-to-reach places. Together, we can improve efficiency and security overall.
10. **Data Fusion and Visualization:** Robotic data collection from diverse sources allows for real-time data fusion and visualization. This offers a thorough perspective of the catastrophe region and facilitates decision-making.

Disaster damage assessment uses mathematical and theoretical ideas from the fields of robotics, sensor fusion, computer vision, structural analysis, and data analytics. Map-making, localization, and path-planning algorithms are crucial for successful navigation in challenging terrain. In conclusion, robots play a significant role in assessing the damage caused by disasters by quickly and accurately delivering information about risks and important areas that require attention. Their deployment provides to a better coordinated and effective disaster response as well as increased human responder safety.

8.11 ROBOTS UTILIZED IN DISASTER RESPONSE SCENARIOS TO PROVIDE SUPPORT

Robots contribute significantly to disaster response scenarios by giving human responders essential assistance. Their capacity to move through dangerous terrain, gather data, and carry out particular duties enhances the efficiency and safety of disaster response operations. Here are a few ways that robots can assist with disaster relief:

1. **Search and Rescue:** Robots using cameras, sensors, and thermal imaging technology can find survivors in wreckage, rubble, or dimly lit regions. They can enter areas that might be dangerous for people and provide rescue teams with up-to-the-minute information.
2. **Debris Removal:** Robots with manipulator arms can help clear impediments and debris to make it simpler for rescue workers to reach the damaged regions and carry out their tasks.
3. **Handling Hazardous goods:** Specialized robots are capable of handling dangerous goods, such as poisons, chemicals, or radioactive substances, lowering the risk to human responders. They have the ability to neutralize, confine, or transport hazardous materials safely.
4. **Communication and Mapping:** Drones offer aerial views of disaster zones and help with mapping by locating entry points, obstacles in the route, and potential dangers.
5. **Medical Assistance:** Medical robots can carry medical supplies to remote or inaccessible regions, perform remote telemedicine, and monitor vital signs. The medical response in disaster-affected areas is improved by this support.
6. **Infrastructure Inspection:** To evaluate damage and structural integrity, robots can survey important infrastructure, including bridges, dams, and power plants. They spot damaged places and avert any following catastrophes.
7. **Data Collection and Analysis:** Robots with various sensors gather information about the environment, gas concentrations, and radiation. Making decisions and identifying risks are made easier with the help of this knowledge.
8. **Fighting Fires:** Firefighting robots can put out fires in dangerous environments by using water cannons or materials that can withstand fire. This lowers the risk to human firefighters.
9. **Delivery of Humanitarian Aid:** Ground robots or drones can provide basic necessities like food, water, and medical supplies to remote or inaccessible regions, assisting impacted communities.
10. **Communication and Coordination:** Robots may communicate with other responders and convey messages and information, improving coordination and assuring effective communication in difficult environments.

The deployment of robots for disaster support incorporates theoretical and mathematical ideas from robotics, control theory, sensor fusion, path planning, and other areas. To ensure secure navigation, precise data collection, and efficient engagement with the environment, algorithms are created.

In conclusion, robots complement human efforts, improve safety, and broaden the capabilities of rescuers in disaster response scenarios. They are crucial resources for reducing the effects of disasters and rescuing lives due to their adaptability and capacity to execute jobs in dangerous environments.

Numerical Problems with Solutions

Problem 1. Search and Rescue Robot Navigation: A search and rescue robot needs to navigate a maze-like structure to locate a simulated survivor. The robot is equipped with LIDAR sensors and can move 0.5 meters per second. It takes 2 seconds for the robot to process LIDAR data and make navigation decisions. How far can the robot travel within a 10-second time frame?

Solution: Time for navigation = 10 seconds – 2 seconds = 8 seconds. Distance traveled = Speed × Time. Distance traveled = 0.5 m/s × 8 s = 4 meters.

Problem 2. Drone Mapping Area of Disaster: A drone is tasked with creating a 3D map of a disaster-affected area using LIDAR technology. The drone flies at an altitude of 50 meters and has a LIDAR scan rate of 1000 points per second. If the drone flies over the area for 5 minutes, how many LIDAR points will it capture?

Solution: Time in seconds = 5 minutes × 60 seconds/minute = 300 seconds. Total LIDAR points = LIDAR scan rate × Time. Total LIDAR points = 1000 points/s × 300 s = 300,000 points.

Problem 3. Hazardous Material Handling Robot Task: A hazardous material handling robot is tasked with moving hazardous waste containers from one location to another. Each container weighs 25 kilograms. The robot's arm exerts a force of 150 newtons to lift the containers. Calculate the work done by the robot to move a container 3 meters horizontally.

Solution: Work done = Force × Distance × Cosine(θ) θ = 0° (since the force is applied horizontally). Work done = 150 N × 3 m × Cosine(0°) = 150 N × 3 m × 1 = 450 J.

Problem 4. Communication Drone Battery Life: A communication drone is deployed to establish communication relays in a disaster-stricken area. The drone's battery has a capacity of 5000 milliampere-hours (mAh). If the drone consumes 200 milliamps of current during its operation, how many hours can the drone operate before its battery is depleted?

Solution: Battery life (hours) = Battery capacity (mAh)/Current consumption (mA). Battery life = 5000 mAh/200 mA = 25 hours.

Problem 5. Medical Robot's Remote Assessment: A medical robot is remotely assessing the vital signs of a survivor using its sensors. The robot measures a heart rate of 80 beats per minute and a body temperature of 37°C. Convert the body temperature to Fahrenheit using the formula: Fahrenheit = Celsius × 9/5 + 32.

Solution: Body temperature in Fahrenheit = 37°C × 9/5 + 32 = 98.6°F.

These numerical problems provide a practical understanding of concepts related to robotics in disaster response, including navigation, data collection, work calculations, and unit conversions.

Short Questions and Answers

Q.1. What is the role of robotics in disaster response?

Ans. Robotics aid in search and rescue, damage assessment, hazardous material handling, communication, and more during disasters.

Q.2. How do robots assist in search and rescue operations?

Ans. Robots can access hazardous areas, locate survivors, and provide real-time information to rescue teams.

Q.3. What type of sensors do disaster response robots use?

Ans. Robots use cameras, LIDAR, thermal sensors, and more to gather data about the environment.

Q.4. How do robots help in hazardous material handling?

Ans. Specialized robots can safely handle and contain hazardous substances, minimizing risks to human responders.

Q.5. What is the role of drones in disaster response?

Ans. Drones provide aerial views, map disaster areas, and establish communication relays.

Q.6. What is the purpose of medical robots in disaster situations?

Ans. Medical robots assist in remote telemedicine, vital sign assessment, and medical supply delivery.

Q.7. How do robots assist in infrastructure inspection after a disaster?

Ans. Robots can inspect critical structures to assess damage and ensure structural integrity.

Q.8. What is the significance of communication and relaying information using robots?

Ans. Robots establish communication links when traditional infrastructure is disrupted, aiding coordination.

Q.9. What are some tasks firefighting robots can perform?

Ans. Firefighting robots combat fires using water cannons or fire-resistant materials in hazardous environments.

Q.10. How do robots contribute to delivering humanitarian aid?

Ans. Robots transport essential supplies like food, water, and medical equipment to remote areas.

Q.11. What is the role of robots in rapid damage assessment?

Ans. Robots quickly evaluate damage and hazards, providing initial information to prioritize response efforts.

Q.12. How do robots assist in subsurface assessment?

Ans. Robots use ground-penetrating radar and ultrasound sensors to assess damage beneath surfaces.

Q.13. What is the advantage of using robots for remote telepresence?

Ans. Robots with cameras and microphones allow experts to assess damage and guide responders remotely.

Q.14. How do robots enhance situational awareness during disasters?

Ans. Robots collect data, analyze it, and relay critical information to responders for informed decision-making.

Q.15. What role do robots play in communicating with survivors?

Ans. Robots can carry communication devices to establish contact with trapped or isolated survivors.

Q.16. How do robots contribute to creating maps of disaster-affected areas?

Ans. Drones equipped with cameras and LIDAR capture images and data for creating accurate 3D maps.

Q.17. What is the significance of robots in handling radioactive materials?

Ans. Robots can safely handle and contain radioactive substances in hazardous environments.

Q.18. How do robots support debris removal efforts?

Ans. Robots with manipulator arms assist in removing debris, enabling better access for human responders.

Q.19. How do robots assist in identifying structural weaknesses?

Ans. Robots equipped with sensors assess vibrations and stress to determine structural integrity.

Q.20. How can robots contribute to improving communication in disaster-stricken areas?

Ans. Robots can establish communication networks using ad hoc systems or satellite links.

Q.211. What types of robots are used for medical assistance in disasters?

Ans. Medical robots include telemedicine platforms, vital sign monitors, and drug delivery robots.

Q.22. How do robots help responders assess hazardous environments?

Ans. Robots equipped with gas and environmental sensors identify dangers and monitor conditions.

Q.23. How do robots assist in analyzing hazardous substances?

Ans. Robots equipped with spectrometers and sensors can identify and analyze chemical substances.

Q.24. What is the role of robots in assessing the structural soundness of buildings?

Ans. Robots equipped with accelerometers and strain gauges evaluate vibrations and deformation.

Q.25. How can robots assist in collecting data during disasters?

Ans. Robots gather data from sensors to provide real-time information about the environment.

Q.26. What are the benefits of using robots for data fusion?

Ans. Robots combine data from multiple sensors for a comprehensive understanding of the environment.

Q.27. How do robots provide remote assistance to responders?

Ans. Robots equipped with cameras and microphones allow experts to guide on-site responders remotely.

Q.28. How do robots handle debris removal tasks?

Ans. Robots equipped with manipulator arms can remove debris to facilitate access for human responders.

Q.29. What tasks can firefighting robots perform in disaster scenarios?

Ans. Firefighting robots use water cannons and fire-resistant materials to combat fires in hazardous areas.

Q.30. How do robots contribute to assessing environmental hazards?

Ans. Robots equipped with sensors monitor gas levels, radiation, and other environmental factors.

Q.31. What is the role of drones in disaster damage assessment?

Ans. Drones provide aerial imagery and 3D mapping for evaluating damage and identifying hazards.

Q.32. How can robots assist in evaluating structural integrity after a disaster?

Ans. Robots equipped with sensors assess stress, deformation, and vibrations in structures.

Q.33. What is the significance of communication relays using robots?

Ans. Robots establish communication links in areas where traditional networks are disrupted.

Q.34. How do robots enhance situational awareness for responders?

Ans. Robots collect and relay data about the environment to improve decision-making.

Q.35. How do robots contribute to delivering medical aid in disasters?

Ans. Medical robots provide remote telemedicine, vital sign monitoring, and drug delivery.

Q.361. How do robots assist in assessing damage in hazardous environments?

Ans. Robots equipped with sensors enter dangerous areas to assess damage and identify risks.

Q.37. How can robots enhance communication between responders?

Ans. Robots establish communication links, ensuring effective coordination among responders.

Q.38. How do robots contribute to delivering humanitarian aid to affected areas?

Ans. Robots transport essential supplies to remote or inaccessible locations.

Q.39. How do robots assist in infrastructure inspection after a disaster?

Ans. Robots inspect critical structures to assess damage and prevent secondary disasters.

Q.40. What is the significance of medical robots in disaster response?

Ans. Medical robots provide remote medical assistance and support to responders in challenging environments.

Q.41. How do robots provide real-time data to responders?

Ans. Robots equipped with sensors gather and relay real-time information about the environment.

Q.42. What is the role of drones in disaster recovery operations?

Ans. Drones capture aerial images and data, aiding in mapping and damage assessment.

Q.43. How do robots handle hazardous materials in disasters?

Ans. Specialized robots safely contain and manage hazardous substances, reducing risks to responders.

Q.44. What types of sensors do robots use for environmental assessment?

Ans. Robots use gas sensors, radiation detectors, and environmental monitors to assess conditions.

Q.45. How do robots assist in remote communication with survivors?

Ans. Robots equipped with communication devices establish contact with isolated or trapped survivors.

Q.46. How do robots aid in subsurface assessment after a disaster?

Ans. Robots use ground-penetrating radar and ultrasound to assess damage beneath surfaces.

Q.47. How do robots assist in medical triage during disasters?

Ans. Medical robots provide remote medical assessments, helping responders prioritize treatment.

Q.48. How do robots help responders assess structural integrity?

Ans. Robots equipped with sensors evaluate vibrations and stress to determine a structure's stability.

Q.49. How do robots assist in hazardous environment exploration?

Ans. Robots enter dangerous areas to assess damage and identify potential hazards.

Q.50. What is the role of robots in creating comprehensive damage assessment reports?

Ans. Robots collect data and provide accurate information for responders to assess the scope of the disaster.

Multiple Choice Questions and Answers

Q.1. What is the primary role of robots in disaster response scenarios?
(*a*) Providing entertainment (*b*) Facilitating transportation
(*c*) Enhancing situational awareness (*d*) Conducting scientific research

Q.2. Which sensor helps robots identify heat signatures of survivors in disaster-stricken areas?
(*a*) LIDAR (*b*) Infrared sensor
(*c*) Gas sensor (*d*) Ultrasound sensor

Q.3. What is the main advantage of using robots for debris removal after a disaster?
(*a*) Robots are less expensive than human labor
(*b*) Robots can work longer hours without rest
(*c*) Robots can access hazardous areas more safely
(*d*) Robots require minimal maintenance

Q.4. How can robots contribute to assessing structural integrity after a disaster?
(*a*) Robots can predict future earthquakes (*b*) Robots can repair damaged structures
(*c*) Robots can evaluate vibrations and stress in structures
(*d*) Robots can analyze the composition of building materials

Q.5. Which technology helps drones create accurate 3D maps of disaster-affected areas?
(*a*) Radar (*b*) Sonar (*c*) LIDAR (*d*) GPS

Q.6. What task can firefighting robots perform during disaster response?
(*a*) Providing medical assistance (*b*) Delivering humanitarian aid
(*c*) Combatting fires in hazardous environments (*d*) Removing debris

Q.7. How do robots contribute to delivering medical assistance in disaster scenarios?
(*a*) Robots perform surgery on patients (*b*) Robots administer vaccinations
(*c*) Robots provide remote medical assessments (*d*) Robots replace human doctors

Q.8. Which type of robot can handle and contain hazardous chemicals in disaster-stricken areas?
(*a*) Cleaning robot (*b*) Medical robot
(*c*) Hazardous material robot (*d*) Search and rescue robot

Q.9. What is the primary purpose of communication relaying using robots in disaster response?
(*a*) Sending emergency alerts to the public (*b*) Enhancing human-robot communication
(*c*) Establishing communication links in areas with disrupted infrastructure
(*d*) Sending video feeds to remote experts

Q.10. How can robots enhance situational awareness for responders?
(*a*) By providing musical entertainment (*b*) By analyzing historical data
(*c*) By collecting and relaying data about the environment
(*d*) By providing navigation assistance

Q.11. What is the main advantage of using drones for mapping disaster-affected areas?
(*a*) Drones are immune to weather conditions (*b*) Drones are highly resistant to damage

(c) Drones can quickly and accurately capture visual and spatial data
(d) Drones can carry heavy loads of supplies

Q.12. How can robots help in accessing hard-to-reach areas during search and rescue operations?
(a) By using secret passageways
(b) By drilling through obstacles
(c) By navigating dangerous environments
(d) By flying over obstacles

Q.13. What technology allows robots to provide real-time visual and audio communication between remote experts and on-site responders?
(a) Virtual reality
(b) Augmented reality
(c) Teleoperation
(d) Holography

Q.14. What is the primary purpose of robots equipped with communication devices in disaster scenarios?
(a) To order supplies
(b) To communicate with other robots
(c) To establish contact with isolated survivors
(d) To play recorded messages

Q.15. How do robots assist in assessing subsurface damage in disaster-stricken areas?
(a) By using x-ray technology
(b) By employing ground-penetrating radar and ultrasound sensors
(c) By drilling into the ground
(d) By using electromagnetic pulses

Q.16. What is the primary role of robots equipped with medical tools in disaster response?
(a) Performing surgeries
(b) Administering medication
(c) Assisting in remote medical assessments
(d) Providing psychological counseling

Q.17. How do robots contribute to data collection during disaster response?
(a) By recording video for entertainment purposes
(b) By gathering data from sensors to provide real-time information
(c) By collecting coins and tokens
(d) By transmitting radio signals

Q.18. What is the role of robots in assessing environmental hazards after a disaster?
(a) Robots predict the occurrence of environmental hazards
(b) Robots actively prevent environmental hazards
(c) Robots monitor and report on gas levels, radiation, and other factors
(d) Robots create hazards intentionally

Q.19. How do robots contribute to remote telemedicine in disaster scenarios?
(a) Robots perform medical procedures on patients
(b) Robots administer medication to patients
(c) Robots deliver supplies to medical facilities
(d) Robots allow medical professionals to assess patients remotely

Q.20. What type of information do robots equipped with sensors relay to responders during disaster response?
(a) Fictional data for entertainment
(b) Data about the nearest coffee shop

(c) Real-time information about the environment, hazards, and conditions
(d) Stock market updates

Q.21. How do robots assist in creating comprehensive maps of disaster-affected areas?
(a) By sketching maps on paper
(b) By using GPS coordinates
(c) By capturing aerial images and data using sensors
(d) By conducting surveys with local residents

Q.22. What is the role of robots equipped with gas sensors in disaster response?
(a) Detecting the presence of extraterrestrial life
(b) Monitoring the air quality for pollution
(c) Identifying the presence of hazardous gases in the environment
(d) Analyzing the composition of rocks and minerals

Q.23. How do robots contribute to assessing structural damage in disaster-affected areas?
(a) By performing acrobatics to examine buildings
(b) By evaluating the emotional state of survivors
(c) By assessing the availability of electricity and water
(d) By using sensors to evaluate vibrations and stress in structures

Q.24. What is the primary role of robots in remote telepresence during disaster response?
(a) Playing video games remotely
(b) Enabling experts to guide on-site responders from a distance
(c) Performing dance routines via live video feed
(d) Providing remote pet grooming services

Q.25. How can robots aid in identifying potential hazards in disaster-stricken environments?
(a) By predicting the weather
(b) By identifying the presence of ghosts and spirits
(c) By using sensors to detect gas leaks, radiation, and other dangers
(d) By performing circus tricks

Q.26. What technology allows drones to create detailed 3D models of disaster-affected areas?
(a) GPS navigation
(b) Sonar
(c) LIDAR
(d) Bluetooth communication

Q.27. What is the primary advantage of using robots equipped with cameras in disaster response?
(a) They can take high-quality selfies
(b) They can capture images of wildlife
(c) They can provide visual data from hard-to-reach or dangerous areas
(d) They can paint landscapes

Q.28. How do robots contribute to medical assessment in disaster situations?
(a) By performing complex surgeries
(b) By providing psychological counseling
(c) By remotely assessing vital signs and injuries
(d) By administering vaccinations

Q.29. What is the main role of robots equipped with manipulator arms in disaster response?

(*a*) Performing magic tricks for entertainment

(*b*) Assisting in human activities by mimicking arm movements

(*c*) Carrying heavy equipment for construction

(*d*) Providing massages to survivors

Q.30. How do robots contribute to communication in disaster-stricken areas with disrupted infrastructure?

(*a*) By broadcasting music concerts (*b*) By sending funny memes to survivors

(*c*) By establishing communication links for responders and survivors

(*d*) By launching fireworks displays

Q.31. What type of data can robots equipped with radiation detectors provide to responders?

(*a*) Internet browsing history (*b*) Local restaurant reviews

(*c*) Information about nearby tourist attractions

(*d*) Data about radiation levels in the environment

Q.32. How do robots contribute to assessing structural stability in disaster-affected areas?

(*a*) By predicting the future weather conditions

(*b*) By analyzing geological formations

(*c*) By evaluating vibrations and stress in buildings

(*d*) By performing archaeological excavations

Q.33. What is the primary role of robots equipped with manipulator arms in disaster recovery operations?

(*a*) Providing musical entertainment

(*b*) Assisting in debris removal and object manipulation

(*c*) Offering dance performances (*d*) Serving meals to survivors

Q.34. How do robots help responders navigate hazardous environments during search and rescue operations?

(*a*) By providing step-by-step directions (*b*) By sending signals to satellites

(*c*) By analyzing historical data

(*d*) By navigating dangerous areas with sensors and cameras

Q.35. What role do robots equipped with communication devices play in disaster response?

(*a*) Sharing jokes with survivors (*b*) Communicating with local wildlife

(*c*) Establishing contact with isolated survivors and relaying information

(*d*) Broadcasting radio stations

Q.36. How do robots assist in damage assessment in disaster scenarios?

(*a*) By predicting the stock market trends (*b*) By playing music for responders

(*c*) By collecting data and providing information about the extent of damage

(*d*) By predicting future weather conditions

Q.37. What is the main purpose of robots equipped with environmental sensors in disaster response?

(*a*) Predicting earthquakes (*b*) Monitoring air pollution levels
(*c*) Detecting the presence of alien life forms (*d*) Assessing water quality in nearby rivers

Q.38. How do robots contribute to delivering supplies to isolated areas during disasters?

(*a*) By teleporting supplies instantly (*b*) By flying using jet engines
(*c*) By navigating hazardous terrain and transporting essential items
(*d*) By digging tunnels to reach survivors

Q.39. What is the primary role of robots equipped with communication devices in disaster recovery?

(*a*) Broadcasting radio programs (*b*) Making phone calls to distant relatives
(*c*) Relaying information between responders and experts
(*d*) Streaming live concerts for entertainment

Q.40. How do robots enhance communication during disaster response in areas with disrupted infrastructure?

(*a*) By sending love letters to survivors
(*b*) By establishing communication links using satellite technology
(*c*) By broadcasting emergency alerts using megaphones
(*d*) By transmitting video game tutorials

Q.41. What is the primary purpose of robots equipped with gas sensors in disaster scenarios?

(*a*) Monitoring the scent of flowers
(*b*) Detecting the presence of hazardous gases in the environment
(*c*) Identifying the best locations for picnics
(*d*) Analyzing the composition of rocks and minerals

Q.42. How do robots enhance communication between experts and on-site responders during disaster recovery?

(*a*) By sending Morse code messages (*b*) By translating text messages into emojis
(*c*) By providing real-time video and audio communication
(*d*) By sending smoke signals

Q.43. What technology allows drones to create accurate maps of disaster-affected areas?

(*a*) Supercharged batteries (*b*) Hypersonic propulsion
(*c*) LIDAR technology (*d*) Quantum computing

Q.44. What is the primary role of robots equipped with communication devices in disaster response?

(*a*) Sending random messages to strangers
(*b*) Providing entertainment by telling jokes
(*c*) Establishing contact with survivors and relaying information
(*d*) Broadcasting radio shows

Q.45. How can robots assist in assessing structural damage after a disaster?

(*a*) By conducting geological surveys

(*b*) By predicting future earthquakes

(*c*) By evaluating vibrations and stress in buildings

(*d*) By analyzing cloud formations

Q.46. What is the primary advantage of using robots equipped with cameras in disaster scenarios?

(*a*) They can take artistic photographs

(*b*) They can capture images of rare animals

(*c*) They can provide visual data from dangerous areas

(*d*) They can write poetry

Q.47. How do robots contribute to delivering medical aid in disaster situations?

(*a*) By performing heart surgeries

(*b*) By conducting dental check-ups

(*c*) By providing remote medical assessments and support

(*d*) By painting medical murals

Q.48. What technology allows drones to create accurate maps of disaster-stricken areas?

(*a*) Wind propulsion

(*b*) Sonar technology

(*c*) LIDAR technology

(*d*) Solar panels

Q.49. How can robots assist in providing communication support in disaster scenarios?

(*a*) By composing symphonies

(*b*) By translating messages into foreign languages

(*c*) By establishing communication links in areas with disrupted infrastructure

(*d*) By creating social media posts

Q.50. What is the main advantage of using robots for hazardous material handling in disasters?

(*a*) Robots are immune to hazardous substances

(*b*) Robots can contain and handle hazardous materials safely

(*c*) Robots can consume hazardous substances without harm

(*d*) Robots can communicate with hazardous materials

Key to Multiple Choice Questions

1. (*c*)	**2.** (*b*)	**3.** (*c*)	**4.** (*c*)	**5.** (*c*)	**6.** (*c*)	**7.** (*c*)	**8.** (*c*)	**9.** (*c*)
10. (*c*)	**11.** (*c*)	**12.** (*c*)	**13.** (*c*)	**14.** (*c*)	**15.** (*b*)	**16.** (*c*)	**17.** (*b*)	**18.** (*c*)
19. (*d*)	**20.** (*c*)	**21.** (*c*)	**22.** (*c*)	**23.** (*d*)	**24.** (*b*)	**25.** (*c*)	**26.** (*c*)	**27.** (*c*)
28. (*c*)	**29.** (*b*)	**30.** (*c*)	**31.** (*d*)	**32.** (*c*)	**33.** (*b*)	**34.** (*d*)	**35.** (*c*)	**36.** (*c*)
37. (*b*)	**38.** (*c*)	**39.** (*c*)	**40.** (*b*)	**41.** (*b*)	**42.** (*c*)	**43.** (*c*)	**44.** (*c*)	**45.** (*c*)
46. (*c*)	**47.** (*c*)	**48.** (*c*)	**49.** (*c*)	**50.** (*b*)				

Project Works

1. **Search and Rescue Robot Description:** Design and build a robot capable of navigating through a simulated disaster environment to locate and rescue "survivors" (represented by objects).

Hints: The search and rescue robot can be equipped with sensors like cameras, thermal sensors, and LIDAR to detect survivors and obstacles. It can use path planning algorithms to navigate the environment and reach the simulated survivors.

2. **Teleoperated Drone for Disaster Assessment Description:** Develop a teleoperated drone that can be controlled remotely to collect real-time images and data from disaster-affected areas.

   Hints: The drone can be equipped with a high-resolution camera and GPS for accurate positioning. Remote operators can control the drone's flight and camera to capture images and videos for damage assessment. The collected data can be transmitted back to the operators for analysis.

3. **Hazardous Material Handling Robot Description:** Create a robot capable of safely handling and containing hazardous materials in a controlled environment.

   Hints: The hazardous material handling robot can have specialized arms and grippers designed to interact with hazardous substances. It can be equipped with sensors to detect the presence of harmful gases or chemicals and use appropriate tools to safely contain and move the materials to a secure location.

4. **Disaster-Resilient Communication Network Description:** Design a communication network using robots and drones to establish communication links in disaster-stricken areas with disrupted infrastructure.

   Hints: The project can involve deploying communication drones equipped with communication relays to establish temporary networks. Ground robots can act as communication nodes, extending the range and coverage of the network, ensuring responders and survivors can communicate effectively.

5. **Remote Medical Assessment Robot Description:** Develop a robot equipped with medical sensors to remotely assess the vital signs and injuries of "survivors" in a disaster scenario.

   Hints: The remote medical assessment robot can have sensors to measure heart rate, temperature, and other vital signs. It can also include a camera for visual assessment. Remote medical professionals can analyze the data collected by the robot to prioritize medical treatment and allocate resources effectively.

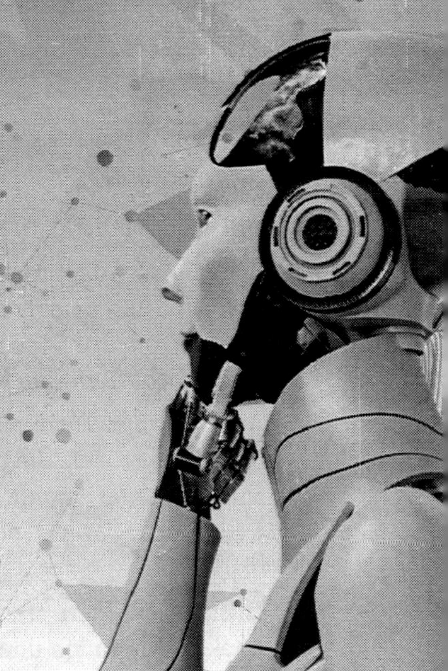

Biologically-Inspired Robotics

9.1 INTRODUCTION

Biologically-inspired robotics, also known as biomimetic or bio-inspired robotics, is a branch of engineering that uses design and development of robots and robotic systems as a source of inspiration from the natural world. The objective is to build machines that resemble or are motivated by the systems, structures, and behaviors of living things. Researchers want to increase robotic efficiency, adaptability, and functionality by modeling nature's structures and principles.

The theory behind biologically inspired robotics is that over millions of years, nature has developed incredibly effective and efficient solutions to a wide range of problems. Researchers can build robots that perform better in complicated and dynamic environments by researching and implementing these methods. Robotics that draws inspiration from biology aims to develop machines that are excellent at a variety of jobs and settings. Researchers hope to open up new possibilities for robotics and automation by studying the natural world and adapting its principles into creative solutions.

9.2 KEY ASPECTS OF BIOLOGICALLY-INSPIRED ROBOTICS

Here are some key aspects of biologically-inspired robotics:

1. **Design Inspiration:** Robots with biological inspiration are built to mimic the physical traits and actions of organic things. This can include characteristics related to locomotion (walking, flying, and swimming), body morphology (limb configurations, skeletal structures), and sensory organs (vision, touch, and hearing).
2. **Locomotive Techniques:** Animals have developed a variety of adaptable and effective means of travelling through various situations. Legged robots, aerial drones, submersible vehicles, and other types of robots have all been influenced by these locomotion techniques. For instance, research into the biomechanics of insects, birds, and fish has resulted in the creation of robots with nimble and effective movement.
3. **Sensor Systems:** Animals have highly developed sensory systems that enable them to understand their surroundings and make decisions based on sensory information. In this area, scientists are

developing robots with sophisticated sensor systems, including computer vision, tactile sensing, and even synthetic whisker systems modeled after rodents.

4. **Behavioral Adaptation:** Robots with biological inspiration frequently try to imitate nature's adaptable and flexible behaviors. This may entail developing robots with the ability to absorb information from their surroundings, adapt to shifting circumstances, and even engage in swarm behaviors akin to those of social insects.
5. **Applications:** There are many different sectors in which biologically inspired robots can be used. These robots can be employed in a variety of tasks, including search and rescue operations, environmental monitoring, agriculture, medical procedures, and extreme environment exploration. For instance, congested or difficult spaces might be navigated by robots modelled like insects.
6. **Difficulties:** There are a number of difficulties in creating robots that are biologically inspired. It takes a thorough understanding of both biology and robotics to convert biological concepts into practical engineering solutions. Furthermore, verifying these robots' dependability, effectiveness, and safety in real-world situations might be challenging.
7. **Highly Interdisciplinary Nature:** Biologically-inspired robotics draws on a variety of disciplines, including biology, robotics, engineering, computer science, biomechanics, and more. For this profession to advance, interactions between experts from these several fields are essential.

9.2.1 Design Inspiration with Technology

Design inspiration in biologically-inspired robotics entails researching the structures, behaviors, and mechanisms found in nature and then translating those insights into technical solutions employing concepts from different scientific fields, including biology, mathematics, and engineering. Let's examine this idea in greater detail using several illustrations:

1. **Locomotion:**

   Biological Inspiring: Terrestrial creatures with complicated terrains, such as insects, have evolved effective loco motional strategies.

   Science, Technology, and Math: Biomimetic robots, like legs, mimic animal and insect movement patterns and limb arrangements. Energy-efficient movements are optimized with the aid of mathematical models derived from physics and biomechanics.

   Example: The "RoboBee" is a miniature flying robot that was designed with bees in mind. It replicates the bee's wing-flapping motion and applies aerodynamics concepts to achieve stable flying.
2. **Biomimetic Sensors:**

   Animals Sensory Systems: Animals have sophisticated sensory systems that allow for perception and navigation.

   Process and Interpret Data: Robots can contain sensors like tactile sensors (inspired by human touch or animal whiskers), acoustic sensors (influenced by bat echolocation), artificial vision systems (cameras), and tactile sensors (inspired by human touch or animal whiskers). These sensors process and interpret data using mathematical algorithms.

   AquaJelly Robot Mimics: As an illustration, the "AquaJelly" robot mimics the swimming motions of jellyfish while moving through water by using a mix of sensors and actuators.

3. **Soft Robotics:**

   Biological Inspiring: Many animals have supple, malleable bodies that enable them to manoeuvre through challenging circumstances.

   Science, Technology, and Math: Soft robotics is the construction of robots that mimics the compliance and flexibility of biological beings by using flexible materials and structures. Understanding deformations and regulating motions are accomplished by mathematical modeling.

   Example: The "Octobot" is a soft robotic octopus that employs pneumatics to contract and extend its soft body in order to produce lifelike movements.

4. **Swarm Robotics:**

   Inspiration from Biology: Social insects like ants and bees display sophisticated group behaviors and self-organization.

   Technology and Mathematics: Swarm robots cooperate autonomously to complete tasks, frequently employing algorithms that are modeled by the behavior of swarms in nature. Interactions and emerging patterns are modeled using mathematics.

   Kilobot: The "Kilobot" is a little robot created for study of swarm robotics. It has the ability to communicate with nearby Kilobots and work together to complete tasks like formation assembly.

5. **Snake Robots:**

   Biological Inspiration: Snakes can easily traverse through tight areas and challenging terrain due to their tremendous adaptability.

   Technology and Mathematics: To simulate the movements of snakes, snake-like robots have segmented bodies and flexible joints. Equations from kinematics and dynamics are utilized to simulate and regulate their motion.

   Modular Snake: As an illustration, the "Modular Snake" robot, which was designed to manoeuvre past obstacles like pipelines and rubble, was motivated by the way snakes move.

6. **Biohybrid Systems:**

   Biohybrids: Biohybrids, which blend biological tissues or cells with artificial components, are inspired by living organisms.

   Technology and Mathematics: Biological cells and robotic structures can be combined to create biohybrids, which frequently require mathematical modeling to comprehend the interactions between biological and artificial elements.

   Illustration: "Cyberplasm" is a biohybrid robot whose movement is controlled by muscle cells. Both biological and mechanical concepts affect how the muscle cells interact with the robotic framework.

Each of these examples draws its design inspiration from how nature resolves particular problems. The transformation of these ideas into practical robots requires the use of technology and mathematics. Robots that mimic the behaviors and capabilities of live things while offering ground-breaking solutions in a variety of sectors are made possible by the combination of mathematical modeling, computational simulations, and engineering principles.

9.2.2 Movement and Locomotion with Technology

Key components of biologically inspired robotics are locomotion and movement, where researchers take cues from the numerous and effective ways that animals move in their natural habitats. Engineers hope to build robots that can successfully navigate a variety of terrain by modeling these movement patterns. Here is an in-depth analysis of movement and locomotion in biologically inspired robots, with examples and a look at the technology and mathematics involved:

1. **Legged Locomotion:**

   Biological Inspiring: To travel on various surfaces, insects, mammals, and other creatures use a vast variety of leg designs.

   Technology and Mathematics: Multi-legged robots are made to adapt to different types of terrain. Equations relating to kinematics and dynamics are used to represent how the legs move and interact with their surroundings.

   Illustration: As an illustration, "Boston Dynamics' Spot" robot, which was modeled after quadrupedal animals, exhibits adaptable legged movement that enables it to move across uneven terrain and through intricate surroundings.

2. **Flying Locomotion:**

   Biological Inspiring: Different flying techniques and styles are displayed by birds, insects, and bats.

   Technology and Mathematics: Different wing morphologies and flight patterns are replicated by flying robots (drones). To ensure stable flight, aerodynamics concepts and control algorithms are applied.

   DelFly Robot: The "DelFly" robot, for instance, imitates the flight of insects like dragonflies. Fluid dynamics research was used to design its flapping-wing system, allowing for quick and energy-efficient flight.

3. **Swimming Locomotion:**

   Biological Inspiring: Fish and other aquatic mammals have effective swimming techniques.

   Technology and Mathematics: To simulate swimming, underwater robots use a variety of propulsion techniques, such as oscillating fins or undulating bodies. Models of fluid dynamics aid in underwater movement optimization.

   RoboTuna Robot: One example is the "RoboTuna" robot, which has potential uses in undersea research since it mimics the streamlined shape and undulating motion of tunas.

4. **Snake-like Locomotion:**

   Biological Inspiration: Snakes navigate difficult conditions by side winding and undulating laterally.

   Technology and Mathematics: To mimic the movement of snakes, snake robots feature segmented bodies and flexible joints. To simulate interactions with surfaces, kinematics and friction models are utilized.

   Example: As an illustration, the "Carnegie Mellon Snakebot" demonstrates the potential for search and rescue operations in catastrophe circumstances by being able to manoeuvre through tight places and debris.

5. **Biohybrid Locomotion:**

   Biological Inspiration: Mollusks and insects both use biological and mechanical elements to move in intricate ways.

Technology and Mathematics: Biohybrid robots combine robotic components with biological tissues or cells. The relationship between live and artificial components can be better understood using mathematical models.

As an illustration, scientists have created biohybrid robots that move by contracting their muscles. These robots have the potential to move more naturally and adaptably.

6. **Climbing and Wall-scaling:**

   Biological Inspiring: Specialised adhesive systems allow insects and animals like geckos to adhere to surfaces and scale barriers.

   Technology and Mathematics: Climbing robots use adhesive substances or grasping techniques that are modelled after natural adhesion. The optimisation of contact forces and adhesion strength is helped by mathematical models.

   Stickybot: As an illustration of the possibilities for wall-scaling robots, "Stickybot" uses specialised feet with adhesive characteristics to emulate the ability of the gecko to scale smooth surfaces.

In each of these instances, technology is essential for converting biological insights into useful robotic systems. Robots can imitate and frequently outperform their biological counterparts' ability to move by combining advanced materials, actuators, sensors, and control algorithms with mathematical models. Biologically inspired robots can excel in a variety of applications, from exploration and disaster response to surveillance and environmental monitoring, by utilizing the principles of movement inherent in nature.

9.2.3 Sensory Systems with Technology

In biologically inspired robotics, sensory systems are essential because they allow robots to perceive and interact with their surroundings. Engineers create sophisticated sensor technologies that enable robots to successfully gather data, make judgments, and navigate by taking inspiration from the sensory abilities of animals. Here is a full discussion of the technology and mathematics involved in biologically inspired robotics' sensory systems, with examples:

1. **Vision and Cameras:**

   Biological Inspiring: Primates in particular have highly evolved vision systems for detecting their surroundings.

   Technology and Mathematics: By recording and processing images, cameras and computer vision algorithms mimic animal vision. Techniques including pattern recognition, object detection, and optical flow analysis are used in image processing.

   RoboBees: As an illustration, "RoboBees" employ microscopic cameras to assess their surroundings. This enables them to avoid obstacles and adhere to certain visual cues.

2. **Tactile Sensing:**

   Bio-Inspiration: Touch receptors in an animal's skin or appendages are used by them to sense pressure, textures, and other physical characteristics.

   Technology and Mathematics: Pressure-sensitive materials and arrays can be used in tactile sensors, which are frequently modelled after biological touch receptors. In order to discover contact patterns and object attributes, mathematics is used to analyse sensor data.

   Takktile Sensors: An illustration of this is the use of flexible materials and arrays of pressure-sensitive cells in "Takktile" sensors, which mimic human touch. Robots can handle items with greater sensitivity and adaptability thanks to these sensors.

3. **Auditory Sensing:**
   Biological Insight: Animals like bats and dolphins utilise echolocation, which involves sending and receiving sound waves, to navigate and find items.
   Technology and Mathematics: Acoustic and ultrasonic sensors emit and detect sound waves to simulate echolocation. Distances can be calculated using time-of-flight calculations and signal processing.
   DolphinBot: As an illustration, "DolphinBot" imitates dolphin echolocation by using underwater acoustic sensors to navigate and find underwater things.
4. **Whisker-based Sensing:**
   Whiskers: Rodents use their whiskers, or vibrissae, to perceive the textures, forms, and obstructions in their environment.
   Mathematics and Technology: Robots are equipped with whisker-like sensors that replicate the tactile sense of rodents. These sensors can provide information on surface characteristics, air currents, and object contact.
   Shrewbot: As an illustration, "Shrewbot" uses fake whiskers to detect impediments and travel in low-light conditions, just like shrews do.
5. **Chemical and Olfactory Sensing:**
   Biological Inspiring: Many animals rely on their sense of smell to find nutrients and identify substances.
   Technology and Mathematics: Chemical sensors are integrated into robots to detect odours and chemicals. These sensors frequently use gas-sensitive materials. Specific compounds can be recognised using signal analysis and pattern recognition.
   Cyborg Nose: As an illustration, the "Cyborg Nose" uses algorithms and gas sensors to detect and identify distinct odours. It has applications in search and rescue and environmental monitoring.
6. **Proprioception and Joint Sensing:**
   Biological Inspiration: Animals can sense their locations and motions because they have proprioceptive sensors in their muscles and joints.
   Technology and Mathematics: To keep track of their own motion, robots employ joint encoders and inertial sensors. Positions and orientations are estimated using kinematic and dynamics equations.
   PR2 Robot: As an illustration, the "PR2" robot can manipulate things and interact with its environment by precisely tracking its own movements using joint encoders and inertial sensors.

Each of these examples uses technology that is intended to mimic the natural sensory modalities. Robots' sensory data collection is interpreted and put to use using mathematical concepts from signal processing to pattern recognition to kinematics and physics. Biologically-inspired robots can move, interact with, and react to their environments in ways that closely mirror the skills of real organisms by combining sophisticated sensing systems inspired by animals.

9.2.4 Behavioral Adaptation with Technology

In biologically inspired robotics, where robots are created to exhibit adaptable and contextually appropriate behaviors like to those seen in organic animals, behavioral adaptation is a core idea. Robots are able to adapt successfully to changing settings, tasks, and circumstances thanks to these adaptive behaviors. Here is a thorough explanation of behavioral adaptation in biologically inspired robots, complete with examples and the underlying technology and mathematics:

1. **Learning and Adaptation:**

   Biological Inspiration: Animals have the capacity to learn from their experiences and modify their behavior in response to shifting environmental factors.

   Technology and Mathematics: Robots can learn from interactions with their environments thanks to machine learning methods like reinforcement learning, neural networks, and genetic algorithms. The optimization and decision-making processes are supported by mathematics.

   Quadrupedal Robot: As an illustration, the quadrupedal robot "ANYmal" employs reinforcement learning to modify its gait patterns to suit various surfaces over time, increasing its stability and effectiveness.

2. **Swarm Behavior and Collective Intelligence:**

   Biological Inspiring: Collective intelligence is demonstrated by the coordinated behaviours of social insects like ants and bees in big groups.

   Technology and Mathematics: Swarm robotics includes creating algorithms that enable networks of robots to communicate and work together autonomously. Complex systems' interactions and emergent behaviours are modelled mathematically.

   RoboBees: As an illustration, "RoboBees" can cooperate in swarms to carry out activities like agricultural pollination or environmental monitoring, adjusting their collective behaviour in response to changing environmental variables.

3. **Contextual Decision-Making:**

   Biological Inspiration: Animals can make choices based on sensory information and environmental signals to accomplish particular objectives.

   Mathematics and Technology: Robots utilize algorithms to make decisions and sensors to comprehend their surroundings. To deal with uncertainty in decision-making, probabilistic reasoning and Bayesian inference are used.

   Aerial Robots: As an illustration, "Aerial Robots" with cameras can identify and navigate around moving objects by using computer vision algorithms, modifying their flight trajectories in real-time.

4. **Morphological Adaptation:**

   Biological Inspiring: Some animals have shape-changing bodily features that enable them to change their bodies to various jobs.

   Science, Technology, and Math: Robots with modular or reconfigurable components can alter their physical configurations to adapt to different jobs or conditions. Equations from mechanics and kinematics are used to model these transformations.

   Robotic Origami: An illustration of "Robotic Origami" is the creation of robots that can fold and unfold their structures to suit various jobs, such as crawling through small places or expanding for bigger ones.

5. **Evolutionary Robotics:**

   Inspiration from Biology: Over time, evolution has moulded biological organisms to better match their ecological niches.

Technology and Mathematics: Through decades of simulations, genetic algorithms are utilised to generate robotic designs and behaviors. The process of selection and adaptation is guided by mathematical optimization approaches.

Evolvability Program: As an illustration, "Evolvability" programs entail developing virtual robots that can carry out particular duties. These robots demonstrate how adaptive features can develop through evolution by progressively increasing their fitness over generations.

6. **Adaptive Control and Feedback Loops:**

   Biological Inspiration: Animals constantly modify their behaviors and movements in response to sensory information in order to maintain equilibrium and accomplish objectives.

   Technology and Mathematics: Robot behavior is regulated using real-time feedback using control methods like PID controllers and adaptive control systems. Mathematical calculations and control theory provide accurate and reliable adjustments.

   Bipedal Robots: As an illustration, "Bipedal Robots" regularly modify their gait patterns depending on sensor data to minimize falls when walking or running.

In each of these cases, adaptive behaviors seen in nature are mimicked by technology. Robots are able to dynamically modify their behaviors and strategies in response to their environments and tasks thanks to mathematical ideas spanning from control theory to machine learning techniques. Researchers want to build robots with biologically inspired design that can function successfully and efficiently in challenging and shifting environments.

9.2.5 Applications with Technology

With a variety of uses, biologically inspired robotics makes use of natural designs and concepts to build robots that are more effective and adaptable at their jobs. Here is a full analysis of some biologically inspired robotics applications, including with the underlying technology, mathematics, and examples:

1. **Search and Rescue:**

   Use: Robots with biological inspiration can explore hazardous and complicated areas, such as catastrophe scenes, to find survivors and gather data.

   Science, Technology, and Math: Robots with vision, touch, and communication sensors can find survivors and gather data for analysis. Search strategy optimisation is made possible by path planning algorithms and machine learning models.

   Snake Robots: As an illustration, "Snake Robots" can discover and assist people in catastrophe circumstances by navigating through debris and cramped areas, increasing the effectiveness of search and rescue operations.

2. **Environmental Monitoring:**

   Use: To collect data for environmental monitoring and research, biologically inspired robots can be used in difficult situations like woods, oceans, or distant locations.

   Technology and Math: Robots with data-collection devices, such as cameras or environmental sensors, can keep track of ecological changes. Data trends and patterns are analyzed using mathematical models.

   BioSwimmer: As an illustration, "BioSwimmer" is an underwater robot designed to look like a fish that gathers information in aquatic settings for jobs like monitoring pollution and researching marine life.

3. **Agriculture and Farming:**

   Use: Animal-inspired robots can help with agricultural chores like crop monitoring, pollination, and harvesting.

   Technology and Mathematics: Ripe crops can be identified by robots with cameras and sensors, while control algorithms and mathematical models optimise their operations.

   Harvesting Robots: One illustration is how "Harvesting Robots" imitate human ability to select delicate fruits and vegetables without breaking them, increasing the effectiveness of agricultural procedures.

4. **Medical Procedures and Assistive Devices:**

   Use: Robots with biological inspiration can help with operations, therapy, and prostheses, improving medical procedures and assisting those with mobility issues.

   Technology and Mathematics: Wearable or implantable gadgets now use robotics technology like soft actuators and sensors. Natural and adaptable movements can be produced with the use of kinematics and biomechanical modelling.

   Robotic Exoskeletons: As an illustration, "Robotic Exoskeletons" help people with mobility problems walk and carry out daily tasks by using mechanical components and control algorithms.

5. **Space Exploration:**

   Application: Adaptive behaviors and effective locomotion are essential for planetary exploration, where biologically inspired robots can be deployed.

   Robots with sophisticated sensors and locomotion systems are being created for navigating difficult terrains in the fields of technology and mathematics. Calculations in mathematics are used in trajectory optimization and navigation algorithms.

   Rovers: As an illustration, "rovers" like NASA's "Curiosity Rover" show off their autonomous navigation and adaptive mobility while operating on Mars' surface, facilitating data collection and scientific research.

6. **Industrial Automation:**

   Use: By adapting to changing tasks and circumstances, biologically inspired robotics can improve automation in manufacturing and assembly operations.

   Technology and Mathematics: Production lines can be optimized by using robots with sensor- and adaptive-control systems. Mathematical modeling is used in work allocation and coordination algorithms.

   Cobots: As an illustration, "Cobots" (collaborative robots) cooperate safely and effectively with human employees in factories by adapting to various assembly jobs.

7. **Underwater Exploration:**

   Use: Marine animal-inspired underwater robots can explore and investigate the depths of oceans and other bodies of water.

   Technology and Mathematics: Robots with underwater sensors and propulsion systems can gather information on geological characteristics and marine life. Their motions are controlled by fluid dynamics and hydrodynamics models.

   AUVs: As an illustration, "AUVs" (Autonomous Underwater Vehicles) use fish-like swimming techniques to explore and gather information in aquatic environments, advancing oceanographic research.

Robotics with biological inspiration applies technology and mathematical ideas to each of these problems to produce novel solutions. Researchers and engineers use inspiration from nature to develop robots that excel in a variety of fields, increasing efficiency, adaptability, and safety in a broad range of tasks and industries.

9.2.6 Challenges in Biologically-Inspired Robotics

In order to fully realize the potential of biomimicry, researchers and engineers must overcome a number of obstacles in the field of biologically inspired robotics. Here is a thorough examination of some of the difficulties in biologically-inspired robotics, along with the requisite technology, mathematics, and examples:

1. **Complexity of Biological Systems:**

   Challenge: Biological organisms are complex and often not fully understood. Translating intricate biological structures and behaviors into functional robotic systems can be challenging.

   Technology and Mathematics: Advanced imaging techniques, computational simulations, and mathematical modeling help researchers understand and replicate biological systems.

   Example: Replicating the agility of insects' flight requires precise understanding of aerodynamics and wing movement, necessitating detailed computational models for accurate emulation.

2. **Integration of Multidisciplinary Knowledge:**

   Challenge: Interdisciplinary cooperation is crucial for biologically-inspired robotics because it calls for knowledge in biology, robotics, engineering, and other fields.

   Technology and Mathematics: To build and create robots, cross-disciplinary teams draw on methods from numerous disciplines. Engineering solutions and biological principles can be reconciled through mathematical modeling.

   Illustration: As an illustration, integrating biological tissues with robotic components requires cooperation between biologists, material scientists, and engineers when creating a robot with a biohybrid muscle actuator.

3. **Energy Efficiency and Power Sources:**

   Problem: It is difficult to recreate the exceptional energy efficiency of many animals in robots that use standard power sources.

   Technology and Mathematics: Energy-harvesting technologies, lightweight materials, and efficient actuators are investigated. Energy-efficient locomotion solutions are created using mathematical optimization techniques.

   Illustration: As an illustration, "RoboBees" are designed to mimic the energy-efficient flight of real bees, which necessitates the development of lightweight materials and effective propulsion systems.

4. **Sensing and Perception Limitations:**

   Obstacle: Since animal senses are highly specialized and tailored to certain surroundings, it is challenging to effectively mimic their perceiving skills.

   Technology and Mathematics: To replicate animal senses, advanced sensor technologies and signal processing techniques are used. The application of machine learning models enhances perceptive capacities.

Illustration: As an illustration, creating sophisticated sound sensors and intricate signal processing algorithms will help robots mimic the sophisticated sonar system of bats for navigation and obstacle avoidance.

5. **Robustness and Adaptability:**

   Problem: It is difficult to mimic animals' resilience and adaptation in robots that might run into unfamiliar situations. Animals thrive in unpredictably changing surroundings.

   Technology and Mathematics: Robots can learn and modify their behaviours depending on real-time input thanks to machine learning, reinforcement learning, and adaptive control algorithms.

   Illustration: As an illustration, complex learning algorithms are needed to build robots that can modify their walking gait to suit diverse surfaces and ambient situations, such as "ANYmal" did.

6. **Ethical and Bioethical Considerations:**

   Difficulty: When biologically inspired robots use living tissues or mimic animal behavior, ethical difficulties arise.

   Technology and Mathematics: To ensure ethical research and application, rules and frameworks are created. Engineers, researchers, and bioethics specialists work together.

   Illustration: As an illustration, designing biohybrid robots raises concerns about how living tissues and organisms should be handled, which sparks debates over the research's ethical implications.

7. **Scaling from Lab to Real World:**

   Obstacle: Complex and chaotic surroundings may prevent real-world performance of robots designed in controlled lab situations.

   Technology and Mathematics: To bridge the gap between controlled environments and real-world scenarios, real-world testing, simulations, and reliable control algorithms are used.

   Illustration: As an illustration, aerial drones that operate flawlessly indoors may encounter difficulties such as wind gusts and obstructions when used outdoors, necessitating adaptation through rigorous testing and algorithm improvement.

Innovating technologically, conducting rigorous scientific research, and working together across disciplines are all necessary to meet these difficulties. Despite the fact that these challenges may be significant, conquering them may result in robotics advances that allow for the creation of robots that are more capable, adaptable, and successful that is inspired by the natural world.

9.2.7 Interdisciplinary Nature of Biologically-Inspired Robotics

The fact that this area gets its inspiration from biology while incorporating numerous technology and mathematical concepts to produce useful robotic systems accounts for the interdisciplinary nature of biologically-inspired robotics. Here is a full analysis of the multidisciplinary nature of technologically and mathematically inspired biological robotics, with examples:

1. **Biology:**

   Contribution: The knowledge of natural systems, behaviors, and structures that biology provides serves as the basis for designing robots.

   Technology and Mathematics: Biomimetic designs are created using insights from biology, which helps engineers better comprehend the mechanics and ideas that underlie animal locomotion, sensory systems, and behaviors.

Illustration: As an illustration, robotic birds and drones that are capable of stable and agile flight are designed using knowledge of the wing structures and flight processes of birds.

2. **Robotics and Engineering:**

   Contribution: The creation of useful robots requires a strong understanding of robotics and engineering.

   Technology and Mathematics: Using technologies like actuators, sensors, materials, and control algorithms, engineers create robotic systems. Robot behaviors and movements are designed using mathematical modeling.

   ANYmal: One example is the quadrupedal robot "ANYmal," which employs engineering principles to emulate the mobility and flexibility of animals like dogs and can thus navigate difficult terrain.

3. **Computer Science and AI:**

   Contribution: Algorithms developed in computer science and artificial intelligence (AI) let robots learn, adapt, and make decisions.

   Science, Technology, Engineering, and Mathematics: Machine learning, reinforcement learning, and computer vision algorithms are used to provide robots the ability to process sensory data, gain experience, and change their behavior on their own.

   Illustration: As an illustration, robots like "RoboBees" use AI to assess visual inputs and modify their flight patterns in order to navigate through challenging environments.

4. **Material Science:**

   Contribution: Material science aids in the creation of adaptable, lightweight materials that are employed in robot construction.

   Technology and Mathematics: Engineers create materials that mimic the flexibility and strength of organic tissues. These materials react reliably under varied circumstances thanks to mathematical modeling.

   Illustration: As an illustration, soft robotics, which uses materials science, produces "Octobots," robots that resemble the compliance and adaptability of living things.

5. **Mathematics and Physics:**

   Contribution: The quantitative tools required to model and simulate robotic systems are provided by mathematics and physics.

   Technology and Mathematics: Kinematics, dynamics, control algorithms, and optimisation procedures are all described by mathematical equations. The creation of mechanisms that mimic natural behaviours is governed by physics principles.

   Illustration: Examples include designing swimming robots that emulate the propulsion systems of aquatic animals using mathematical simulations of fluid dynamics.

6. **Bioethics and Philosophy:**

   Contribution: The ethical issues underlying biologically inspired robotics are influenced by bioethics and philosophy.

   Technology and Mathematics: Researches and engineers are guided by ethical frameworks and discussions while dealing with living tissues or simulating animal behavior.

   Illustration: For instance, standards and laws for creating biohybrid robots that employ living tissues are influenced by ethical discussions concerning the use of animals in research.

7. **Ecology and Environmental Science:**

   Contribution: Ecology and environmental science shed light on how living things interact with their surroundings.

   Technology and Mathematics: Having an understanding of ecological linkages aids in building robots that can interact with natural ecosystems when doing jobs like environmental monitoring.

   Illustration: As an example, to guarantee that aquatic ecosystems are not significantly harmed, robots employed to monitor marine habitats use knowledge from marine ecology.

Because of the interdisciplinary cooperation between these disciplines, scientists and engineers can build biologically inspired robots that not only replicate natural systems but also progress a variety of scientific subjects. By working together, we can make sure that biologically inspired robotics continues to be a vibrant, cutting-edge field that gains from the combined knowledge of many other fields.

9.3 INSPIRATION FROM NATURE TO DEVELOP ROBOTS THAT MIMIC BIOLOGICAL SYSTEMS

Biomimicry, also known as biologically inspired design, is the practice of creating robots that mimic biological processes by taking design cues from nature. In this method, the concepts, structures, behaviors, and functions of live organisms are studied, and these elements are subsequently replicated in robotic systems. Here is how scientists are using nature as a model to build robots that replicate biological systems:

1. **Observation and Study:** Researchers first observe and investigate many creatures, including animals, in their natural habitats. They examine how these organisms move, engage with their environment, detect their surroundings, and carry out different tasks.
2. **Identifying Key Features:** Researchers pinpoint the essential traits, mechanisms, and behaviors that allow organisms to efficiently carry out particular activities. These characteristics can include physical traits like wings, limbs, or appendages as well as sensory traits like vision, touch, and hearing as well as locomotion techniques.
3. **Understanding Basic Concepts:** To better understand the fundamental concepts of biology, physics, and mechanics that underlie the behaviors we experience. This knowledge facilitates the conversion of natural systems into useful robotic designs.
4. **Converting to Robotics:** Based on their observations and comprehension, researchers create robotic systems that reproduce the traits and behaviors they have found. Designing mechanical parts, sensors, actuators, and control algorithms that resemble natural systems is required for this.
5. **Biomimetic Design:** Scientists use biomimetic design concepts to build robots that closely mimic the structure and behavior of the biological animals that served as their inspiration. This encompasses copying designs, elements, and motions.
6. **Materials and Actuators:** Scientists frequently investigate brand-new substances that resemble the mechanical characteristics of biological tissues. They create soft, flexible materials that allow robots to safely interact with their surroundings, imitate animal movements, and adjust to various settings.
7. **Sensor Integration:** To mimic the sensory capacities of animals, researchers build sophisticated sensors into robots, such as cameras, tactile sensors, and audio sensors. Robots can detect their surroundings and interact with them thanks to these sensors.

8. **Control Algorithms:** Researchers create control algorithms that are motivated by biological brain networks and feedback mechanisms in order to mimic adaptive behaviors. Robots can process sensory inputs and modify their actions in real-time thanks to these algorithms.
9. **Learning and Adaptation:** To help robots learn from their experiences, adapt to new circumstances, and get better over time, researchers use machine learning and artificial intelligence approaches.
10. **Testing and Refinement:** Robots that have been developed are tested first in controlled settings and then in real-world conditions. Data on their performance and behavior are gathered, and the design and algorithms are subsequently improved repeatedly in response to the findings.
11. **Interdisciplinary:** Creating robots that replicate biological processes necessitates cooperation between researchers in a variety of fields, including biology, engineering, computer science, material science, and mathematics. To assure the success of the biomimetic approach, experts from these domains collaborate.
12. **Ethical Considerations:** When taking cues from nature, especially when including living tissues or closely mimicking animal behaviours, researchers also take ethical considerations into account. Responsible research and application are governed by ethical frameworks.

Examples of these ecologically inspired robots include drones that imitate bird flight, robotic fish for underwater exploration, soft robots that replicate the movements of marine animals, and biohybrid robots that incorporate both living and artificial tissues. Researchers want to develop robots that are more adaptable, effective, and capable of carrying out tasks in challenging and dynamic circumstances. They do this by taking inspiration from nature.

9.4 POTENTIAL APPLICATIONS OF ROBOTS THAT MIMIC BIOLOGICAL SYSTEMS

Robots that replicate biological systems have numerous potential uses in a variety of sectors. These applications use the effectiveness, adaptability, and specialized skills of living things to tackle challenging issues. Here are some potential uses for robots that draw inspiration from biological systems, along with the corresponding technical and mathematical requirements:

1. **Search and Rescue:**

   Use: Animal-inspired robots that can navigate collapsed buildings can find survivors and gather data in disaster-stricken places.

   Science, Technology, and Math: Biomimetic robots with sensors for vision and mobility can move through trash and adapt to difficult terrain. Path planning and real-time changes are guided by mathematics.

   Illustration: Drones that resemble bird flight can access difficult-to-reach regions, while snake-like robots with flexible bodies can navigate through confined tunnels to seek for survivors.

2. **Medical Procedures and Surgeries:**

   Use: Robots that are modeled after how people and animals move can help with operations and medical procedures, improving accuracy and lowering invasiveness.

   Technology and Mathematics: Biomimetic joint mechanics in robotic arms allow them to mimic the motions of human limbs. To ensure precise and secure interactions, mathematics models kinematics and dynamics

Illustration: As an illustration, robotic surgical assistants imitate the dexterity of human hands to help surgeons carry out complex procedures with greater accuracy.

3. **Environmental Monitoring and Exploration:**

   Use: Biomimetic robots may investigate challenging areas, such distant forests or deep oceans, to gather information and monitor ecosystems.

   Technology and Mathematics: Robots with sensors and adaptable locomotion systems can go across difficult terrain while gathering data. The interpretation of sensor data and energy-efficient mobility are improved by mathematical models.

   Illustration: Bird- or insect-inspired aerial drones can fly across forests or distant locations while gathering environmental data and keeping an eye on ecosystem changes.

4. **Agriculture and Farming:**

   Use: Robots that mimic animal behaviors can help with chores like planting, pollinating, and harvesting, increasing agricultural productivity.

   Technology and Mathematics: Robots that mimic the movements of animals like bees for pollination or birds for seed distribution include tools or appendages that are specially designed. The best motion planning and control are guided by mathematics.

   Illustration: Robotic pollinators can mimic the actions of bees to help pollinate plants in greenhouses, orchards, and agricultural fields.

5. **Underwater Exploration:**

   Use: Robots modeled after marine animals can explore the ocean's depths while gathering information and researching marine biodiversity.

   Science, Technology, and Math: Biomimetic underwater robots mimic the hydrodynamics and swimming movements of marine animals. In underwater conditions, mathematical models direct propulsion and manoeuvring.

   Illustration: Sensor-equipped robotic fish may explore underwater terrain, gathering information on marine ecosystems and performing research in locations that are inaccessible to human divers.

6. **Disaster Response and Hazardous Environments:**

   Use: To undertake inspections and maintenance in risky areas like nuclear power plants or chemical spills, biomimetic robots can be used.

   Science, Technology, and Math: Robots with adaptive locomotion mechanisms may travel across dangerous terrains and avoid obstacles. Algorithms for navigation and risk assessment are guided by mathematics.

   Illustration: Robots that resemble snakes, for instance, can check and maintain machinery in small spaces or places where radiation exposure may be present.

7. **Human Assistance and Rehabilitation:**

   Use: Robots that mimic human movement can help people with mobility issues or with rehabilitation.

   Technology and Mathematics: Robotic exoskeletons that imitate human joint movements let people walk and carry out other tasks. Control algorithms and mathematical modelling guarantee efficient and safe support.

Illustration: As an illustration, wearable exoskeletons can help stroke victims recover their strength and mobility during rehabilitation.

Integration of technology and mathematics is essential in each of these applications. Robots can perform a variety of activities and work in a wide range of industries thanks to the integration of robotics technology, sensors, control algorithms, and mathematical models. Researchers want to build versatile, adaptable robots that can take on challenging tasks by taking their cues from nature.

Numerical Problems with Solutions

Problem 1. Locomotion Efficiency Comparison: Compare the energy consumption of a legged robot and a wheeled robot to cover a distance of 100 meters. The legged robot requires 5 Joules per step and takes 50 steps to cover the distance. The wheeled robot requires 2 Joules per meter.

Solution: Energy consumption of the legged robot = 5 Joules/step * 50 steps = 250 Joules. Energy consumption of the wheeled robot = 2 Joules/m * 100 meters = 200 Joules.

The wheeled robot consumes less energy to cover the same distance.

Problem 2. Flapping Wing Frequency: A robotic bird with flapping wings has a flapping frequency of 10 Hz. Calculate the time period and the angular frequency (in radians per second) of the wing motion.

Solution: Flapping frequency (f) = 10 Hz. Time period (T) = 1/f = 1/10 = 0.1 seconds. Angular frequency (ω) = 2πf = 2π * 10 = 20π radians/second.

Problem 3. Aspect Ratio Calculation: You are designing a bird-inspired drone with a wingspan of 80 cm. If the wing area is 1200 cm^2, calculate the aspect ratio of the wings.

Solution: Wingspan (b) = 80 cm Wing area (A) = 1200 cm^2.

Aspect ratio (AR) = $b^2/A = 80^2/1200 = 5.33$

Problem 4. Snake Robot Path Length: A snake-like robot is navigating a curvy path. It has moved 2 meters forward, 1 meter backward, 3 meters forward, and then 2.5 meters backward. Calculate the total distance covered by the robot.

Solution: Total distance covered = 2 + 1 + 3 + 2.5 = 8.5 meters.

Problem 5. Biohybrid Robot Motion: A biohybrid robot with living muscle tissues contracts at a speed of 2 cm/s. If it contracts for 5 seconds, calculate the total distance the robot moves.

Solution: Speed = 2 cm/s, Time (t) = 5 seconds

Distance = Speed * Time = 2 cm/s * 5 s = 10 cm.

Problem 6. Sensor Range Calculation: You are designing a robotic fish with ultrasonic sensors for obstacle detection. The sensors can detect obstacles within a range of 3 meters. If the fish is navigating through a narrow passage, how close can it get to the walls while still detecting obstacles?

Solution: Sensor range = 3 meters.

The fish can get as close as half of the sensor range to the walls: Minimum distance = 3 meters/2 = 1.5 meters

Problem 7. Biomimetic Arm Length: You are designing a biomimetic robot arm that mimics the arm of a primate. If the primate's arm has a length of 60 cm and the robot arm is designed to be 80% of the primate's arm length, calculate the length of the robot arm.

Solution: Primate arm length = 60 cm, Robot arm length = 80% of 60 cm = 0.8 * 60 cm = 48 cm.

Problem 8. Exoskeleton Actuator Torque: An exoskeleton assisting a human's knee joint requires a torque of 30 Nm to aid in movement. If the lever arm (distance from the joint) is 0.1 meters, calculate the force needed to produce this torque.

Solution: Torque (τ) = 30 Nm, Lever arm (r) = 0.1 meters.

Force (F) = Torque/Lever arm = 30 Nm/0.1 m = 300 N.

Problem 9. Reinforcement Learning Iterations: A robot is learning to navigate a maze using reinforcement learning. It takes an average of 20 iterations to reach the goal. If the robot performs 5 iterations per minute, how long does it take to learn the optimal path?

Solution: Iterations to learn = 20, Iterations per minute = 5.

Time to learn = Iterations to learn/Iterations per minute = 20/5 = 4 minutes

Problem 10. Snake Robot Speed: A snake-like robot with 10 segments each of length 15 cm moves forward at a speed of 2 cm/s. Calculate the speed of the robot's head.

Solution: Number of segments = 10, Length of each segment = 15 cm, Speed of robot = 2 cm/s.

Speed of head = Speed of robot – (Number of segments – 1) * Length of each segment

Speed of head = 2 cm/s – (10 – 1) * 15 cm = –133 cm/s

The head is moving backward at a speed of 133 cm/s.

Problem 11. Biomimetic Wing Design: You are designing a robotic bird inspired by real birds' wing shapes. The wingspan of your robotic bird is 50 cm. If you want to replicate the aspect ratio (AR) of an eagle, which has an AR of 7.5, calculate the length and width of the wing.

Solution: Given: Wingspan of the robotic bird (b) = 50 cm, Aspect Ratio (AR) = 7.5.

Aspect Ratio (AR) = Wing Length (l)/Wing Width (w) $7.5 = l/w$

Solving for l : l = AR * w $l = 7.5 * w$

Now, we know that the wingspan (b) is equal to twice the wing length (l): $b = 2 * l$

Substitute the value of l from the AR equation: $50 = 2 * (7.5 * w)$ $w = 50/15 = 3.33$ cm

Now, calculate the length: $l = 7.5 * 3.33 = 24.975$ cm

So, the length of the wing is approximately 24.975 cm, and the width is approximately 3.33 cm.

Problem 12. Energy Consumption Comparison: Compare the energy consumption of a wheeled robot (Robot A) and a legged robot (Robot B) over a 1-hour period. Robot A requires 100 Watts of power and travels at a speed of 1 m/s. Robot B requires 120 Watts of power and travels at a speed of 0.8 m/s. Which robot consumes less energy per unit distance?

Solution: Energy consumption per unit distance (ECD) can be calculated using the formula: ECD = Power/Speed

For Robot A: Power = 100 Watts, Speed = 1 m/s ECD (Robot A)= 100/1 = 100 Ws/m (Watt-seconds per meter)

For Robot B: Power = 120 Watts, Speed = 0.8 m/s ECD (Robot B) = 120/0.8 = 150 Ws/m.

Since 100 Ws/m < 150 Ws/m, Robot A consumes less energy per unit distance compared to Robot B.

Problem 13. Snake-Like Robot Length Calculation: You are designing a snake-like robot with 10

segments, each segment having a length of 20 cm. Calculate the total length of the robot when all segments are connected end-to-end.

Solution: Given: Number of segments = 10, Length of each segment = 20 cm.

Total length = Number of segments * Length of each segment Total length

= 10 * 20 = 200 cm.

So, the total length of the snake-like robot is 200 cm.

Short Questions and Answers

Q.1. What is biologically-inspired robotics?

Ans. Biologically-inspired robotics involves designing robots that mimic the behaviors, structures, and mechanisms found in living organisms.

Q.2. What is the goal of biologically-inspired robotics?

Ans. The goal is to create robots with enhanced adaptability, efficiency, and functionality by drawing inspiration from biological systems.

Q.3. Give an example of locomotion inspired by nature.

Ans. Snake-like robots imitate the lateral undulation of snakes for navigating confined spaces.

Q.4. What's the purpose of biohybrid robots?

Ans. Biohybrid robots combine living tissues or cells with artificial components to achieve more lifelike and adaptive behaviors.

Q.5. How are drones inspired by birds used?

Ans. Drones inspired by bird flight can navigate complex environments and gather aerial data.

Q.6. What is soft robotics?

Ans. Soft robotics focuses on creating robots with flexible, deformable materials that mimic the compliance of biological tissues.

Q.7. How can swarm behavior be applied in robotics?

Ans. Swarm robots can work together in a coordinated manner, mimicking social insect behaviors, to perform tasks like exploration or environmental monitoring.

Q.8. What's the significance of tactile sensing in biologically-inspired robotics?

Ans. Tactile sensing allows robots to detect textures, pressure, and interactions with the environment, similar to how animals use touch.

Q.9. How do robots mimic echolocation?

Ans. Robots can emit and detect sound waves to navigate and sense objects underwater, similar to how bats and dolphins use echolocation.

Q.10. What is the role of control algorithms in biologically-inspired robotics?

Ans. Control algorithms replicate animal behaviors and enable robots to adjust their actions based on sensory inputs and objectives.

Q.11. How does bioethics relate to biologically-inspired robotics?

Ans. Bioethics considers the ethical implications of using living tissues or replicating animal behaviors in robots.

Q.12. What's the advantage of using modular components in robotic design?

Ans. Modular components allow robots to adapt to different tasks by rearranging or reconfiguring their parts.

Q.13. What's an example of morphological adaptation in robots?

Ans. Robots that can change shape or configuration to adapt to different tasks, similar to animals with adaptable bodies.

Q.14. How does reinforcement learning apply to biologically-inspired robotics? –

Ans. Reinforcement learning enables robots to learn from interactions with their environment and improve their behaviors over time.

Q.15. How are aerial robots inspired by insects used?

Ans. Aerial robots mimicking insect flight can navigate tight spaces and perform tasks like surveillance or exploration.

Q.16. What's the purpose of studying animal locomotion in robotics?

Ans. Studying animal locomotion helps engineers design robots that can navigate various terrains and environments effectively.

Q.17. What's the key benefit of biomimetic design in robotics?

Ans. Biomimetic design allows robots to replicate natural functionalities and excel in specific tasks, benefiting from evolutionary optimization.

Q.18. How can robots with soft actuators be advantageous?

Ans. Soft actuators allow robots to interact with delicate environments and objects, making them suitable for tasks like medical procedures.

Q.19. What is proprioception and why is it important in robotics?

Ans. Proprioception is the ability to sense the position and movement of one's body. Robots use this to adapt their movements and maintain stability.

Q.20. How do robots mimic insect behaviors like swarming?

Ans. Robots can be programmed to follow simple rules and interact with neighbors to mimic emergent swarm behaviors seen in insects.

Q.21. How does biomimicry contribute to sustainable design in robotics?

Ans. Biomimicry promotes the use of energy-efficient designs and materials inspired by nature, leading to more sustainable robots.

Q.22. What's the significance of using lightweight materials in biomimetic robotics? –

Ans. Lightweight materials help robots achieve energy-efficient movement and reduce their environmental impact.

Q.23. How does studying animal vision influence robotic camera systems?

Ans. Insights from animal vision systems inform the design of robotic cameras and computer vision algorithms for image processing.

Q.24. How are robots that mimic animal behaviors used in education?

Ans. Educational robots inspire curiosity and help students understand complex concepts in biology, engineering, and robotics.

Q.25. What's the advantage of creating robots with adaptive behaviors?

Ans. Robots with adaptive behaviors can respond to changing environments and tasks, making them versatile and useful in dynamic scenarios.

Q.26. How are robots inspired by fish used in underwater exploration?

Ans. Fish-inspired robots can swim efficiently in underwater environments, collecting data and exploring areas that are difficult for traditional vehicles to reach.

Q.27. How does the study of muscle and joint mechanics influence robotics?

Ans. Understanding muscle and joint mechanics informs the design of robots with lifelike movements and accurate control.

Q.28. How do robots mimic animal sensing mechanisms like touch and proprioception? –

Ans. Robots integrate tactile sensors and joint encoders to replicate touch and proprioception, allowing them to interact with the environment and adapt their motions.

Q.29. How are drones inspired by birds used for environmental monitoring?

Ans. Bird-inspired drones can fly through forests and other challenging environments to collect data on biodiversity and habitat health.

Q.30. What's the potential impact of biologically-inspired robots in space exploration? –

Ans. Biologically-inspired robots can navigate complex planetary terrains and gather data in environments where human presence is challenging.

Q.31. How do researchers address ethical concerns in developing biohybrid robots?

Ans. Researchers collaborate with ethicists and regulators to ensure that biohybrid robots involving living tissues are developed responsibly and ethically.

Q.32. What's the significance of studying animal behaviors for robotics?

Ans. Animal behaviors provide insights into efficient locomotion, sensing, and adaptation strategies that can be replicated in robots.

Q.33. How can robots inspired by birds be useful in agriculture?

Ans. Bird-inspired robots can assist in tasks like pest control or crop monitoring, improving efficiency and reducing the need for chemical interventions.

Q.34. What's the role of kinematics in designing biomimetic robotic movements?

Ans. Kinematics helps model the motion and orientation of robot components, ensuring accurate replication of animal movements.

Q.35. How are robots with adaptive control used in industrial settings?

Ans. Adaptive control allows robots to optimize their movements based on changing tasks and environments, enhancing efficiency in manufacturing processes.

Q.36. How does the concept of swarm intelligence apply to robotic systems?

Ans. Swarm intelligence involves designing robots that work collectively, sharing information and tasks, similar to how social insects cooperate.

Q.37. What's the advantage of mimicking animal locomotion in underwater robots?

Ans. Underwater robots inspired by fish and marine animals can achieve efficient swimming and maneuvering in underwater environments.

Q.38. How are robots inspired by insects used for agricultural tasks?

Ans. Insect-inspired robots can perform tasks like pollination, crop monitoring, and pest control, enhancing agricultural productivity.

Q.39. How do robots with soft gripping mechanisms mimic animal manipulation abilities?

Ans. Soft grippers imitate the delicate manipulation skills of animals like octopuses, allowing robots to handle fragile objects without damage.

Q.40. How can robots mimic animal behaviors in performing complex tasks like assembly?

Ans. Robots can replicate the coordinated movements and interactions seen in animal swarms to optimize assembly processes in manufacturing.

Q.41. How does the study of biomimicry encourage sustainable technology development?

Ans. Biomimicry promotes designs that minimize energy consumption, waste generation, and resource usage, leading to more sustainable robots.

Q.42. What's the significance of studying insect flight for aerial drones?

Ans. Insect flight mechanisms provide insights into agile and energy-efficient flight, which can be applied to the design of drones for various purposes.

Q.43. How do robots inspired by animals adapt to changing environments?

Ans. Robots use adaptive control algorithms to process sensory inputs and adjust their actions, ensuring optimal performance in varying conditions.

Q.44. How are robots inspired by animals used in scientific research?

Ans. Robots replicate animal behaviors and physiology to study animal cognition, biomechanics, and ecological interactions in controlled environments.

Q.45. What's the potential of soft robots in medical applications?

Ans. Soft robots can interact safely with human tissues and organs, making them suitable for tasks like minimally invasive surgeries and rehabilitation.

Q.46. How does studying animal sensory systems influence robotic perception?

Ans. Insights from animal sensory systems inform the development of sensors and algorithms that enable robots to perceive their environment.

Q.47. How do robots imitate animals for exploration of extreme environments?

Ans. Robots can replicate behaviors of extremophile organisms to explore harsh environments like deep-sea hydrothermal vents or polar regions.

Q.48. How can robots mimic animal behaviors for environmental conservation?

Ans. Robots can replicate behaviors of animals in fragile ecosystems to study wildlife behavior, monitor habitats, and contribute to conservation efforts.

Q.49. What's the potential of robotic prosthetics inspired by human movement?

Ans. Robotic prosthetics that mimic natural joint movements and muscle contractions can enhance the mobility and quality of life for amputees.

Q.50. How do swarm robots communicate and coordinate their actions?

Ans. Swarm robots use local communication and simple rules to interact with neighboring robots, creating emergent behaviors and coordinated actions.

Multiple Choice Questions and Answers

Q.1. What is biologically-inspired robotics?

(*a*) Robots that mimic biological systems
(*b*) Robots made from biological materials
(*c*) Robots that study biology
(*d*) Robots with advanced technology

Q.2. What is the primary goal of biologically-inspired robotics?

(*a*) To replicate animal behavior exactly
(*b*) To create robots with unique behaviors
(*c*) To design robots that outperform animals
(*d*) To achieve adaptability and efficiency

Q.3. Which of the following is an example of biomimetic design?

(*a*) Creating robots without sensors
(*b*) Designing robots with wheels only
(*c*) Replicating animal behaviors in robots
(*d*) Developing robots without control algorithms

Q.4. What is the advantage of soft robotics?

(*a*) Hard robotics is outdated
(*b*) Soft robots can fly longer distances
(*c*) Soft robots can interact with delicate environments
(*d*) Soft robots are more expensive

Q.5. How do robots with adaptive behaviors respond to changing environments?

(*a*) They shut down automatically
(*b*) They remain stationary
(*c*) They adjust their actions based on sensory inputs
(*d*) They always follow a predetermined path

Q.6. Which field contributes to the ethical considerations in biologically-inspired robotics?

(*a*) Material science
(*b*) Engineering
(*c*) Computer science
(*d*) Bioethics

Q.7. What is the purpose of studying animal locomotion in robotics?

(*a*) To prove robots are better than animals
(*b*) To mimic animal movements exactly
(*c*) To create robots that can navigate various terrains
(*d*) To replace animals with robots in the wild

Q.8. Which area of robotics focuses on replicating animal vision systems?

(*a*) Soft robotics
(*b*) Control algorithms
(*c*) Computer vision
(*d*) Kinematics

Q.9. How do swarm robots achieve coordinated behaviors?

(*a*) By following a central leader
(*b*) By using global positioning systems (GPS)
(*c*) By local communication and simple rules
(*d*) By moving randomly

Q.10. How does reinforcement learning apply to biologically inspired robotics?

(*a*) Robots are rewarded with candy (*b*) Robots learn by watching TV
(*c*) Robots learn from interactions with their environment
(*d*) Robots follow predetermined paths

Q.11. What is the significance of studying animal behaviors for robotics?
(*a*) It provides entertainment for researchers (*b*) It helps in creating robotic animals
(*c*) It informs efficient locomotion and adaptation strategies
(*d*) It replaces the need for sensors in robots

Q.12. What is proprioception in robotics?
(*a*) The ability to mimic animal behaviors (*b*) The sense of taste in robots
(*c*) The ability to sense position and movement of the robot's body
(*d*) The ability to communicate with other robots

Q.13. How do biohybrid robots combine living tissues and artificial components?
(*a*) By creating robots that can reproduce (*b*) By using only living tissues
(*c*) By integrating biology textbooks with robots
(*d*) By combining engineered materials and living cells

Q.14. What is the role of material science in biologically-inspired robotics?
(*a*) To study animal behaviors (*b*) To create robots with hard materials only
(*c*) To develop lightweight and flexible materials for robots
(*d*) To create robots without any materials

Q.15. What is the significance of swarm behavior in robotics?
(*a*) It allows robots to work individually (*b*) It mimics solitary animal behaviors
(*c*) It enables robots to work together in a coordinated manner
(*d*) It prevents robots from communicating with each other

Q.16. How do robots mimic echolocation?
(*a*) By emitting light (*b*) By using GPS
(*c*) By emitting and detecting sound waves (*d*) By using infrared sensors

Q.17. What's the primary goal of studying animal sensory systems in robotics?
(*a*) To replace animal senses with robotic sensors
(*b*) To create robots with extra senses (*c*) To develop advanced computers
(*d*) To design robots that can perceive and interact with their environments

Q.18. How do robots with adaptive control adjust their actions?
(*a*) By shutting down (*b*) By following a predetermined path
(*c*) By processing sensory inputs and feedback (*d*) By moving randomly

Q.19. What's the potential of robots inspired by animals in space exploration?
(*a*) They can replace astronauts (*b*) They can travel faster than light
(*c*) They can navigate planetary terrains and gather data
(*d*) They can communicate with extraterrestrial life

Q.20. How do robots with modular components enhance adaptability?

(*a*) They can change their appearance (*b*) They can change their programming
(*c*) They can rearrange or reconfigure parts for different tasks
(*d*) They can transform into different types of animals

Key to Multiple Choice Questions

1. (*a*)	**2.** (*d*)	**3.** (*c*)	**4.** (*c*)	**5.** (*c*)	**6.** (*d*)	**7.** (*c*)	**8.** (*c*)	**9.** (*c*)
10. (*c*)	**11.** (*c*)	**12.** (*c*)	**13.** (*d*)	**14.** (*c*)	**15.** (*c*)	**16.** (*c*)	**17.** (*d*)	**18.** (*c*)
19. (*c*)	**20.** (*c*)							

Projects Works

1. **Biohybrid Muscle-Powered Robot:** Create a robot that uses living muscle tissues (biohybrid) to generate movement and demonstrate lifelike locomotion.

   Hints: Design a robot with skeletal muscle cells cultured on flexible scaffolds. Apply electrical stimulation to induce muscle contractions, resulting in robotic movement.

2. **Insect-Inspired Swarm Robotics:** Develop a swarm of small robots that mimic insect behaviors like foraging, navigation, and cooperative tasks.

   Hints: Design robots with simple algorithms for local communication and task sharing. Test the swarm's ability to collectively perform tasks like exploration or object retrieval.

3. **Biomimetic Fish Robot for Underwater Exploration:** Design a robotic fish that replicates the swimming patterns of real fish to explore underwater environments.

   Hints: Develop a fish-inspired robotic tail and fin mechanism for propulsion. Implement sensors to mimic fish sensory behaviors, allowing the robot to navigate underwater.

4. **Soft Robotic Gripper Inspired by Octopus Tentacles:** Create a soft robotic gripper that imitates the dexterity and flexibility of octopus tentacles for delicate object manipulation.

   Hints: Design a gripper with flexible materials and multiple degrees of freedom. Utilize pneumatics to control the gripper's movements, enabling it to grasp objects of various shapes.

5. **Bird-Inspired Aerial Surveillance Drone:** Build a drone that replicates bird flight to achieve agile and efficient aerial surveillance capabilities.

   Hints: Develop a drone with flapping wings or wing morphing mechanisms to mimic bird flight. Integrate cameras for visual surveillance and control algorithms for stable flight.

6. **Gecko-Inspired Climbing Robot:** Create a climbing robot that mimics the adhesive capabilities of gecko feet to navigate vertical surfaces.

   Hints: Design a robot with specialized feet featuring microstructured adhesive surfaces. Explore van der Waals forces or other adhesion mechanisms for wall-climbing abilities.

7. **Robot-Inspired Biohybrid Plant Interaction:** Develop a robot that interacts with plants by mimicking the movements of insects for pollination or plant care.

   Hints: Design a robot with appendages that replicate insect motions. Equip it with sensors to detect plant stimuli and enable interactions like pollination or pest removal.

8. **Bat-Inspired Sonar Navigation System:** Create a navigation system for a robot that mimics the echolocation abilities of bats for obstacle detection and avoidance.

Hints: Develop an acoustic sensor system capable of emitting and receiving sound waves. Implement signal processing algorithms to interpret the echoes and enable navigation.

9. **Biomechanics-Inspired Robotic Prosthetics:** Design a prosthetic limb that replicates human biomechanics and movement for amputees.

   Hints: Create a prosthetic limb with joint mechanisms that imitate natural human movement. Develop control algorithms that respond to residual muscle signals or external sensors.

10. **Turtle-Inspired Autonomous Underwater Vehicle (AUV):** Build an AUV that emulates the swimming and navigation behaviors of sea turtles for ocean exploration.

    Hints: Design a streamlined AUV with a flapping fin mechanism to replicate turtle swimming. Integrate sensors and navigation algorithms to enable autonomous exploration.

10

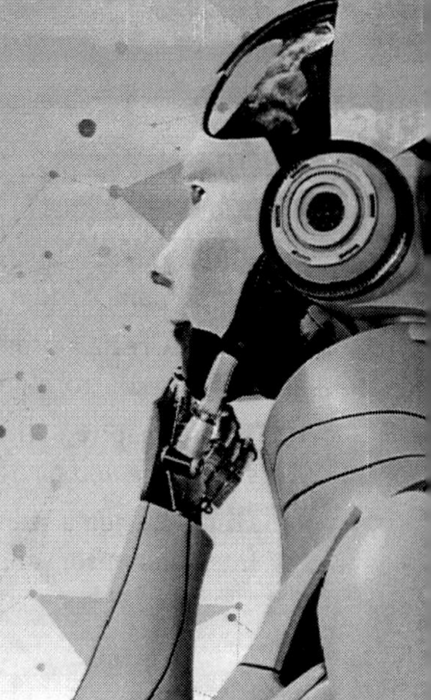

Robotics in Entertainment and Service Industries

10.1 INTRODUCTION

Robotics has made great progress in transforming the leisure and service sectors. In order to improve and automate a variety of duties, activities, and experiences within these industries, this technological breakthrough involves the design, creation, and use of robots. Robotics offers a wide range of uses that increase productivity, novelty, and patron engagement in theme parks, theatres, restaurants, hotels, and even personal settings. Robotics have given rise to previously futuristic types of entertainment in the entertainment sector. Animated characters interact with tourists in theme parks and other attractions, creating engaging and immersive experiences. The distinction between reality and fiction is muddled by these robots' ability to mimic human movement, voice, and gestures. The development of robotic actors, musicians, and entertainers has also pushed the frontiers of what is possible on stage and screen. Robots have been incorporated into a variety of positions in the service sector to improve workflow and client experience. Robots are used in restaurants and cafés as servers or bartenders to improve the eating experience through automated interactions and accurate order delivery. To streamline guest services and reduce human resource restrictions, hotels deploy robots for jobs including room service delivery, concierge services, and housekeeping.

Through smart companions and technology, robots are also entering our personal environments. Routine jobs like vacuuming, mopping, and even keeping an eye on home security can be handled by domestic robots. More and more social robots are being developed for emotional support and companionship, particularly in medical settings where they might help with patient care and rehabilitation. A variety of technologies, such as artificial intelligence, machine learning, computer vision, and sensor technologies, are driving the integration of robotics into various industries. These developments increase the utility and adaptability of robots by enabling them to sense their surroundings, make wise judgments, and adapt to changing settings. While incorporating robotics into the entertainment and service sectors has many advantages, there are drawbacks as well. Some of the issues that need to be properly considered include controlling potential job displacement, managing ethical concerns, and ensuring the safety of both humans and robots. Robotics is revolutionizing the entertainment and service sectors by bringing new approaches to audience engagement, improving customer experiences, and streamlining operations. The convergence of robots and these industries will probably result in even more stunning and creative applications as technology develops, influencing how entertainment and service interactions are conducted in the future.

10.2 USE OF ROBOTS IN THE ENTERTAINMENT SECTOR

Over time, the employment of robots in the entertainment industry has changed tremendously, opening up new possibilities for innovation, interactivity, and immersion. Robotics has improved a number of entertainment mediums and captured audiences in fresh ways, from stage performers to mechanical characters at amusement parks. Here are some important fields in which robots are having an impact.

10.2.1 Theme Parks and Attractions

Theme parks and attractions are one of the earliest and most well-known uses of robots in entertainment. Animatronic characters engage with guests and provide remarkable encounters by bringing fantastical realms to life. The distinction between fiction and reality is effectively blurred by these robots' ability to mimic human-like gestures, attitudes, and movements. Animatronics, which range from lifelike pirates to talking dinosaurs, are now a crucial part of theme parks' immersive ambiance. Robots are used in theme parks and attractions to show how technology and mathematics can be combined to produce engaging and immersive entertainment.

Let's explore how these elements come together, along with examples:

1. **Animatronics:** Lifelike robotic characters that carry out predetermined movements, behaviors, and interactions are known as animatronics. They significantly rely on mathematics, computer programming, and mechanical engineering to coordinate movements and produce believable behaviors. Animated creatures move naturally and realistically thanks to the use of servos, actuators, sensors, and exact timing algorithms.

   An illustration of this is the "Pirates of the Caribbean" attraction at Disneyland, which involves animatronics pirates, mermaids, and other characters in a daring adventure. Through complex mechanical designs and programming, these robots replicate activities like rowing, drinking, and swordplay.

2. **Robotic Ride Systems:** To create engaging and dynamic experiences, theme park attractions frequently use robotics. These rides which range from interactive roller coasters to motion simulators, use robotics, motion control, and physics to produce exciting and enjoyable attractions.

   For instance, the "Transformers: The Ride 3D" at Universal Studios immerses visitors in a battle between the Autobots and Deceptions using motion simulators and 3D images. The robotic movements are precisely timed to match the action on screen, which heightens the sense of immersion in the Transformers world.

3. **Interactive Characters:** This can react to visitors' actions, gestures, and even voice are made possible by sophisticated robotics and sensors. To read and react to a variety of stimuli, these characters need complex programming and mathematics.

   As an illustration, Disney's "Talking Mickey Mouse" is an interactive character that converses with park visitors using speech synthesis and facial recognition technologies. The robot uses sophisticated picture recognition and natural language processing algorithms to track and respond to specific visitors.

4. **Intricate Control Systems:** Intricate show control systems are used by theme parks behind the scenes to coordinate the activities and motions of numerous robots and other elements. To produce seamless performances, these systems use programming, logic design, and mathematical algorithms.

   A complicated choreography of animatronics, fountains, lights, and pyrotechnics is used in the nightly displays at theme parks like Disney's "Fantasmic!" The timing of each component is carefully controlled by the show control system, resulting in an enthralling display for the audience.

5. **Physical Interactions:** Robots can be included into virtual reality (VR) and augmented reality (AR) experiences to give physical interactions and feedback, boosting immersion. Robotics, computer vision, and spatial tracking technologies are all combined in these experiences.

   Using "Star Wars: Secrets of the Empire" by the VOID as an illustration, users interact with real-world objects that are synchronized with the virtual environment while wearing VR headsets and hap tic vests. Participants can physically interact with virtual features thanks to robots and sensors, which heighten the sense of immersion in the "Star Wars" world.

In some instances, the robotic elements of theme park attractions are designed, engineered, and orchestrated with the use of technology and mathematics. The fusion of technology and mathematics yields the engaging and immersive experiences that visitors to theme parks and attractions enjoy, from generating lifelike movements to synchronizing actions and replies.

10.2.2 Robotic Performers

Additionally, robotics has impacted the performing arts. Robots are being created to imitate human performers, musicians, and dancers. They are able to do intricate dance routines, play musical instruments with amazing dexterity, and even take part in theatrical productions. These robotic performers broaden the range of what can be accomplished on stage by fusing technology and creativity to produce one-of-a-kind displays. A fascinating intersection of technology, mathematics, and creative artistic expression may be seen in robotic performers in the entertainment industry. These robots are created to mimic human actions, gestures, and facial expressions, and in order to function convincingly and realistically, they need to combine engineering, coding, and mathematics principles.

Let's delve into how technology and mathematics contribute to robotic performers, along with examples:

1. **Biomechanics and Kinematics:** Robotic actors must accurately imitate human movements. Engineers may create robots that move realistically and naturally by using biomechanics concepts to understand how human joints and muscles function. The movements of the robots are modeled and managed using kinematics, a field of mathematics.

   The humanoid robot "Atlas" from Boston Dynamics performs intricate movements including walking, jumping, and even back flips. Engineers have developed a robot that can perform dynamic actions with astounding dexterity by utilizing biomechanical and kinematic concepts.

2. **Motion Capture and Control:** Using motion capture technology, human movements are captured and converted into digital information. Robots are then programmed to imitate those movements using this data. Modern control algorithms guarantee that the robot moves smoothly.

   For instance, Disney's "Stuntronics" are robotic stunt artists who do accurate airborne manoeuvres using motion capture data. These machines can mimic the moves of human acrobats by swinging, flipping, and flying through the air.

3. **Human-Robot Communication:** Human communication and psychology principles are needed to produce convincing interactions between robotic actors and live audiences. Robots can react to cues from human audiences thanks to technology like natural language processing and facial recognition.

   For instance, Hanson Robotics' "Sophia" is a social humanoid robot that interacts and converses with people. Sophia provides a realistic and engaging presence by fusing AI, facial recognition, and expressive motions.
4. **Expressive Robotics:** Robotic actors need devices that can imitate facial expressions, hand movements, and emotional states. This calls for complex technical layouts and emotional-translating control systems.

   A humanoid robot named "RoboThespian" is an example of this. It is capable of expressing emotions, giving monologues, and even delivering Shakespearean soliloquies. It uses a combination of mechanical parts and preprogrammed responses to generate its expressive face and body movements.
5. **Computer Vision and Sensing:** Robotic performers frequently employ computer vision to interact with items and people as well as to sense their surroundings. Robots are equipped with sensing systems that allow them to navigate and react appropriately, such as depth sensors and cameras.

   As an illustration, human dancers and the robot "KUKA LBR iiwa" have collaborated in theatrical performances. It makes real-time adjustments to its own motions to stay in tune with the dancers' movements using vision systems.
6. **Machine Learning and AI:** Artificial intelligence and machine learning help robotic performers adapt. Robots can learn from interactions and develop their performances over time thanks to these technologies.

   As an illustration, the ABB Robotics robot "YuMi" has led orchestras and even performed with renowned artists. Because of its machine learning capabilities, it can adjust to various conductors' approaches and stay in time with the orchestra.

In these instances, the fusion of technology and mathematics is crucial in the development of robotic entertainers that astound and delight audiences. Robotic performers demonstrate the wonderful symbiosis between engineering, mathematics, and creative innovation by mimicking human movements and engaging with people in lifelike ways.

10.2.3 Virtual Reality (VR) and Augmented Reality (AR) Experiences

In conjunction with virtual and augmented reality technologies, robots are being integrated to enhance immersive experiences. Robots can serve as physical extensions of virtual characters, providing hap tic feedback or responding to user interactions. This fusion of robotics with VR and AR creates a multi-sensory experience that engages audiences on a deeper level. The use of robots in conjunction with Virtual Reality (VR) and Augmented Reality (AR) experiences adds a layer of physical interaction and immersion to virtual worlds. This integration involves a blend of technology, mathematics, and creative design to create captivating and interactive entertainment.

Let's explore how these elements come together, along with examples:

1. **Telepresence Robots:** Remotely controlled robotic platforms featuring cameras, screens, and sensors are known as telepresence robots. They improve remote participation and collaboration by allowing users to digitally tour physical areas and interact with the environment.

For instance, "Double Robotics" sells telepresence robots that let customers virtually attend events and travel. Users can engage with people and situations in real-time, independent of where they are physically located, by controlling the robot's motions.

2. **Haptic Feedback Devices:** Users may "feel" virtual objects and environments thanks to haptic devices, which give them tactile feedback. To imitate sensations like texture, pressure, and vibration, this technology uses mechanics, sensors, and mathematics.

   As an illustration, the "HaptX Gloves" replicate touch in virtual environments via haptic feedback. Users can interact with and feel virtual things thanks to the sensors and actuators integrated into these gloves, which simulate the feeling of touch.

3. **Robotic Interfaces for VR/AR:** VR/AR experiences can be improved by using robots as interfaces. The distance between the digital and physical worlds can be closed by these gadgets, which can imitate actual interactions with virtual items and settings.

   As an illustration, Microsoft's "HoloLens" augmented reality headset enables users to interact with holographic objects in their immediate surroundings. Users may easily control and interact with virtual items by combining hand motions and voice instructions.

   Robotic performers for mixed reality experiences can be incorporated into AR experiences to combine real-world and virtual components. These robots can interact with both real-world and virtual characters and objects and respond to user input.

   Example: The augmented reality (AR) gaming robot "MekaMon" communicates with virtual animals in the real world. Through a mobile app, users may direct the robot's motions while participating in challenges and conflicts that combine real-world action with virtual game play.

5. **VR-Assisted Robot Control and Training:** Before performing activities in the real world, robots can be programmed and controlled in virtual surroundings. To assure exact replication of robot behavior, this method uses programming, physical simulations, and mathematical modeling.

   As an illustration, "Shadow Robot Company" trains operators to operate sophisticated robotic systems using VR. In order to prepare for jobs in real-world settings, users can practice controlling robot arms in a virtual environment.

   Robot-Assisted VR/AR Experiences: Robots can be used as physical companions or guides in VR/AR experiences, increasing user immersion and engagement.

   For instance, "Buddy" is a social robot created to engage kids and offer educational opportunities. Buddy may lead kids through interactive learning exercises via augmented reality applications, encouraging a mix of real-world and digital involvement.

In these instances, the incorporation of robotics with VR and AR experiences demonstrates the synergy of technology, mathematics, and creativity. Robotics is used to create new and interesting types of entertainment by bringing concrete interactions, real-world connections, and a sense of presence to virtual and augmented environments.

10.2.4 Interactive Exhibits and Installations

Interactive robotic installations are being used in museums, galleries, and exhibitions to engage people in aesthetic and educational experiences. People can engage with robots in these installations to learn about science, history, and technology in an interesting and entertaining way. These installations frequently encourage hands-on investigation. Visitors can engage in entertaining and instructive experiences thanks to interactive displays and installations that use robots in the entertainment industry. To produce interactive and immersive experiences, these installations frequently integrate technology, mathematics, and creative design.

Let's explore how these elements converge, along with examples:

1. **Kinetic Sculptures:** These dynamic, moving works of art are made using robotic mechanisms. In order to create their symmetrical and hypnotic movements, these artworks use technical concepts, mechanical design, and mathematical computations.

   Theo Jansen, for instance, created a collection of wind-driven kinetic sculptures called "Strandbeests" that imitate the gait of animals. These sculptures can move smoothly across beaches thanks to intricate mechanical systems and links.

2. **Robotic Art Installations:** In interactive art compositions, robots are incorporated into robotic art installations. These installations combine artistic vision, programming, and sensors to produce mesmerizing and frequently interactive visual experiences.

   Example: Kollision's "Umbrella Swarm" installation has robotic umbrellas that react to the weather by opening and closing in unison. The umbrellas move in unison, producing a dynamic and visually arresting spectacle.

3. **Educational Robots and Workshops:** Through interactive exhibitions' hands-on activities and workshops, visitors may learn about robotics and technology. These exhibits engage visitors in learning experiences through educational content, coding challenges, and mathematical ideas.

   For instance, visitors at "LEGO Mind storms" exhibits can assemble and program robots from LEGO parts. These exhibitions provide visitors with a fun and engaging introduction to robotics, engineering, and programming.

4. **Interactive Exhibits:** Interactive exhibits may include robotic games and puzzles that test visitors' problem-solving abilities. To win these games, you need to be able to reason logically, use algorithms, and grasp math.

   A robot named "CubeStormer II" can solve Rubik's Cubes in a flash. Visitors may observe the robot's deft motions as it applies algorithms to analyze and resolve the issues.

5. **Artwork Produced by Robots:** Robots can be programmed to produce visual art, frequently with the assistance of human collaborators. For distinctive and unforeseen outcomes, these installations combine technology with creative expression and mathematical algorithms.

   Example: The robot artist "e-David" uses several painting techniques and image analysis to produce unique works of art. Based on how it perceives the composition of the piece of art, the robot dynamically modifies its brushstrokes.

   Robots can perform in interactive performances that are dynamic and responsive. Interactive installations can include robots as performers. In order to engage audiences, these performances use sensors, algorithms, and in-person interactions.

   Examples are the robotic swings in "Pendulum Choir" which move in time with the music. The installation gives viewers a multi-sensory experience by fusing motion control with timed choreography.

In these instances, the design and implementation of interactive exhibits and installations utilizing robots demonstrate the partnership of technology, mathematics, and artistic creativity. These exhibits provide visitors the chance to interact with technology, investigate mathematical ideas, and enjoy entertainment in fresh and engaging ways.

10.2.5 Robot-Assisted Live Shows

Some live performances include robots as essential elements of the presentation. For instance, drones are employed to produce complex aerial displays that coordinate with musical performances and light displays, bringing a dynamic visual component to festivals and concerts. Live performances with robot assistance are an example of how technology, mathematics, and creative performance are combined in the entertainment industry. Robots frequently appear on these performances alongside human performers as dynamic performers, team members, or even co-stars. These shows' technological and mathematical underpinnings allow for seamless coordination, exact movements, and interactive interactions.

Let's explore how these elements come together, along with examples:

1. Robot-assisted live performances necessitate exact synchronization between robots and human performers. Programming, algorithms, and mathematical timing are all used in this coordination to make sure that movements and activities are smooth and on-time.

   In theatrical productions, the "RoboThespian" robot has been utilized to interact with live performers. The robot interacts with human co-stars in dynamic situations created by its coordinated movements and written conversation.

2. **Robotics and Dance Collaboration:** Algorithms that map robotic movements to the rhythm and choreography of the performance are used to incorporate robots into dance performances. To ensure synchronization, this collaboration needs programming and mathematical calculations.

   In the "DanceDraw" initiative, human dancers and a robot work together to produce live artwork. A dynamic performance that combines dance and visual art is produced as a result of the robot's movements being coordinated with the dancers' actions.

3. **Music and Robotic Instruments:** In live performances, robots can be programmed to play musical instruments. To understand musical notation, manage instruments, and sync with other performers, these robots need algorithms.

   An illustration of this is the "Robot Orchestra" initiative, which shows robots performing various musical instruments. Orchestral performances are made possible by the programming of each robot to read music notation, control its instrument, and respond to conductor signals.

4. **Interactive Audience Engagement:** Shows with robot assistance can include audience engagements in which the robots react in real time to audience inputs. Sensors, algorithms, and mathematical models are used in this interaction to interpret and react to audience inputs.

   A live performance called "Puppet Parade" features robotic puppets interacting with the audience. The audience's motions and the movements of the robots are coordinated by puppeteers to provide a fun and interactive experience.

5. **Dynamic Visual and Lighting Effects:** Live performances can include robotic components in their visual and lighting presentations. These components are frequently designed to change and move in time with the performance, creating a lively visual aspect.

   Example: To generate elaborate visual effects that improve the overall experience, robotic arms outfitted with LED lights are employed during concerts and musical performances. These robotic arms move in carefully orchestrated patterns that match the music.

6. Interactive storytelling with robots as characters is a possibility in robot-assisted live performances. These robots react to audience input and add to the story with pre-written dialogue and actions.

An interactive and immersive narrative experience is produced when the "RoboThespian" is employed in live storytelling performances where it engages with the audience and performs tale sequences.

In these instances, the incorporation of robots into live entertainment performances exemplifies the blending of technology, mathematics, and creative expression. Robots can collaborate, perform, and interact with audiences in creative and compelling ways because to the programming, algorithms, and mathematical computations those underlie these presentations.

10.2.6 Robotic Characters in Film and TV

Robots now play both production and character roles in films and television shows. These technologies are crucial in producing aesthetically attractive and engrossing on-screen experiences, from famous characters like R2-D2 and C-3PO in "Star Wars" to cutting-edge visual effects made possible by robotics. Robotic characters in films and television show how technology, mathematics, and imaginative storytelling can all coexist. These characters contribute to immersive storylines and present distinctive viewpoints on the interaction between humans and robots. They are frequently brought to life by a combination of robotics, visual effects, and acting.

Let's explore how these elements come together, along with examples:

1. **Robotics and Animatronics:** To produce convincing motions and expressions, robotic characters in movies and television shows frequently use robotics and animatronics. In order to imitate human-like actions, these technologies use mechanical engineering, programming, and mathematical calculations.

   As an illustration, in the film "Jurassic Park," mechanical dinosaurs were used to generate believable and dramatic interactions with the human protagonists. These mechanical animals had intricate replicating systems, which gave them a central role in the story.

2. **Visual Effects and CGI:** Using visual effects and computer-generated imagery (CGI), robotic characters can be smoothly incorporated into real-world settings and performers. These methods make use of sophisticated computing tools, math, and physics simulations.

   As an illustration, the Marvel Cinematic Universe's "Iron Man" is created using a blend of practical effects, CGI, and motion capture. Through the use of technology and special effects, his robotic outfit seems to be an extension of the actor's body.

3. **Motion Capture and Performance Capture:** These technologies capture the gestures and facial expressions of actors, which are subsequently used to create robotic characters. For these technologies to accurately translate human motions onto robotic forms, mathematical algorithms are needed.

   In the movie "Avatar," for instance, the alien Na'vi characters were produced via performance capture, in which the expressive motions of the performers were recorded and then converted into the movements of the characters.

4. **Emotional Expression and AI:** Through programmed behaviors and artificial intelligence, robotic figures can exhibit emotions and personalities. These characters need algorithms to control how they react to certain circumstances and engage with other characters.

   An illustration of this is the humanoid robot "C-3PO" from the "Star Wars" series, who expresses a variety of emotions through his interactions and reactions. The demeanour of the character is memorable in part due to his AI-driven reactions.

5. **Human-Robot Relationships in Storytelling:** Robotic characters frequently examine the dynamics of human-robot relationships, posing issues of empathy, identity, and morality. These stories combine storytelling with emotion, technology, and philosophy.

   Example: The film "Ex Machina" delves into the relationship between a human and an advanced AI robot. The story explores themes of consciousness, manipulation, and trust through intricate character interactions.

6. **Transhumanism and Futuristic Themes:** These can be embodied by robotic figures, obfuscating the distinction between humans and machines. These stories look at how technology might alter society and the way people live.

   In the TV show "Westworld," robotic hosts interact with human visitors in a theme park. The program examines moral conundrums, consciousness, and the effects of developing sentient robots.

These examples show how technology, arithmetic, and imaginative storytelling interact to create robotic characters for cinema and television. These people enrich immersive storytelling, spark stimulating debates, and demonstrate how technology has the power to alter the entertainment environment.

10.2.7 Robot Competitions and Events

Robotics contests and events, including drone racing, sumo wrestling, and robot battles, have gained popularity as forms of entertainment. These competitions, where robots display their skills and cutting-edge designs, engage crowds because of their excitement and competitiveness. Robotics contests and events are examples of how technology, arithmetic, and a competitive spirit can coexist in the entertainment industry. These competitions offer a stage for demonstrating engineering creativity, problem-solving abilities, and robotics advancements.

Let's explore how these elements come together, along with examples:

1. **Robot Wars and Battle Competitions:** Creating robots for robot wars and battle competitions involves competing in combat situations. To build sturdy and powerful fighting devices, participants use engineering principles, mechanical design, and mathematical computations.

   For instance, teams of engineers and hobbyists construct combat robots for the TV program "BattleBots" so they can engage in arena warfare. For these robots to endure impact, damage, and tactical manoeuvring, meticulous design is necessary.

2. **Sumo Robot Contests:** In sumo robot contests, competitors are required to construct robots that can expel rivals from a sumo ring. Engineers use mechanics, sensor technologies, and mathematical techniques to build robots that can recognize adversaries and execute manoeuvres.

   As an illustration, the sumo robot competition on "RoboGames" (formerly known as RoboGames and the ROBOlympics) requires robots to independently navigate the sumo ring and push their opponents outside of the ring's limits.

3. **Robot Soccer Competitions:** In robot soccer competitions, robots that can play soccer autonomously or remotely compete. In order to construct robots capable of moving around the field, identifying the ball, and making tactical judgements, participants use arithmetic, programming, and sensor technology.

   The international robotics competition "RoboCup" features a number of leagues, including robot football. Through competitive events, RoboCup hopes to promote robotics and artificial intelligence research.

4. **Drone Racing Contests:** In drone racing contests, competitors must manoeuvre drones quickly via complicated courses. To guide the drones precisely and effectively, participants use control algorithms, mathematics, and physics.

   For instance, the "Drone Racing League" (DRL) sponsors expert drone racing tournaments in which pilots control drones through challenging and aesthetically pleasing courses. The accuracy necessary for effective racing is aided by the application of maths and technology.

5. **Autonomous Vehicle Challenges:** Designing cars that can follow predetermined routes without human intervention is an autonomous vehicle problem. To build autonomous vehicles that can sense their surroundings and make judgments, participants use sensor fusion, machine learning, and mathematical models.

   Examples include the "DARPA Grand Challenge" and "Roborace" competitions, which put teams under pressure to create autonomous cars that can successfully navigate difficult courses without the assistance of humans. The developments in self-driving technology are on display at these events.

6. **Robotic Creativity and Art Competitions:** Robot competitions aren't just about fights and racing; certain occasions highlight the artistic and creative uses of robotics. Participants build robots that make original pieces of art by fusing technology, mathematics, and aesthetic vision.

   Example: In the "RobotArt" competition, robots use a variety of tools and techniques to produce original pieces of art. Robots that can paint, sketch, and produce visual art in a variety of genres are controlled by participants.

These instances show how technology, mathematics, and a desire to compete are all integrated into the world of robot competitions and events. These activities not only amuse spectators but also increase engineering, problem-solving, and robotics knowledge while showing robotics' potential in a variety of contexts.

10.2.8 Robot-Assisted Sports

Some sports have started including robots as players or participants, however this trend is still relatively new. Robot football contests, for instance, pit teams of robots against one another in games, showing the sports applications of robotics and AI. A interesting nexus of technology, mathematics, and competitive enjoyment can be found in robot-assisted sports. These sports produce dynamic and interesting competitions by fusing engineering creativity, programming know-how, and mathematical techniques.

Let's explore how these elements come together, along with examples:

1. **Robot Soccer:** Creating autonomous or remotely controlled robots that can play soccer is known as robot soccer. Participants in this sport must use math, sensor technology, and programming to navigate the field, identify the ball, and plan their moves.

   Example: As part of its larger goal to enhance robotics and AI research, "RoboCup" holds a robot football competition. Teams from all over the world compete to create robots that can use coordinated motions and decision-making algorithms to play football.

2. **Drone Racing:** Pilots must manoeuvre drones through complex, time-sensitive circuits in drone racing. To accurately pilot the drones and manoeuvre around challenging obstacles, competitors use on mathematics, physics, and control algorithms.

   An such is the "Drone Racing League" (DRL), which hosts expert drone racing contests in which competitors pilot drones through difficult courses in breathtaking scenery. Success depends on precise piloting and control techniques.

3. **Robot Sumo Wrestling:** Robot sumo wrestling entails creating robots that participate in a ring designed in the form of a sumo match with the goal of shoving rivals outside the ring. In order to build robots that can recognize adversaries and make effective manoeuvres, engineers combine mechanical design, sensor technologies, and mathematical techniques.

   Using robotic sumo wrestlers from various categories, the well-known "All Japan Robot-Sumo Tournament" features robotic sumo wrestlers competing to outmanoeuvre and force opponents outside the ring.

4. **Autonomous Racing:** In autonomous racing, competitors must create vehicles that can navigate racetracks without assistance from humans. In order to give the vehicles in this sport the ability to navigate and make judgments in real-time, sensor fusion, machine learning, and mathematical models are needed.

   Using "Roborace" as an illustration, autonomous racing vehicles compete in exciting race settings. Teams create AI technologies and software algorithms to help automobiles navigate challenging tracks without colliding.

5. **Robot Combat:** In robot combat, robots are built and then pitted against one another in an arena. Participants build robots that can resist hits and effectively incapacitate opponents using mechanical engineering, structural design, and programming.

   Example: The popular TV program "BattleBots" features teams creating combat robots to engage in head-to-head combat. The robots need to have sturdy structure, functional weaponry, and tactical manoeuvres.

6. **Robot Golf:** The goal of robot golf is to push engineers to create robots that can accurately hit golf balls and navigate golf courses. To obtain precise shots in this sport, control algorithms, sensor technologies, and mathematics are used.

   For instance, competitions like the "Japan Robot Golf Association Cup" show off robot golfers navigating courses, figuring out distances, and hitting golf balls at predetermined goals.

   In these instances, the world of robot-assisted sports clearly demonstrates the interaction of technology, mathematics, and competitive spirit. By showcasing the adaptability and potential of robots in a variety of athletic scenarios, these sports not only enthral spectators but also push the boundaries of robotics and technical innovation.

Robotics in the entertainment industry is enabling new engagement and creative levels. It converts inert experiences into participatory and interactive ones, giving viewers a distinctive and memorable way to engage with technology and art. The opportunities for using robotics into entertainment are likely to grow much further as technology develops.

10.3 USE OF ROBOTS IN THE HOSPITALITY

The usage of robots in the hospitality business has drawn a lot of attention recently as technology has continued to transform many industries. Robots are being used in the hospitality industry to boost customer service, increase operational effectiveness, and provide customers unique experiences.

Here's how robots are being utilized in different aspects of the hospitality industry:

1. **Customer Service:** Robots are being used to help with check-ins, check-outs, and concierge services in hotels and restaurants. These machines may tell visitors on the location and points of interest nearby and even suggest places to eat and have fun.

Robots can conduct check-ins at the Henn na Hotel in Japan, and they can even control the lights and air conditioning in the rooms based on the preferences of the guests.

2. **Room Service and Delivery:** Meals, beverages, and other products are being delivered to guest rooms via robotic delivery systems. These robots efficiently deliver orders by moving through the hotel's hallways and lifts.

   To transport food, drinks, and utilities straight to guest rooms, several hotels use robots with compartments. This minimizes the need for human intervention during delivery.

3. **Cleaning and Housekeeping:** Robots are being created to carry out standard cleaning jobs like mopping floors and washing windows. These autonomous robots can keep rooms and public areas clean by themselves.

   As an illustration, some hotels utilize the "Roomba" robotic hoover cleaner to keep the hallways and rooms tidy. For activities like hotel housekeeping, more sophisticated robots are being developed.

4. **Entertainment and Interaction:** Robots are being utilized to entertain and interact with visitors through fun activities including games, shows, and storytelling. These robots engage visitors and help create a positive hotel experience.

   As an illustration, several hotels have robotic bartenders that can prepare drinks and serve customers while conversing with them, increasing the social atmosphere.

5. **Language Translation:** Robots that can translate between different languages can enable staff members and visitors who speak different languages communicate more effectively. Smoother interactions and higher visitor satisfaction are made possible by these robots.

   As an illustration, SoftBank's "Pepper" robot can understand and respond in several languages while assisting visitors and responding to their inquiries.

6. **Event Assistance:** Robots may offer information, direct people to specific spots, and even act as interactive tour guides during exhibitions and conferences.

   As an illustration, several conference venues employ robots to help visitors with directions, scheduling details, and other conference-related questions.

7. **Contactless Services:** In light of the Covid-19 pandemic, robots are being used to deliver contactless services, decreasing the necessity for direct human connection and lowering the danger of transmission.

   Towels, toiletries, and other supplies are delivered to guest rooms by robots in some hotels, which eliminate the need for in-person encounters with personnel.

Robotics integration in the hotel sector strives to improve visitor experiences, streamline operations, and add a futuristic and inventive element to the guest journey. While robots have many advantages, it's crucial to strike the correct balance between automation and human interaction to efficiently meet each guest's specific demands.

10.4 USE OF ROBOTS IN THE CUSTOMER SERVICE

The way businesses connect with their clients is changing as a result of the usage of robots in customer care. These robots, which are frequently propelled by artificial intelligence (AI) and cutting-edge technology, are used to deliver effective, dependable, and customised customer service across a variety of industries.

Here's how robots are being utilized in customer service:

1. **Virtual Assistants:** AI-powered solutions that can answer client questions and offer information instantly include virtual assistants. They can react to a wide range of inquiries, from simple FAQs to more complicated ones, and are accessible around-the-clock.
   As an illustration, numerous websites and online platforms employ chatbots to instantly respond to client inquiries and help with tasks like order tracking, problem-solving, and product recommendations.
2. **Client Inquiry Management:** By managing and classifying a high amount of client queries, robots can either direct them to the right human agents or immediately respond to them based on predetermined criteria.
   Using an example, AI-powered systems can categorise incoming calls in call centers and route clients to the proper departments or self-service choices, cutting down on wait times and boosting productivity.
3. **Personalised Offers:** By examining consumer behavior and interests, robots can make offers and recommendations for certain products. By adapting recommendations to specific needs, this improves the consumer experience.
   Example: To make product recommendations that are in line with a customer's interests, e-commerce companies frequently use AI algorithms to analyze browsing and purchase history.
4. **Voice Assistants and Phone Support:** Using skills such as voice recognition and natural language processing, voice-activated assistants can handle customer support requests over the phone.
   For instance, voice assistants like Google Assistant and Amazon's Alexa may help users with things like making appointments, checking the status of orders, and getting product information.
5. **Self-Service Kiosks and Terminals:** Robotic kiosks and terminals can assist consumers with self-service tasks like making food orders, checking in for flights, or obtaining information.
   Self-service kiosks are used in airports for check-in and luggage drop-off, which decreases wait times and gives passengers more control over their travel experience.
6. **Multilingual Support:** By overcoming communication gaps and enhancing accessibility, robots with language translation capabilities can offer support to clients who speak different languages. Robots that can translate between languages might help travelers in airports or tourists in public places, promoting communication between speakers of various languages.
7. **Customer Feedback and Surveys:** To determine customer satisfaction and pinpoint areas for development, robots can collect customer feedback and perform surveys.
   As an illustration, chatbots can encourage customers to participate in post-interaction surveys, gathering insightful data that enables firms to improve their customer care tactics.

Robotic customer service has many advantages, including cost savings, availability around-the-clock, scalability, and uniformity in responses. Striking a balance between automation and human engagement is crucial, though, as some complex or emotionally charged topics could call for the sensitivity and nuanced understanding only human agents can offer. Enhancing the client experience should be the top priority when integrating robots into customer service, while keeping the human touch where essential.

10.5 REAL-WORLD APPLICATIONS OF ROBOTICS IN MANUFACTURING

Robotics has transformed the manufacturing sector by boosting productivity, accuracy, and efficiency across numerous industries. Robots are used in manufacturing for complex activities that require precision and repetition in addition to assembly lines. Here are a few examples of real-world manufacturing uses for robots:

1. **Production and Assembly:** Robots are frequently employed to accurately connect components together to create items. They are capable of handling jobs that call for consistent accuracy, reducing mistakes, and guaranteeing uniform quality in bulk manufacturing.

   For instance, robots are used in the automobile industry to quickly and accurately assemble car parts like doors, engines, and chassis.

2. **Welding:** Robots are frequently employed for welding operations, which call for accuracy, repetitive motions, and the capacity to function in potentially dangerous conditions like high temperatures and gases.

   To produce strong and reliable welds, robots can execute arc welding, spot welding, and laser welding during the manufacture of metal items.

3. **Material Handling:** Robots are excellent at transporting components between production steps, loading and unloading materials, and carrying finished goods to storage facilities.

   Robots can efficiently manage inventory, sort parcels, and carry supplies in factories and warehouses, which reduces manual labor and increases productivity.

4. **Inspection and Quality Control:** To check products for flaws and guarantee that only high-quality goods are supplied to customers, robots outfitted with sophisticated sensors and cameras are utilized.

   Robots can do visual inspections to find circuit board flaws in the production of electronics, ensuring the finished product satisfies exacting quality standards.

5. **Painting and Coating:** To ensure even coverage and minimize paint and coating waste, robots are used for precise painting and coating applications.

   Example: To ensure consistency and adherence to stringent safety rules, coatings are applied to aircraft components by robots in the aerospace sector.

6. **Palletizing and Packing:** Robots can automate the stacking of finished goods on pallets for shipment as well as the packaging of goods into boxes, crates or other containers.

   As an illustration, food and beverage producers utilize robots to package goods, lowering the possibility of contamination and guaranteeing precise portioning.

7. **CNC Machining:** To carry out intricate and accurate machining tasks like milling, drilling, and grinding, robots are linked into computer numerical control (CNC) machines.

   Example: In CNC machining centers, robotic arms are used to perform intricate operations on metal, plastic, and other materials, resulting in high-quality finished products.

8. **Electronics Manufacturing:** Precision and miniaturization are essential for creating electronic products like circuit boards and Smartphone.

   An illustration is the precise placement of microscopic electronic components onto circuit boards using surface-mount technology (SMT) robots.

9. **3D printing and Additive Manufacturing:** By building up layers of material, robotic 3D printers build complex constructions and prototypes, providing creative options for individualized designs and quick prototyping.

   As an illustration, aerospace businesses use robotic 3D printing to produce complex and lightweight aero plane components, increasing fuel economy.

The use of collaborative robots (cobots), which can safely operate alongside human workers, is one of the evolving robotics applications in industry. By improving productivity, quality, and safety while also advancing automation and innovation in manufacturing processes, these technologies have revolutionized entire sectors.

10.6 REAL-WORLD APPLICATIONS OF ROBOTICS IN LOGISTICS

Robotics has drastically changed how commodities are handled, processed, and transported inside of warehouses and distribution centers, having a tremendous impact on the logistics sector. Robotics in logistics serves to increase productivity, accuracy, and supply chain management as a whole. Here are a few examples of robotics in logistics in the real world:

1. AGVs and AMRs, or autonomous mobile robots, are two examples. AMRs and AGVs are robots that can navigate warehouses and distribution centers on their own thanks to sensors and navigational systems onboard. To specific places, they can convey cargo, pallets, and containers.
   Example: To shorten walking distances and speed up order fulfillment, Amazon's fulfillment centers use Kiva robots (formerly known as Amazon Robotics) to move product shelves to human pickers.
2. **Goods-to-Person Systems:** By bringing product storage units or shelves to human pickers, robotic systems reduce the need for pickers to go throughout the warehouse.
   For instance, "AutoStore" is a small-scale robotic storage system that automates the movement of product bins to pickers in order to maximize storage and speed up order processing.
3. **Robotic Picking and Packing:** To automate the fulfillment process, robots are made to recognize, choose, and pack things from shelves or bins.
   As an illustration, "Pick and Place" robots employ gripping mechanisms and computer vision to remove various items from bins and put them into shipping containers.
4. **Sorting and Distribution:** To facilitate effective distribution procedures, robots are employed to sort and categorize objects according to size, weight, and destination.
   An illustration would be automated robotic sortation systems in logistics hubs.
5. **Last-Mile Delivery Robots:** In metropolitan areas, robotic delivery vehicles drive along streets and sidewalks to bring packages to consumers' doorsteps.
   For instance, in some places, businesses like Starship Technologies and Amazon's Scout utilise small wheeled robots to deliver products to consumers on their own.
6. **Drones for Delivery:** Drones are used to deliver tiny products to far-off or difficult-to-reach locations, cutting down on delivery times and costs.
   As an illustration, businesses like Zipline and Wing, a division of Google's parent firm Alphabet, use drones to carry consumer products and medical supplies to remote areas.
7. **Automated Warehousing:** In a warehouse, robots may load and unload trucks, store and retrieve products, and manage inventory, among other functions.
   Example: The "Stretch" robot from robotics company Boston Dynamics is made to handle material handling jobs like truck unloading, palletizing, and inventory management.
8. **Collaborative Robots (Cobots) in Logistics:** Cobots assist human workers in jobs that call for both robotic precision and human judgment.
   As an illustration, in some warehouses, cobots help human employees by supplying equipment, moving large objects, or carrying out jobs that call for dexterity and attention to detail.

By increasing speed, precision, and efficiency throughout the supply chain, these real-world applications show how robots is altering logistics. Robotics integration in logistics contributes to the expansion of e-commerce and international trade by streamlining processes, reducing labor-intensive jobs, and optimizing resource use.

10.7 REAL-WORLD APPLICATIONS OF ROBOTICS IN OTHER INDUSTRIAL SECTORS

Numerous industrial areas have seen a substantial impact from robotics, which has transformed operations, increased productivity, and allowed for creative approaches to challenging jobs. Here are some examples of actual robotics applications in several industrial fields:

1. **Healthcare:**

   Surgical Robots: With more accuracy and less invasiveness, surgical robots let doctors carry out minimally intrusive treatments.

   Robots that assist with rehabilitation and physical therapy can lead patients through activities and track their development.

   Telemedicine Robots: Robots with cameras and displays allow for patient and doctor consultations over long distances.

2. **Agriculture:**

   Harvesting Robots: To increase efficiency in agricultural harvesting, robots are made to pick fruits and vegetables with little to no harm.

   Precision Farming: Robots fitted with sensors and cameras keep an eye on the health of the crops, improve irrigation, and apply fertilizer or pesticides where they are needed.

3. **Construction:**

   Bricklaying Robots: Bricks may be laid quickly and precisely by robots, which lowers labor costs and expedites building.

   Autonomous Vehicles: Site preparation, material transportation, and excavation can all be done using self-driving construction machinery.

4. **Energy and Utilities:**

   Inspection Robots: Robots with cameras and sensors are able to inspect and repair vital infrastructure, including pipelines, power lines, and wind turbines.

   Robots are employed for jobs in hazardous settings to assist in the decommissioning of nuclear sites.

5. **Retail:**

   Robots that manage inventory may autonomously roam retail aisles while scanning shelves to monitor inventory levels and determine when restocking is necessary.

   Customer Assistance Robots: These machines assist customers by directing them to certain products and responding to their inquiries.

6. **Mining:**

   Autonomous Hauling Trucks: These vehicles deliver extracted materials within mines while enhancing efficiency and safety.

   Remote-Controlled Drilling: To lower the risk to human workers, robots control drilling equipment from a distance.

7. **Space Exploration:**

   Planetary Rovers: Robots that explore far-off planets, like the Mars rovers, gather information and photos to use in scientific study.

   Robots can be used to maintain, refill, or enhance satellites while they are in space.

8. **Education:**

   Educational Robots: Students are introduced to STEM disciplines, programming, and problem-solving using robots as teaching aids.

   Robotics Kits: By allowing students to construct and program robots, educational kits promote experiential learning.

9. **Automotive Manufacturing:**

   Robotic Welding: Robots work quickly and precisely to complete complex welding jobs on car assembly lines.

   Paint Robots: To provide a consistent finish, robots evenly apply paint to automobiles.

10. **Aerospace:**

    Repair of Aircraft Robots: Robots check, fix and clean up after aircraft as part of their maintenance duties.

    Unmanned Airborne Vehicles (UAVs): Drones are utilized in aerospace applications for airborne surveys, inspections, and data collection.

These instances show how robots is being used in a variety of ways to alleviate problems, boost productivity, and stimulate innovation across numerous industrial sectors. Robotics technology is constantly improving, and this has the potential to change and improve industries in ways that were previously thought to be unthinkable.

Numerical Problems with Solutions

Problem 1. A theme park has 10 robot characters for a parade. Each robot requires 3 hours of maintenance daily. How many hours of maintenance are needed for all the robots in a week?
Solution: 210 hours (10 robots * 3 hours/robot * 7 days).

Problem 2. A robot entertainer can perform 4 different dance routines. If each routine lasts 5 minutes and the robot performs them back-to-back, how long will the entire performance be?
Solution: 20 minutes (4 routines * 5 minutes/routine).

Problem 3. An interactive robot exhibit has 50 visitors every hour. If the exhibit operates for 8 hours each day, how many visitors experience the exhibit in a week?
Solution: 2800 visitors (50 visitors/hour * 8 hours/day * 7 days).

Problem 4. A robotic puppet show has 3 robots, each performing a 15-minute show. If all robots run simultaneously, how long will the entire show last?
Solution: 15 minutes (The longest show duration among the robots).

Problem 5. A virtual reality tour guide robot covers a 1-kilometer path in 10 minutes. What is its average speed in meters per second?
Solution: 0.1667 m/s (1000 meters/600 seconds).

Problem 6. An escape room has a robot providing hints to participants. If the robot gives 5 hints every 15 minutes, how many hints does it give in 2 hours?

Solution: 40 hints (5 hints/15 minutes * 8 intervals).

Problem 7. A robot dance group performs a routine in 3 minutes. If they perform it every 20 minutes, how many times can they perform in 2 hours?

Solution: 6 performances (120 minutes/20 minutes/performance).

Problem 8. A drone racing competition consists of 10 rounds. If each round takes 2 minutes to complete, how long will the entire competition last?

Solution: 20 minutes (10 rounds * 2 minutes/round).

Problem 9. An autonomous robotic barista serves 15 customers per hour. How many customers can it serve in a 6-hour shift?

Solution: 90 customers (15 customers/hour * 6 hours).

Problem 10. A robotic puppet show charges $10 per ticket. If they sell 150 tickets for a performance, how much revenue is generated?

Solution: $1500 ($10/ticket * 150 tickets).

Problem 11. An AR-based interactive exhibit has 2 robots guiding visitors. Each robot interacts with 20 visitors per hour. How many visitors can both robots interact with in a 6-hour operating period?

Solution: 240 visitors (2 robots * 20 visitors/hour * 6 hours).

Problem 12. A robotic magic show has a total of 8 tricks. If each trick takes 4 minutes to perform, how long is the entire show?

Solution: 32 minutes (8 tricks * 4 minutes/trick).

Problem 13. An entertainment event has 5 robots performing simultaneously. If each robot generates 3 hours of battery life, how many hours of entertainment can they provide in total?

Solution: 15 hours (5 robots * 3 hours/robot).

Problem 14. A robotic character takes 10 seconds to move from one end of the stage to the other. If the stage is 20 meters long, what is the robot's speed in meters per second?

Solution: 2 m/s (20 meters/10 seconds).

Problem 15. A robotic orchestra has 15 members, each playing a 3-minute piece. If they perform one after the other, how long will the entire performance be?

Solution: 45 minutes (15 members * 3 minutes/member).

Problem 16. An augmented reality exhibit has a robot interacting with 40 visitors per hour. If the exhibit operates for 5 hours, how many visitors experience the exhibit?

Solution: 200 visitors (40 visitors/hour * 5 hours).

Problem 17. A robotic concession stand serves an average of 25 customers per hour. If it operates for 9 hours, how many customers does it serve in total?

Solution: 225 customers (25 customers/hour * 9 hours).

Problem 18. A robotic dance workshop has 2 robots practicing a routine for 45 minutes each. If they practice simultaneously, how long is the combined practice session?

Solution: 45 minutes (The longest practice duration among the robots).

Problem 19. An entertainment event has 3 robot performers. If each robot costs $500 to maintain per month, what is the total maintenance cost for 6 months?

Solution: $9000 ($500/robot * 3 robots * 6 months).

Problem 20. An interactive robot exhibit attracts 1200 visitors in a week. If each visitor spends an average of 10 minutes at the exhibit, how many hours of engagement does the exhibit provide in total?

Solution: 200 hours (1200 visitors * 10 minutes/visitor/60 minutes/hour).

Short Questions and Answers

Q.1. What are some examples of robotics in the entertainment sector?

Ans. Robotic performers, virtual reality experiences, interactive exhibits, and robotic characters in films and TV.

Q.2. How do robots enhance theme park attractions?

Ans. Robots can act as characters, guides, and performers, enhancing the immersive experience for visitors.

Q.3. What's an example of a robotic performer in entertainment?

Ans. Disney's animatronic characters like the singing dolls in "It's a Small World" ride.

Q.4. How do robots contribute to virtual reality experiences?

Ans. Robots can simulate physical sensations in virtual environments, enhancing immersion.

Q.5. What's an example of a robot-assisted live show?

Ans. Robots performing choreographed dances alongside human dancers.

Q.6. How do robots participate in drone racing?

Ans. Robots can be programmed to pilot drones through complex racing courses.

Q.7. What's the role of robots in interactive exhibits?

Ans. Robots can engage visitors in hands-on learning experiences and demonstrations.

Q.8. How do robots enhance customer service in the entertainment sector?

Ans. Robots can provide information, guide visitors, and offer unique interactions.

Q.9. What's an example of a robotic character in a film?

Ans. R2-D2 from "Star Wars" is a famous robotic character.

Q.10. How do robots simulate emotions in entertainment?

Ans. Robots are programmed to mimic expressions and behaviors to convey emotions.

Q.11. What's the significance of robots in theme park attractions?

Ans. Robots create immersive experiences, enhancing visitor engagement and enjoyment.

Q.12. How are drones used in filming and photography?

Ans. Drones capture aerial shots and unique perspectives in films and TV shows.

Q.13. What's the role of robots in stage performances?

Ans. Robots can be programmed to dance, move, and interact with human performers.

Q.14. How do robots contribute to augmented reality experiences?

Ans. Robots can interact with users in augmented reality, enhancing the sense of presence.

Q.15. What's an example of a robotic puppet show?

Ans. "Puppet Parade" features interactive robots performing alongside human puppeteers.

Q.16. How do robots enhance interactivity in museum exhibits?

Ans. Robots can guide visitors, answer questions, and provide interactive demonstrations.

Q.17. What's the role of robots in haunted house attractions?

Ans. Robots can provide unexpected scares and enhance the thrill for visitors.

Q.18. How do robots contribute to art installations?

Ans. Robots can create dynamic movements and visual effects in artistic compositions.

Q.19. What's an example of a robot-made artwork?

Ans. "e-David" is a robot artist that paints using algorithms and perception.

Q.20. How do robots simulate human-like movements in performances?

Ans. Robots use advanced mechanical systems and algorithms for precise motions.

Q.21. How do robots assist in theater productions?

Ans. Robots like "RoboThespian" can perform scripted dialogues and actions on stage.

Q.22. How do robots interact with virtual reality environments?

Ans. Robots can be programmed to react to users' movements and actions in VR.

Q.23. What's the advantage of using robots in character animation?

Ans. Robots can replicate consistent movements and expressions, enhancing animation quality.

Q.24. How do robots assist in conducting orchestras?

Ans. Robots equipped with motion control systems can mimic conducting movements.

Q.25. How do robots contribute to live music performances?

Ans. Robots can play instruments, adding unique elements to musical compositions.

Q.26. What's the concept of robot-assisted storytelling?

Ans. Robots can act as characters and storytellers, engaging audiences in interactive narratives.

Q.27. How do robots enhance the "wow" factor in entertainment?

Ans. Robots introduce novel and futuristic elements that captivate audiences.

Q.28. What's the role of robots in theme park parades?

Ans. Robots can be part of parade floats, adding dynamic movements and excitement.

Q.29. How are robots used in interactive installations for educational purposes?

Ans. Robots engage visitors in learning activities and STEM concepts through hands-on experiences.

Q.30. What's the impact of robots on audience engagement in entertainment?
Ans. Robots create memorable and interactive experiences that capture viewers' attention.

Q.31. How do robots contribute to immersive storytelling in films?
Ans. Robots can be used to portray characters with unique abilities and interactions.

Q.32. What's the role of robots in dance performances?
Ans. Robots can be programmed to execute intricate dance routines alongside human performers.

Q.33. How do robots simulate animal behaviors in attractions?
Ans. Robots mimic natural movements and behaviors, enhancing realism.

Q.34. How are robots used in promotional events?
Ans. Robots can engage attendees, distribute promotional materials, and create a buzz.

Q.35. What's the significance of humanoid robots in entertainment?
Ans. Humanoid robots create relatable interactions and can perform human-like actions.

Q.36. How do robots participate in competitive games and sports events?
Ans. Robots can compete in robotic soccer tournaments, drone races, and more.

Q.37. How do robots enhance magic shows and illusions?
Ans. Robots can perform intricate and precise actions that add to the illusion.

Q.38. How are robots integrated into circus performances?
Ans. Robots can be programmed to perform acrobatics and stunts in circus acts.

Q.39. What's the role of robots in creating dynamic lighting effects?
Ans. Robots with LED lights can create stunning visual displays synchronized with music.

Q.40. How are robots used in entertainment events for children?
Ans. Robots can interact with children, tell stories, and engage them in fun activities.

Q.41. How do robots add an element of surprise in entertainment?
Ans. Robots can appear unexpectedly and perform actions that amaze and entertain.

Q.42. What's the advantage of using robots for repetitive tasks in entertainment?
Ans. Robots can perform tasks consistently without getting fatigued, ensuring reliable performances.

Q.43. How do robots contribute to futuristic themes in entertainment?
Ans. Robots embody futuristic concepts and technology, aligning with sci-fi narratives.

Q.44. How are robots used in entertainment events to capture data?
Ans. Robots can gather audience feedback and preferences for future improvements.

Q.45. What's the role of robots in creating multi-sensory experiences?
Ans. Robots can incorporate tactile, visual, and auditory elements to engage multiple senses.

Q.46. How are robots integrated into escape room experiences?
Ans. Robots can provide clues, challenges, and surprises, enhancing the immersive adventure.

Q.47. How do robots contribute to the "wow" factor in theme park attractions?
Ans. Robots can perform awe-inspiring actions that leave visitors amazed and talking about the experience.

Q.48. How are robots used in events that combine entertainment and education?

Ans. Robots can engage audiences in educational content through interactive demonstrations.

Q.49. How do robots create memorable moments in live shows?

Ans. Robots can perform unique actions that stick in the audience's memory long after the performance.

Q.50. How are robots integrated into music festivals and concerts?

Ans. Robots can provide visual and interactive elements that enhance the overall concert experience.

Multiple Choice Questions and Answers

Q.1. Which of the following is an example of a robot-assisted live show?

(*a*) Virtual reality experience
(*b*) Autonomous vehicle demonstration
(*c*) Robot dance performance
(*d*) Drone racing competition

Q.2. In the entertainment sector, robots can enhance which of the following aspects?

(*a*) Sound engineering
(*b*) Set design
(*c*) Audience engagement
(*d*) Costumed actors

Q.3. What role do robots play in interactive museum exhibits?

(*a*) Cleaning
(*b*) Ticket sales
(*c*) Guiding visitors
(*d*) Repairing artifacts

Q.4. How are robots used in theme park attractions?

(*a*) Selling merchandise
(*b*) Providing security
(*c*) Enhancing guest experience
(*d*) Repairing rides

Q.5. Robotic characters in films and TV shows often require:

(*a*) Long breaks
(*b*) Makeup artists
(*c*) Script rehearsals
(*d*) Precise movements

Q.6. What's a common application of robots in drone racing?

(*a*) Piloting the drones
(*b*) Selling tickets
(*c*) Providing commentary
(*d*) Designing the race course

Q.7. How do robots contribute to virtual reality experiences?

(*a*) They provide haptic feedback.
(*b*) They generate 3D models.
(*c*) They create visual effects.
(*d*) They simulate real-world movements.

Q.8. Robotic puppets in interactive exhibits can:

(*a*) Sing songs
(*b*) Provide weather forecasts
(*c*) Hand out brochures
(*d*) Engage in conversations

Q.9. What's the advantage of using robots in stage performances?

(*a*) Lower production costs
(*b*) Guaranteed applause
(*c*) Precise and repeatable movements
(*d*) No need for rehearsals

Q.10. How do robots contribute to the "wow" factor in entertainment?

(*a*) By using loud sound effects (*b*) By performing unexpected actions
(*c*) By using bright lighting (*d*) By wearing flashy costumes

Q.11. Robots designed for theme park parades often feature:

(*a*) Realistic animal behavior (*b*) Colorful lighting effects
(*c*) Remote-controlled movements (*d*) Ability to perform magic tricks

Q.12. In the context of entertainment, what's the significance of humanoid robots?

(*a*) They are skilled musicians (*b*) They can perform complex calculations
(*c*) They can interact in human-like ways
(*d*) They are made from environmentally friendly materials

Q.13. How do robots simulate emotions in entertainment?

(*a*) By analyzing weather data
(*b*) Through complex mathematical algorithms
(*c*) By mimicking facial expressions and behaviors
(*d*) Through random movements

Q.14. How do drones enhance filming and photography?

(*a*) By replacing cameras entirely (*b*) By providing aerial perspectives
(*c*) By acting as lighting equipment (*d*) By animating characters

Q.15. Robots used in retail for customer service can:

(*a*) Cook gourmet meals (*b*) Analyze stock market trends
(*c*) Assist customers with inquiries (*d*) Write novels

Q.16. How do robots contribute to magic shows and illusions?

(*a*) By predicting future events (*b*) By creating holograms
(*c*) By performing complex card tricks (*d*) By executing precise actions

Q.17. What's the role of robots in creating multi-sensory experiences?

(*a*) Creating virtual reality content
(*b*) Engaging audiences through touch, sight, and sound
(*c*) Generating complex mathematical patterns
(*d*) Handling food and drink services

Q.18. How are robots used in promotional events?

(*a*) By providing financial advice (*b*) By distributing brochures
(*c*) By performing dance routines (*d*) By performing acrobatics

Q.19. How do robots simulate animal behaviors in attractions?

(*a*) By mimicking bird songs (*b*) By analyzing soil samples
(*c*) By studying marine life (*d*) By imitating movements and sounds

Q.20. What's the advantage of using robots for repetitive tasks in entertainment?

(*a*) They require no electricity (*b*) They can work without instructions
(*c*) They don't need maintenance (*d*) They ensure consistency and reliability

Q.21. How do robots contribute to last-mile delivery in entertainment?

(*a*) By providing musical performances
(*b*) By delivering packages to customers' doorsteps
(*c*) By painting murals on buildings
(*d*) By analyzing traffic patterns

Q.22. What's the role of robots in creating dynamic lighting effects?

(*a*) Creating thunder and lightning
(*b*) Producing holographic images
(*c*) Enhancing visual aesthetics
(*d*) Creating wind effects

Q.23. How do robots enhance customer experiences in the service industry?

(*a*) By predicting future events
(*b*) By automating all human tasks
(*c*) By providing personalized assistance
(*d*) By performing stand-up comedy

Q.24. What's the advantage of using robots for repetitive tasks in the service industry?

(*a*) They never make mistakes
(*b*) They require no training
(*c*) They are always available
(*d*) They never need breaks

Q.25. How do robots simulate human-like movements in performances?

(*a*) Through random actions
(*b*) By analyzing historical data
(*c*) Through advanced mechanical systems and algorithms
(*d*) By playing recorded sounds

Q.26. Robots designed for surgical procedures:

(*a*) Operate without any human input
(*b*) Can perform surgery autonomously
(*c*) Assist surgeons with precise movements
(*d*) Provide medical diagnoses

Q.27. How do robots assist in agricultural tasks?

(*a*) By planting seeds using catapults
(*b*) By transporting livestock
(*c*) By analyzing weather patterns
(*d*) By picking fruits and vegetables

Q.28. What's the role of robots in space exploration?

(*a*) They build spacecraft
(*b*) They communicate with aliens
(*c*) They repair satellites
(*d*) They gather data and images on distant planets

Q.29. How do robots contribute to rehabilitation in healthcare?

(*a*) By performing medical procedures
(*b*) By providing emotional support
(*c*) By assisting patients with exercises
(*d*) By diagnosing illnesses

Q.30. In the entertainment industry, robots can be used to enhance which of the following?

(*a*) Fashion design
(*b*) Audience engagement
(*c*) Culinary arts
(*d*) Political debates

Q.31. What's the significance of robots in interactive storytelling?

(*a*) They replace human authors
(*b*) They add visual effects to stories
(*c*) They engage audiences through interactive choices
(*d*) They predict the future

Q.32. How do robots participate in dance performances?

(*a*) By controlling lighting effects
(*b*) By operating sound systems
(*c*) By mimicking human movements
(*d*) By managing ticket sales

Q.33. Robots can enhance which of the following in the entertainment industry?

(*a*) Predicting the weather
(*b*) Providing legal advice
(*c*) Audience interaction and engagement
(*d*) Predicting stock market trends

Q.34. What's the role of robots in creating immersive experiences in entertainment?

(*a*) They play live music
(*b*) They analyze economic trends
(*c*) They provide fashion advice
(*d*) They engage all the senses to transport audiences

Q.35. Robots used in service industries can assist with:

(*a*) Astronomical observations
(*b*) Diagnosing medical conditions
(*c*) Negotiating international treaties
(*d*) Customer inquiries and information

Q.36. How do robots simulate animal behaviors in entertainment attractions?

(*a*) By predicting weather patterns
(*b*) By creating virtual reality simulations
(*c*) By mimicking movements and sounds
(*d*) By analyzing DNA samples

Q.37. What's the role of robots in escape room experiences?

(*a*) Selling tickets
(*b*) Providing commentary
(*c*) Offering food and drinks
(*d*) Providing clues and challenges

Q.38. How do robots enhance interactivity in entertainment events?

(*a*) By providing background music
(*b*) By analyzing audience demographics
(*c*) By engaging audiences in activities and choices
(*d*) By providing catering services

Q.39. What's the significance of using robots in musical performances?

(*a*) They create custom stage designs
(*b*) They write lyrics for songs
(*c*) They provide real-time audience feedback
(*d*) They play instruments and add unique elements

Q.40. Robots used in haunted house attractions can provide:

(*a*) Legal advice
(*b*) Tax preparation services
(*c*) Unexpected scares and thrills
(*d*) Musical performances

Q.41. How do robots contribute to dynamic lighting effects in entertainment?

(*a*) By predicting weather patterns
(*b*) By analyzing stock market trends
(*c*) By creating custom light shows
(*d*) By generating mathematical equations

Q.42. What's the role of robots in creating immersive experiences in VR and AR?

(*a*) Providing financial advice
(*b*) Designing costumes
(*c*) Enhancing the sense of presence and realism
(*d*) Creating visual effects for movies

Q.43. How do robots contribute to engaging children in entertainment?

(*a*) By providing legal advice
(*b*) By offering cooking lessons
(*c*) By performing magic tricks
(*d*) By telling stories and leading activities

Q.44. What's the significance of using robots in theme park parades?

(*a*) They predict the weather
(*b*) They provide catering services
(*c*) They add dynamic movements and excitement
(*d*) They manage ticket sales

Q.45. How do robots contribute to creating memorable moments in live shows?

(*a*) By providing medical advice
(*b*) By predicting future events
(*c*) By performing unexpected actions
(*d*) By designing set decorations

Q.46. How are robots used in robotic puppet shows?

(*a*) By designing costumes
(*b*) By controlling lighting effects
(*c*) By providing catering services
(*d*) By interacting with puppets and audiences

Q.47. What's the role of robots in combining education with entertainment?

(*a*) Predicting stock market trends
(*b*) Distributing promotional materials
(*c*) Engaging audiences in hands-on learning
(*d*) Providing live music performances

Q.48. How do robots contribute to adding a futuristic element to entertainment?

(*a*) By predicting future events
(*b*) By analyzing historical data
(*c*) By using advanced technology and robotics
(*d*) By performing traditional dance routines

Q.49. How do robots enhance customer service in the entertainment sector?

(*a*) By designing costumes
(*b*) By analyzing economic trends
(*c*) By providing personalized interactions and information
(*d*) By predicting weather patterns

Q.50. What's the significance of robots in theme park attractions?

(*a*) They replace human actors
(*b*) They enhance guest experience and engagement
(*c*) They predict stock market trends
(*d*) They provide catering services

Key to Multiple Choice Questions

1. (*c*)	**2.** (*c*)	**3.** (*c*)	**4.** (*c*)	**5.** (*d*)	**6.** (*a*)	**7.** (*a*)	**8.** (*d*)	**9.** (*c*)
10. (*b*)	**11.** (*b*)	**12.** (*c*)	**13.** (*c*)	**14.** (*b*)	**15.** (*c*)	**16.** (*d*)	**17.** (*b*)	**18.** (*b*)
19. (*d*)	**20.** (*d*)	**21.** (*b*)	**22.** (*c*)	**23.** (*c*)	**24.** (*d*)	**25.** (*c*)	**26.** (*c*)	**27.** (*d*)
28. (*d*)	**29.** (*c*)	**30.** (*b*)	**31.** (*c*)	**32.** (*c*)	**33.** (*c*)	**34.** (*d*)	**35.** (*d*)	**36.** (*c*)
37. (*d*)	**38.** (*c*)	**39.** (*d*)	**40.** (*c*)	**41.** (*c*)	**42.** (*c*)	**43.** (*d*)	**44.** (*c*)	**45.** (*c*)
46. (*d*)	**47.** (*c*)	**48.** (*c*)	**49.** (*c*)	**50.** (*b*)				

Project Works

1. **Robotic Puppet Show:** Design and build a robotic puppet that can interact with audiences, tell stories, and perform movements based on user input.
2. **Virtual Reality Tour Guide Robot:** Create a robot that navigates a virtual reality environment, guiding users through interactive tours and providing information about different attractions.
3. **Augmented Reality Interactive Exhibit:** Develop an AR-based interactive exhibit that uses robots to engage visitors, explain concepts, and provide an immersive learning experience.
4. **Robot-Assisted Escape Room:** Design a robot that serves as a puzzle master in an escape room, providing hints and clues to participants as they try to solve challenges.
5. **Robotic Dance Performance:** Create a team of synchronized robots capable of performing complex dance routines, enhancing live entertainment events.
6. **Autonomous Drone Racing League:** Organize an autonomous drone racing competition where participants program drones to navigate a challenging race course.
7. **Robot-Assisted Magic Show:** Develop a robot that can execute magic tricks, enhancing the visual appeal and surprise factor of the show.
8. **Interactive Robotic Characters in Theme Parks:** Build robots that can interact with visitors in theme parks, guiding them, posing for photos, and providing information about attractions.
9. **Robot-Assisted Art Installation:** Create a robotic art installation that incorporates movement, lights, and sound to engage viewers and convey a specific artistic concept.
10. **Collaborative Robot for Cooking Demonstrations:** Design a collaborative robot that can assist chefs in cooking demonstrations, showcasing recipes and techniques.
11. **Autonomous Robotic Barista:** Build a robot capable of preparing and serving coffee or beverages in a café setting.
12. **Robotic Live Music Performance:** Develop a robot ensemble capable of playing musical instruments and performing live music on stage.
13. **Robot-Assisted Theater Production:** Create robots that can handle set changes, assist with prop movements, and contribute to the overall production of a play.
14. **Robot-Enhanced Interactive Storytelling:** Design robots that play the roles of characters in interactive storytelling experiences, responding to user choices and actions.
15. **Robot-Operated Haunted House:** Construct a haunted house attraction where robots create unexpected scares and interactions for visitors.
16. **Augmented Reality Costume Character:** Develop an AR system that overlays virtual costumes onto real performers, enhancing their appearances in live events.
17. **Robotic Art Restoration:** Create robots that can delicately restore and repair artworks in museums, ensuring their preservation.
18. **Autonomous Robotic Concession Stand:** Design a robot-operated concession stand that serves snacks, beverages, and merchandise at entertainment venues.
19. **Robot-Assisted Dance Workshop:** Organize a dance workshop where participants collaborate with robots to learn and perform choreographed routines.
20. **Robot-Mediated Virtual Reality Experience:** Develop a robot that physically interacts with users in a virtual reality environment, enhancing the sense of presence and immersion.

11

Automation and the Workforce: Emerging Trends

11.1 INTRODUCTION

Technology has drastically changed many facets of our lives in recent years, and the workplace is no exception. Automation, which is being fueled by advancements in robots, artificial intelligence (AI), machine learning, and other cutting-edge technologies, is changing the employment landscape and upending preconceived ideas about what constitutes work. Opportunities and problems for people, corporations, and larger society are both brought forth by this shift. Automation technologies have the potential to significantly improve efficiency, productivity, and even job quality in the workforce. Previously labor- and time-intensive tasks can now be completed with unparalleled accuracy and speed. New job positions have been created as a result, especially in the areas of technology development, data analysis, and process optimization. By removing or minimizing the need for employees to engage in risky activities, automation has also improved workplace safety. The fast development of automation has, however, led to worries about how it may affect jobs that have historically been done by people. There is a chance that machines will take over routine and repetitive work, which are particularly vulnerable to automation. This has sparked questions about the potential for particular worker categories to be replaced as well as the necessity of reskilling and upskilling programs to guarantee that people stay relevant in the changing labor market.

Automation and the workforce have a complicated and varied interaction. Beyond just economic factors, this interaction also touches on ethical, social, and policy-related issues. As automation develops, it is essential to look at the new trends influencing the workforce of the future and the tactics that may be used to maximize its advantages while minimizing any potential drawbacks. The various aspects of automation and its effects on the workforce are covered in this series of articles. We delve into the ongoing transition and analyze ways that people, businesses, and governments might adapt and succeed in the automated economy, from analyzing the roles that automation plays in various industries to talking about the skills that will be in demand in the automated economy. We will learn about the newest trends in automation and the workforce as we progress through this chapter, and we'll attempt to create a thorough knowledge of the opportunities and difficulties that lie ahead. We can collaboratively create a future in which automation and human potential coexist peacefully by staying up to date on the most recent advancements and actively participating in critical conversations.

11.2 NEW TRENDS IN AUTOMATION AND THE WORKFORCE

These new trends demonstrate the various ways that automation is changing the workforce and various businesses. Automation offers chances for more efficiency and creativity, but it also has ramifications for jobs, skills, ethics, and societal well-being that must be carefully considered. To effectively utilize automation in the years to come, it will be essential to carefully navigate these tendencies.

11.2.1 Collaborative Robotics (Cobots)

Due to safety concerns, conventional industrial robots were frequently kept apart from human workers. However, collaborative robots, also known as cobots, are made to work alongside people, increasing productivity and efficiency while preserving a secure workplace. This tendency is especially noticeable in the manufacturing and warehousing industries, where cobots help with jobs that call for strength and precision, enhancing overall production operations. Cobots, also known as collaborative robots, are a subtype of industrial robots created to cooperate with human workers in a shared workspace. They can work with people effectively and safely thanks to their sophisticated sensing, control, and communication systems. Robotics, sensors, and control algorithms—all of which are grounded by mathematical ideas—come together to make up the technology underpinning cobots.

1. **Robotic Manipulation:** Like typical industrial robots, cobots have robotic arms, but with modifications that improve their safety and ability to interact with people. These arms frequently have several joints, allowing for intricate positioning and movements. The robot's motion is modelled and controlled using kinematic and dynamical mathematical ideas. To determine the joint angles needed to attain a given end-effector position, inverse kinematics is applied, and dynamic models guarantee smooth and precise motion execution.
2. **Sensing and Perception:** Sensors are essential for cobots' interaction with their surroundings and human partners. Cameras, force/torque sensors, lidar, depth sensors, and other sensors are included in this group. It takes the application of mathematical methods like image processing, computer vision, and sensor fusion to decipher sensor data and extract useful information. Computer vision algorithms, for instance, can assist the cobot in recognizing items, people, and their poses, enabling accurate and safe collaboration.
3. **Force and Contact Control:** Cobots must be able to sense forces applied to them during interactions with people or things and respond accordingly. Robot end effector force is regulated using force control algorithms, resulting in delicate and compliant contact. To establish a seamless and secure human-robot interface, control theory, impedance control, and force feedback mathematics are used.
4. **Motion Planning and Collision Avoidance:** Cobots must plan their movements to avoid obstacles and humans in order to interact successfully. Motion planning algorithms create collision-free pathways by using mathematical methods from computational geometry and optimization. Through the use of trajectory planning, the robot is kept from making sudden or jerky movements that would endanger its human coworkers.
5. **Safety Monitoring and Control:** The design of cobots must prioritize worker safety. The control system incorporates mathematical models of safety zones, speed limits, and robot behavior. This enables the cobot to modify its behavior when a person gets too close by slowing down, stopping, or stopping. Calculating safe inter-relationship lengths, speeds, and forces during interaction requires the application of mathematical approaches.

6. **Human-Robot Interaction:** Mathematical understanding and modeling of human behavior and purpose are important for human-robot interaction. Probabilistic modeling and predictive analytics are used by cobots to anticipate and react to human intentions and actions. For instance, based on a collaborator's present trajectory, mathematical models might be employed to predict where they are likely to move next.
7. **Control and Feedback:** Cobots depend on control loops that modify their behavior in response to feedback from sensors. To maintain desired locations, forces, or speeds, control theory mathematical principles, such as proportional-integral-derivative (PID) control and state feedback control, are used.

In conclusion, collaborative robotics (cobots) rely on an intricate balancing act of several technologies and mathematical concepts. These ideas cover a variety of topics, including kinematics, dynamics, computer vision, control theory, optimization, and probability theory. These mathematical underpinnings enable cobots to effectively and safely cooperate with people, creating new opportunities for increased productivity and adaptability across a range of industries.

11.2.2 Intelligent Process Automation (IPA)

In order to automate complex commercial processes including cognitive and decision-making tasks, IPA combines robotic process automation (RPA) with AI and machine learning. With mundane jobs being automated, this trend is revolutionizing sectors like banking, healthcare, and customer service while freeing up human workers to work on more value-added tasks. Intelligent Process Automation (IPA) is the process of streamlining and improving complicated business processes through the use of automation technologies like Robotic Process Automation (RPA) and Artificial Intelligence (AI). Through AI and machine learning, IPA seeks to not only automate repetitive work but also improve cognitive and decision-making abilities.

The technology underlying IPA consists of a variety of parts, each of which relies on mathematical concepts to work.

1. **Robotic Process Automation (RPA):** RPA automates repetitive, rule-based operations by imitating human behavior in computerized systems. RPA bots interact with user interfaces to do activities including form processing, data entry, and data extraction. Although RPA's fundamental functionality doesn't largely rely on complex mathematics, it does include scripting and logical sequencing.
2. **Artificial Intelligence (AI) and Machine Learning (ML):** IPA's capacity to manage challenging data-driven activities depends heavily on AI and ML. For the algorithms and models in these technologies, a strong mathematical foundation is necessary.

   Machine Learning Algorithms: To develop and train machine learning models, mathematical ideas from the fields of statistics, calculus, and linear algebra are used. IPA frequently uses machine learning problems like classification, regression, clustering, and natural language processing (NLP).

   Deep Learning: A branch of computer learning, deep learning employs layers of neural networks to extract complicated patterns from data. Deep neural networks must be trained and optimized using calculus and linear algebra.

   Natural Language Processing (NLP): NLP techniques call for mathematical tools like probabilistic models (like Hidden Markov Models) and linear algebra (for word embeddings) in order to analyze and comprehend text.

3. **Process Mapping and Optimization:** Modeling and improving business processes involves the use of mathematics. To analyze process flows, spot bottlenecks, and optimize resource allocation, methods from operations research, optimization theory, and queuing theory are employed.
4. **Data Analysis and Decision Support:** A common IPA task is to analyze huge datasets in order to gain knowledge and assist with decision-making. In order to derive meaningful information from data, statistical analysis, hypothesis testing, and data visualization approaches rely on mathematics.
5. **Cognitive Automation:** Cognitive automation involves automating cognitive processes including pattern recognition, comprehension of natural language, and judgement. This is dependent on mathematical models, which provide computers the ability to comprehend intricate facts and information.
6. **Process Automation Workflow:** Using logic and control theory, multiple automation components are orchestrated in an end-to-end workflow. The order of tasks is decided using logic-based rules by workflow automation platforms and business process modelling tools.
7. **Predictive Analytics:** Predictive analytics projects future results using previous data and statistical methods. Regression analysis and other analytical methods are used to forecast trends and guide decision-making.
8. **Robustness and Dependability:** The robustness and dependability of IPA systems are guaranteed by mathematics. Systems that can gracefully accept exceptions and errors are designed using concepts from reliability engineering, such as failure analysis and fault tolerance.

In conclusion, Intelligent Process Automation (IPA) is a complex strategy that optimizes business processes by fusing automation technology, artificial intelligence, and machine learning. Numerous mathematical concepts, like as statistics, linear algebra, calculus, and optimization theory, are necessary for the effective use of IPA. These mathematical underpinnings allow IPA systems to automate processes, analyze data, make defensible judgments, and optimize workflows, ultimately resulting in more productivity and better commercial outcomes.

11.2.3 Hyper-Automation

Hyper-automation is the creation of end-to-end automated processes by combining several automation technologies, such as RPA, AI, and analytics. It seeks to improve efficiency and cut costs by streamlining procedures throughout an organization. This tendency could have an influence on a variety of sectors and jobs. Hyper-automation is the creation of end-to-end automated processes that are used throughout an organization by combining several automation technologies, such as Robotic Process Automation (RPA), Artificial Intelligence (AI), machine learning, and advanced analytics. By combining the capabilities of numerous technologies, this complete strategy seeks to maximize efficiency, decrease manual involvement, and improve decision-making.

Hyper-automation technology is made up of a variety of interconnected parts, each having their own mathematical foundations.

1. **Robotic Process Automation (RPA):** RPA automates repetitive, rule-based processes, laying the groundwork for hyper-automation. Bots that replicate human interactions with computer systems are programmed using mathematical ideas such as logical sequencing and conditional statements.

2. **Artificial Intelligence (AI) and Machine Learning:** By allowing the system to undertake more difficult and cognitive tasks, AI and machine learning are essential to hyper-automation.

   Supervised and Unsupervised Learning: By combining supervised and unsupervised learning algorithms, hyper-automation systems may learn from previous data. In these algorithms, feature extraction, model training, and pattern recognition are all accomplished using statistical techniques and linear algebra.

   Natural Language Processing (NLP): NLP approaches use statistical models and mathematical methodologies for sentiment analysis, language comprehension, and text synthesis. NLP operations involve text and language analysis.

   Reinforcement Learning: Reinforcement learning, a subfield of machine learning that employs mathematical optimization to train agents to make sequential decisions and discover the most effective tactics through trial and error, is sometimes incorporated into hyper-automation systems.

3. **Advanced Analytics and Data Integration:** To make intelligent decisions, hyper-automation depends on data from many sources. Advanced analytics techniques use mathematical techniques from statistics and data science for data pretreatment, modification, and aggregation.

4. **Process Mapping and Optimization:** Complex business processes are modeled, simulated, and optimized using mathematics. To locate bottlenecks, improve resource distribution, and forecast results, concepts from operations research, optimization theory, and simulation are used.

5. **Workflow Orchestration:** Workflow orchestration is a component of hyper-automation, which coordinates task flows across various automation systems. To make sure that tasks are completed in the proper order and dependencies are successfully managed, workflow management solutions make use of logic and control theory.

6. **Predictive and Prescriptive Analytics:** Hyper-automation makes use of predictive analytics to project upcoming patterns and results. Predictive models are created using mathematical methods like regression and time series analysis. Prescriptive analytics, which employ optimization and decision-making algorithms, also uses data and constraints to help prescribe the best course of action.

7. **Cognitive Automation and Decision Support:** Mathematical knowledge of complex data and context enables cognitive automation. AI-driven decision support systems analyze data using mathematical models to offer useful information to human decision-makers.

8. **AI-Powered Automation Strategy:** AI-powered automation methods are created using mathematics. In order to do this, data must be analyzed to identify jobs that may be automated, resources must be allocated optimally, and potential benefits must be calculated.

9. **Performance Monitoring and Adaptive Systems:** Feedback loops and adaptive techniques are frequently used in hyper-automation systems. The system adjusts to shifting situations and maintains desired performance levels thanks to control theory and feedback control techniques.

To sum up, hyper-automation is an all-encompassing strategy that integrates a range of automation technologies and mathematical principles to produce end-to-end automated operations. Hyper-automation systems may optimize operations, make educated judgments, and adapt to dynamic settings by utilizing mathematical principles from a variety of domains, including AI, machine learning, statistics, optimization, and control theory.

11.2.4 Remote Work and Digital Collaboration Tools

The COVID-19 epidemic has pushed the use of digital technologies for collaboration and remote work. Virtual assistants, workflow automation, and AI-powered analytics are just a few examples of the automation technologies that have been crucial in sustaining business operations and fostering remote team collaboration. In today's connected and globalised society, remote work and digital collaboration tools have become indispensable for enabling teams to work together seamlessly despite geographical distances.

These tools make use of a variety of technologies to facilitate work automation, communication, and collaboration. Some mathematical principles underlie these technologies' functionality and optimization even if mathematics isn't usually directly involved.

1. **Video Conferencing and Collaboration Platforms:** Platforms enabling real-time video, audio, and text collaboration include Zoom, Microsoft Teams, and Slack. These applications use algorithms for video compression, transmission, and network optimization to ensure smooth communication, even though their basic operation doesn't heavily rely on math.
2. **Tools for Managing Tasks and Projects:** Sites like Trello, Asana, and Jira facilitate the management of tasks, projects, and workflows. Although not inherently mathematical, these tools frequently include fundamental scheduling and organization ideas, which are considered to be applications of mathematics.
3. **Cloud Computing and File Sharing:** Documents and files can be accessed remotely through cloud storage services like Google Drive, Dropbox, and Microsoft OneDrive. Concepts from networking, data storage, and data management are essential to cloud computing.
4. **Workflow Automation:** Users can connect different apps and automate repetitive operations using workflow automation solutions like Integromat and Zapier. These technologies make use of mathematical logic-related ideas like conditional logic and trigger-based actions.
5. Data analysis and visualization are performed using programs like Microsoft Excel, Google Sheets, and Tableau. These tools execute fundamental mathematical operations, but when analyzing data and producing visualizations, more sophisticated statistical and mathematical techniques are used.
6. **Artificial Intelligence and Machine Learning:** Some collaboration solutions make use of AI and ML for automated responses, chatbots, and predictive analytics. These functionalities are fundamentally based on the mathematical ideas of algorithms, model training, and statistical analysis.
7. **Cyber Security and Encryption:** Tools for remote work and collaboration frequently come with security features like encryption, which uses mathematical techniques to protect data and communications.
8. **Network Optimization:** To optimize data routing, minimize latency, and guarantee secure connections, virtual private networks (VPNs) and other remote access technologies use mathematical methods.
9. **Time Tracking and Analytics:** To compute and display productivity indicators, time tracking applications and analytics dashboards frequently use fundamental mathematical ideas like arithmetic and statistics.
10. Machine Translation Natural language processing (NLP) approaches, which rely on probabilistic models and statistical methodologies for language understanding and translation, are used by collaboration platforms with language translation features.

11. **Scheduling Optimization Algorithms:** To plan tasks, distribute resources, and manage workloads, some collaboration solutions, particularly those pertaining to project management and resource allocation, use optimization algorithms.

In conclusion, even though not all digital collaboration platforms directly include complicated mathematics, a number of mathematical ideas underpin their operation and optimization. These tools' efficiency and efficacy are boosted by ideas from networking, data analysis, encryption, artificial intelligence, and optimization, allowing for smooth remote work and cross-border collaboration.

11.2.5 Reskilling and Upskilling

Reskilling and upskilling programs have become crucial for ensuring that the workforce remains relevant and adaptable as automation transforms the nature of employment. Companies and educational institutions are investing in initiatives that give workers the knowledge and abilities they need to collaborate with automation technologies and perform more challenging tasks. To prepare the workforce for the shifting demands of automation and developing technologies, reskilling and upskilling are essential measures. In order for employees to stay relevant and flexible in the changing labor market, these tactics entail giving them new skills and information. Although technology and mathematics are not fundamental to reskilling and upskilling, they play significant roles in providing effective training and instruction.

1. **Online Learning Platforms:** These offer a variety of courses and programs for upskilling and reskilling, such as Coursera, edX, and Khan Academy. Technology is used by these platforms to distribute instructional content, such as movies, tests, interactive activities, and assignments.
2. **E-Learning Management Systems:** Organizations can build and manage online courses using e-learning management systems (LMS), including Moodle and Blackboard. These systems frequently include tools for delivering information, conducting assessments, assigning grades, and communicating; these tools call for elementary programming and database administration.
3. **Augmented Reality (AR) and Virtual Reality (VR) Technologies:** These are being employed more and more for immersive learning environments. In addition to virtual labs and possibilities for hands-on learning, they also provide simulations. The design of is supported by mathematics, notably in geometry and spatial thinking.
4. **Artificial Intelligence (AI) and Personalization:** Platforms powered by AI can tailor learning experiences based on unique learning preferences and styles as well as strengths and shortcomings. Algorithms evaluate the progress of learners and modify the content as necessary. Natural language processing is used in machine learning approaches to deliver recommendations and comments that are specifically tailored to the user.
5. **Data Analytics and Learning Insights:** Learning platforms frequently gather and analyze data on user progress, engagement, and results. To learn more about the efficacy of training programs, data analytics approaches such as statistics and data visualization are used.
6. **Gamification and Serious Games:** To increase motivation and engagement, gamification involves integrating game aspects into instructional content. Game-based situations are used in serious games to teach lessons. Challenges, scoring systems, and feedback loops are all designed using mathematical principles in both gamification and serious games.
7. **Simulation and Modeling:** For training purposes, simulation technologies are employed to simulate real-world circumstances. Simulations can assist students in honing their technical

abilities in a safe setting. Accurate simulations reflecting real-world dynamics are produced using mathematics.

8. **Learning Algorithms and Adaptive Assessments:** Learning algorithms evaluate students' performance and replies to offer tailored evaluations and suggestions for more study. These algorithms make use of ideas from pattern recognition and machine learning.
9. **Programming and Coding Skills:** These abilities are becoming more and more valuable as automation and technology improve. Programming languages and coding techniques, which have mathematical roots in logic and algorithms, are taught through courses offered by learning platforms and coding bootcamps.
10. **Critical Thinking and Problem-Solving:** Although not directly related to technology, improving one's ability to think critically and solve problems is an essential part of reskilling and upskilling. These abilities call for analytical thinking and reasoning, which incorporate logical thought processes and mathematical ideas.

In conclusion, technology and mathematics play important roles in providing effective training and education, even though they are not the only emphasis of reskilling and upskilling. The workforce needs to be prepared for an automated and digitized future, and online learning platforms, AI-driven personalization, simulation technologies, data analytics, and programming skills all rely on these concepts.

11.2.6 Augmented Analytics

Through augmented analytics, which uses AI and machine learning to automate insights production, data preparation, and pattern detection, automation is increasing data analysis and decision-making. Because of this, business users can make decisions based on data without needing to have highly developed analytical abilities. Data preparation, analysis, and interpretation are all improved by augmented analytics, a technology-driven method that blends artificial intelligence (AI), machine learning, and data analytics.

This strategy strives to provide non-technical users with advanced analytics capabilities, empowering them to more successfully make data-driven decisions. Although not all aspects of augmented analytics directly use mathematics, many of their components do so in order to function.

1. **Data Preparation and Cleaning:** To detect and manage missing values, outliers, and inconsistencies in datasets, automated data preparation solutions use algorithms. Data cleaning and preprocessing methods from statistics are used, such as mean imputation and standard deviation.
2. **Natural Language Processing (NLP):** NLP enables users to query data in natural language. Tokenization, sentiment analysis, and named entity recognition are a few examples of techniques that use statistical models and mathematical methods to analyze and comprehend language.
3. **Auto ML (Automated Machine Learning):** Auto ML platforms automate feature engineering, adjusting hyper parameters, and model selection. These systems use algorithms for optimization, frequently using methods from evolutionary or Bayesian algorithms.
4. **Data Exploration and Visualization:** Interactive visualizations produced by augmented analytics technologies are frequently used by users to examine and comprehend data. These visualizations use mathematical ideas for scale, representation, and data modification.
5. **Pattern Recognition and Anomaly Detection:** To find patterns and anomalies in data, machine learning methods are used. To find data linkages and outliers, techniques like clustering, classification, and anomaly detection rely on mathematical models.

6. **Predictive and Prescriptive Analytics:** Augmented analytics solutions provide insights that are both predictive and prescriptive. Based on past data, predictive models utilize mathematical algorithms to predict future results. To recommend the best actions, prescriptive analytics uses optimization algorithms.
7. **Feature Selection and Dimensionality Reduction:** Principal component analysis (PCA) and feature selection methods, both of which are part of machine learning, lower the dimensionality of data while preserving important information. These techniques are based on matrix operations and linear algebra.
8. **Time Series Analysis:** To model and forecast trends for time-series data, techniques like autoregressive integrated moving average (ARIMA) and exponential smoothing are used. These techniques rely on time-series modeling's mathematical underpinnings.
9. **Statistical Analysis and Hypothesis Testing:** Using augmented analytics, significant links in data can be found and hypotheses can be confirmed using statistical testing. T-tests and ANOVA are two statistical and mathematical reasoning techniques.
10. **Data-driven Insights and Recommendations:** Collaborative filtering and matrix factorization techniques are the foundation of recommendation systems, which are frequently employed in e-commerce and content platforms. Linear algebra and optimization techniques are used in these methods.
11. Data storytelling is the process of developing narratives based on data insights. Effective data storytelling frequently calls for knowledge of statistical significance, causality, and relationships, despite not being directly quantitative.

To sum up, augmented analytics uses a mix of AI, machine learning, and data analytics to give users better insights and decision-making abilities. Even though not every facet of augmented analytics is substantially technical, many of its elements use mathematical ideas, algorithms, and models. Non-technical individuals can gain valuable insights from complicated data sets as a result of this synergy and use those insights to make wise decisions.

11.2.7 Autonomous Vehicles and Drones

The use of autonomous vehicles and drones for delivery and transportation is changing in the automotive and logistics sectors. While these technologies may improve productivity and decrease human error, they also raise concerns about job displacement in industries like delivery and driving.

Drones and autonomous cars are examples of cutting-edge automation technology that could revolutionize logistics, transportation, and other industries. For navigation and interaction with their surroundings, these technologies rely on a combination of sensors, algorithms, and control systems. Their automation and safe and effective operation are both made possible by mathematics.

1. **Sensor Fusion:** GPS, cameras, lidar, radar, ultrasonic sensors, and lidar are just a few of the sensors that autonomous vehicles and drones are equipped with. The goal of sensor fusion approaches is to build a thorough picture of the environment around the vehicle by merging data from several sensors. For the purpose of accurately combining and interpreting sensor data, mathematics—particularly statistics and signal processing—is used.
2. **Perception and Object Detection:** To interpret sensor data and identify objects, pedestrians, cars, and barriers in real-time, mathematical algorithms, such as computer vision techniques and deep learning models, are used. To identify and categorize items, image processing, feature extraction, and neural networks are used.

3. **Path Planning and Trajectory Optimization:** Drones and autonomous vehicles need to map out the best routes to take to get where they're going while avoiding hazards. To find secure and effective routes, robotics and optimization techniques like dynamic programming and A* search algorithms are used.
4. **Control Systems:** Maintaining stability and control of autonomous vehicles and drones requires a solid understanding of control theory. Control algorithms that modify vehicle dynamics and react to changing circumstances are created using calculus-related ideas and differential equations.
5. **Localization and Mapping:** Simultaneous Localization and Mapping (SLAM) is a technique for predicting the position of a moving object within a map of its environment. To obtain precise localization and estimation, mathematical models for odometry, geometry, and probability are utilized.
6. **Path Following and Trajectory Tracking:** Autonomous vehicles and drones accurately follow preset courses and trajectories thanks to mathematical methods for feedback control. For trajectory tracking, proportional-integral-derivative (PID) control and other control methods are used.
7. **Machine Learning for Decision-Making:** Complex judgments are made in dynamic contexts using machine learning models, such as reinforcement learning and probabilistic models. These models use probabilistic reasoning and mathematical optimization concepts to learn from past data and adapt to changing conditions.
8. **Collision Avoidance:** To avoid collisions with other cars and objects, collision avoidance systems compute the safest distances, speeds, and manoeuvres using mathematical algorithms. Effective collision avoidance systems depend on prediction models and real-time data processing.
9. **Localization and Navigation Algorithms:** Accurate GPS-based localization and navigation rely on mathematical methods like triangulation and trigonometry. Kalman filters and other estimating techniques help to increase the precision of placement.
10. **Sensor Calibration:** Mathematical transformations and models are used to calibrate sensors for accuracy and consistency. For instance, sensor readings for lidar must be calibrated using geometric and algebraic ideas.

In conclusion, in order to function independently, autonomous vehicles and drones rely on a combination of sensor technology, control systems, path planning, and data analysis. These technologies' design, navigation, and safety are all based on mathematics, which includes a wide range of disciplines like geometry, calculus, statistics, and control theory. This ensures that the technologies work effectively in a variety of complicated and dynamic contexts.

11.2.8 AI-Powered Recruitment and Talent Management

By automating the candidate sourcing, screening, and matching processes, AI-driven tools are revolutionizing the hiring process. AI is also used to analyze employee performance data, which enables businesses to make better choices regarding hiring, training, and talent management.

Artificial intelligence (AI) and data analytics are used in AI-powered recruitment and talent management to improve the process of finding, assessing, and managing talent within organizations. The recruitment lifecycle is streamlined by these technologies in a number of ways, from candidate sourcing to employee performance evaluation. Not all parts of these technologies actually use mathematics, but many of its features rely on mathematical ideas and techniques.

1. **Candidate Sourcing and Matching:** To match the best candidates to available positions, AI algorithms examine candidate profiles and job specifications. To comprehend and match keywords and context, these algorithms employ natural language processing (NLP) methods, which include probabilistic models and semantic analysis.
2. **Screening and Resume Parsing:** AI-powered systems glean pertinent data from resumes and applications. To identify abilities, expertise, and qualifications, these systems rely on NLP and machine learning algorithms, frequently using statistical models and classification approaches.
3. **Predictive Analytics for Recruiting Success:** Predictive analytics models anticipate which candidates will succeed in a specific function based on historical recruiting data. To create predictive models based on past trends, regression analysis and machine learning mathematical approaches are used.
4. **Assessment of Behavioral and Skill Characteristics:** AI platforms can evaluate candidates' behavioral and skill characteristics through online tests. These evaluations frequently involve data analysis and psychometric testing, which call for mathematical measurement and statistical analytic principles.
5. **Video Interviews and Analysis:** AI-powered video interview systems can use machine vision and NLP to analyse candidates' verbal and nonverbal signs. These platforms might use sentiment analysis and facial recognition algorithms, which rely on linguistic and mathematical models for picture processing.
6. **Detection and Mitigation of Prejudice:** AI algorithms are used to identify and address unintentional bias in hiring procedures. Identification of bias trends in hiring decisions is aided by statistical approaches and probabilistic models.
7. **Analysis of Employee Performance:** Talent management tools analyze employee performance data using AI to identify top performers and areas for development. Performance metrics are evaluated using data analytic techniques, such as regression analysis and data visualization.
8. **Succession Planning and Skill Gap Analysis:** AI-powered systems can find candidates for upcoming leadership roles based on performance and skill data. A mathematical comparison of desired skills and existing abilities in the workforce is used in skill gap analysis.
9. **Recommendations for On Boarding and Training:** AI algorithms can provide tailored on boarding and training programs for new hires based on their profiles and abilities. Machine learning models are frequently used in these suggestions to recommend the best learning paths.
10. **Workforce Analytics:** AI-driven analytics technologies assist businesses in understanding the trends and patterns in their workforce. Correlations are found and data-driven judgements are made using mathematical approaches including clustering, regression, and time series analysis.
11. **Natural Language Processing for Feedback Analysis:** AI systems examine sentiment in employee feedback to determine engagement and satisfaction. The use of mathematical language models and probabilistic techniques is a key component of NLP techniques like sentiment analysis and topic modeling.

In conclusion, AI-powered recruitment and talent management solutions make use of a range of technologies to speed up the hiring process and improve personnel management. Although not all parts require sophisticated mathematics, statistical, natural language processing, and machine learning mathematical ideas are essential for creating precise algorithms, models, and analyses that improve the efficacy and efficiency of these technologies.

11.2.9 Smart Manufacturing and Industry 4.0

In order to implement Industry 4.0, manufacturing processes must incorporate automation, IoT, cloud computing, and AI. This development makes it possible for "smart factories" to adjust to shifting customer needs in real-time, improving both production efficiency and quality.

In order to build more effective, adaptable, and data-driven production systems, smart manufacturing, which is frequently related with Industry 4.0, refers to the integration of digital technology, automation, data analytics, and IoT (Internet of Things) into the manufacturing process. Design, optimization, and use of these technologies all heavily rely on mathematics.

1. **Internet of Things (IoT) and Sensors:** To collect data in real-time, IoT devices and sensors are integrated into manufacturing processes and equipment. Signal processing, statistical analysis, and data fusion are examples of mathematical techniques that are used to process sensor data and derive insightful conclusions.
2. **Data Analytics and Big Data:** Smart industrial technologies create enormous volumes of data. To draw conclusions from the data, spot patterns, and improve processes, data analytics techniques such as statistical analysis, machine learning, and data visualisation are utilised.
3. Proactive maintenance is made possible by predictive maintenance, which employs machine learning algorithms to forecast when equipment is likely to break. Based on past data, methods like regression analysis and time-series forecasting are used to make predictions about equipment breakdowns.
4. **Digital Twins:** A digital twin is a representation of a physical system, process, or product that exists digitally. Accurate digital copies that can be utilized for testing, optimization, and predictive analysis are produced using mathematical modeling techniques like simulation and finite element analysis.
5. **Cyber-Physical Systems:** Cyber-physical systems combine computer algorithms with networking and physical processes. To keep the physical and digital components in sync and coordinated, real-time algorithms, embedded systems, and control theory are used.
6. **Cloud and Edge Computing:** These two types of computing help handle and store the massive amounts of data produced by smart manufacturing. To control data processing and storage, resource allocation models, distributed algorithms, and optimization approaches are used.
7. **Robotics and Automation:** The foundation of smart manufacturing is the use of industrial robots and automated systems. Kinematics, dynamics, and control theory are all used in the design of robotic movements in order to ensure precision and optimize robotic trajectories.
8. **Supply Chain Optimization:** Inventory management, demand forecasting, transportation routing, and scheduling mathematical models are used to optimize supply chains. Supply chain activities are made more efficient by using optimization algorithms, linear programming, and queuing theory.
9. **Process Simulation and Optimization:** Manufacturing processes are simulated and optimized mathematically to find areas for improvement. Finding the ideal settings and combinations for optimum performance is made easier by optimization algorithms.
10. **Quality Control and Statistical Process Control (SPC):** Statistical tools like hypothesis testing, control charts, and analysis of variance (ANOVA) are used in smart manufacturing for quality control. These techniques make sure items satisfy quality requirements and identify deviations.

11. **Human-Machine Interaction:** In smart manufacturing systems, user interfaces and human-machine interfaces (HMIs) rely on usability design and human-computer interaction (HCI) principles. These elements may require user behavior analysis and usability testing, although not being fundamentally mathematical.

In conclusion, a combination of different technologies and mathematical ideas is required for smart manufacturing and Industry 4.0 technologies. Mathematics underpins the design, operation, and optimization of smart manufacturing systems, leading to increased efficiency, quality, and flexibility in the manufacturing process. This includes data analytics and optimization, control theory, and simulation.

11.2.10 Ethical and Social Considerations

As automation spreads, ethical questions about AI bias, job loss, and the potential for technology to exacerbate already-existing injustices grow increasingly important. A crucial trend in the automation landscape is the balancing of technology advancement with social responsibility.

Ethical and social concerns are becoming more and more crucial as automation technologies develop. These factors take into account how automation may affect society, people, and the workforce. Some mathematical concepts are applicable to comprehending and resolving these ethical and social difficulties, even though mathematics is not directly involved in all aspects.

1. **Prejudice and Fairness in Algorithms:** Automated systems that are powered by algorithms have the potential to unintentionally inject prejudice and injustice into the decision-making process. The measurement and quantification of bias, as well as the creation of just and equitable algorithms, all depend on mathematics.
2. **Data Protection and Privacy:** Automation frequently depends on the collection and processing of significant amounts of personal data. Encryption, secure data exchange, and anonymization are used in conjunction with cryptography and mathematical approaches to guarantee data privacy.
3. **Privacy and Surveillance Issues:** Automated security measures like facial recognition raise questions about civil liberties and personal privacy. Algorithms and mathematical models can be used to predict how monitoring might affect individual liberties.
4. **Explain Ability and Transparency:** Complicated machine learning models might be challenging to comprehend. Mathematical methods are used in techniques like interpretability algorithms and model visualization to describe how AI systems make decisions.
5. **Accountability and Responsibility:** It can be difficult to assign blame for automated systems' deeds. The roles and duties of different stakeholders, such as developers, users, and organizations, are defined with the aid of concepts from logic and ethics.
6. **Work Displacement and Economic Impact:** The work market may shift as a result of automation. The probable effects on employment rates, wages, and general economic dynamics are forecasted with the aid of mathematical models and economic theories.
7. **Ethical Decision-Making and Moral Dilemmas:** Formalising ethical principles and weighing the repercussions of various options are two aspects of mathematical models for ethical decision-making. Moral conundrums brought forth by automated systems can be resolved with the aid of ethical frameworks and mathematical reasoning.
8. **Bias Detection and Mitigation:** Detecting bias in automated systems and reducing it need for mathematical and statistical methods. Data and algorithmic discriminatory patterns can be found using statistical methods.

9. **Consent and Autonomy:** Automated systems may affect people's capacity for giving informed consent or exercising autonomy. Principles of autonomy are frequently found in legal frameworks and ethical standards, and they may be backed by mathematical principles of logic and reasoning.
10. **Digital Divide and Access Inequality:** Automation technologies have the potential to make disparities in education and technology access even worse. The prospective effects of automation on underserved populations can be evaluated using mathematical models, which can also assist guide governmental decisions.
11. **Social Impact Assessments:** To predict how automation may affect many societal aspects such as healthcare, education, and public services, social impact assessments can be carried out using mathematical modeling and data analytic approaches.

In conclusion, ethical and social issues in automation demand a multidisciplinary strategy that combines ethical concepts and mathematical thinking. For automation technologies to have a good impact on society, mathematical concepts are essential for assessing bias, analyzing impact, deriving meaning from algorithms, and developing just and equitable systems.

11.2.11 Human-AI Collaboration

Automation is increasingly concentrating on augmenting human talents as opposed to replacing people. Designing systems where AI supports humans in decision-making, problem-solving, and creative tasks, leading to synergistic results, is referred to as human-AI collaboration.

The term "human-AI collaboration" describes the interaction of human and artificial intelligence in a variety of tasks, such as problem-solving and decision-making. In this partnership, AI will be used to improve both human and machine performance. Although not all parts directly require maths, the creation and improvement of the technologies that enable this collaboration are guided by mathematical ideas.

1. **Natural Language Processing (NLP):** NLP makes it possible for humans and AI systems to connect and communicate. There are mathematical underpinnings for methods like sentiment analysis, named entity recognition, and text synthesis that use probabilistic models and linguistic algorithms.
2. **Sentiment Analysis and Emotion Detection:** AI can deduce emotions and sentiments from human language and facial expressions. Based on textual or visual signals, emotional states are categorized using statistical models and pattern recognition algorithms.
3. **Recommender Systems:** Recommender systems driven by AI make recommendations for goods, content, or behaviors based on user preferences and behavior. To forecast user preferences, these systems use mathematical techniques like collaborative filtering and matrix factorization.
4. **Decision Support Systems:** By analyzing massive datasets and offering insights, AI systems can help humans make well-informed decisions. The creation of suggestions involves the use of statistical analysis, probability theory, and machine learning algorithms.
5. **Data Interpretation and Visualization:** By creating visualizations and summaries, AI systems can assist humans in understanding complex data. Mathematical concepts from geometry, statistics, and visualization design are used in data visualization strategies.
6. **Automated content generation:** AI is capable of producing content like articles, reports, and code. Natural language generation (NLG) methods use grammatical rules and mathematical language models to create coherent text.

7. **Medical Diagnostics and Image Analysis:** AI helps with image analysis and diagnosis in industries like medicine. Applying mathematical techniques like deep learning, pattern recognition, and image processing
8. **Cognitive Augmentation:** AI can improve human cognitive skills like pattern identification and memory recall. Mathematical models are used in techniques like brain augmentation to enhance cognitive abilities.
9. **Usability and Human-AI Interaction:** Creating user-friendly interfaces for human-AI interaction requires consideration of usability and user experience (UX) design principles. Usability testing and human behavior analysis can benefit from mathematical methods even though they are not mathematical in nature.
10. **Game Playing and Strategy:** AI programs can work with people in games and in situations requiring strategic thinking. AI and humans can collaboratively design tactics with the use of game theory and optimization tools.
11. **Support for Ethical Decision-Making:** AI can inform humans about decisions' potential ethical ramifications. The formalization of ethical principles and the evaluation of outcomes are key components of mathematical models for ethical decision-making.
12. **Learning and Skill Enhancement:** Educational systems with AI-powered technology can adjust to students' specific learning preferences and rates. Machine learning algorithms use mathematical techniques for personalization to analyze performance data and modify content distribution.

In conclusion, human-AI collaboration makes use of AI technologies to improve human talents and judgment. Many mathematical principles are used in the development and optimization of AI technologies that enable this collaboration, ranging from natural language processing and image analysis to statistical modeling and ethical decision-making support, even though not all aspects involve complex mathematics.

11.2.12 Customization and Personalization

By examining client information and preferences, automation is allowing companies to create highly customized goods and services. In e-commerce, marketing, and healthcare, automation allows experiences to be tailored to specific user demands.

Customizing and personalizing products, services, and experiences to suit customer preferences and requirements. Organizations can deliver personalized services effectively and at scale thanks to automation technologies, which are frequently powered by AI and data analytics. Although not all parts explicitly require mathematics, the algorithms and models that drive these technologies are based on mathematical ideas.

1. **Data Gathering and Profiling:** Automation systems gather information from a range of sources, such as user behavior, preferences, and previous interactions. The creation of thorough user profiles is aided by mathematical data collection methods like sampling and data aggregation.
2. **Recommender Systems:** To make product, service, or content recommendations, recommender systems examine user data. To match user preferences with objects in a database, mathematical procedures called collaborative filtering and content-based filtering are used.
3. **Machine Learning Models:** Machine learning models, such as decision trees, neural networks, and clustering algorithms, are frequently used in personalization. These models use regression, classification, and mathematical optimization to create predictions based on data patterns.

4. **Natural Language Processing (NLP):** NLP is used to evaluate and comprehend user-generated text, allowing for customized responses and suggestions. Probabilistic models and language algorithms are used in techniques like topic modeling and sentiment analysis.
5. **Dynamic Content Generation:** Automation can create content that is dynamically targeted to certain consumers. In order to do this, mathematical models are used to alter the text, pictures, and layouts according on user information.
6. **A/B Testing and Experimentation:** In A/B testing, various versions of content or features are tried to see which performs better with users. Mathematical statistics play a role in this process. Decision-making is aided by statistical significance calculations and hypothesis testing.
7. **Customer Segmentation:** By grouping customers with comparable traits, mathematical clustering techniques enable focused marketing and personalization tactics for particular segments.
8. **Contextualization:** By taking factors like location and device type into account, personalization is frequently improved. Context-based customization benefits from mathematical techniques like responsive design and geographical analysis.
9. **Time Series Analysis:** Time series analysis aids in the modeling of trends and patterns across time when personalization involves temporal data. This is crucial for recommendations that are specific to the user, such as in the distribution of material or financial services.
10. **Privacy-Preserving Personalization:** Encryption and anonymization methods are necessary to protect user privacy while enhancing experiences. User data is secured using mathematical techniques and cryptography.
11. **Optimization Algorithms:** Algorithms for mathematical optimization assist in selecting the optimal arrangement of personalized components, such as product recommendations, to increase user engagement or pleasure.
12. **Reinforcement Learning:** In some circumstances, such as in adaptive educational platforms, reinforcement learning is employed for personalization. The learning process is guided by mathematical optimization approaches that modify how the knowledge is delivered.
13. **User Experience (UX) Design:** Although UX design is not exclusively mathematical, it frequently entails usability testing and user behavior analysis, both of which can benefit from mathematical methodologies.

In conclusion, automation technologies that use AI, data analytics, and mathematical algorithms to customize experiences for specific people are what enable customization and personalization. The algorithms, models, and statistical methods used by organizations to effectively deliver personalized offers across a variety of disciplines, from e-commerce to healthcare, are all based on mathematical concepts.

11.3 ROBOTIC PROCESS AUTOMATION (RPA)

The disruptive technology known as robotic process automation (RPA) automates routine, rule-based processes by deploying software robots or "bots." RPA enables businesses to automate processes, boost productivity, cut down on errors, and free up human employees to work on projects with higher added value. Increased operational agility, financial savings, and technology breakthroughs are a few of the drivers that have fueled the growth of RPA. Here are some crucial elements of the development of RPA:

1. **Technological Developments:** RPA's capabilities have been greatly improved by developments in AI, machine learning, and cognitive technologies. Bots can now carry out increasingly difficult jobs, including those requiring judgement calls, natural language processing, and data analysis.
2. **Cost Effectiveness:** Compared to conventional IT initiatives, RPA provides a cost-effective approach for automating activities. Bots may be swiftly introduced without requiring significant changes to current systems, which lowers development and deployment costs.
3. **Scalability:** RPA makes it simple for businesses to scale their automation initiatives. It is a versatile option for enterprises of all sizes because extra bots can be deployed to accommodate rising demands.
4. **Lessening of Errors:** RPA lessens human errors related to repetitious jobs. Bots adhere to predetermined rules and routines, reducing the possibility of errors brought on by fatigue or human error.
5. **Quick Implementation:** RPA can be implemented without a lot of coding or integration work. Numerous RPA platforms include intuitive user interfaces that let non-technical users create, test, and implement automation workflows.
6. **Integration with Legacy Systems:** RPA can connect legacy systems to cutting-edge technology. In order to facilitate data sharing and process automation throughout the organization, bots can communicate with a wide range of programs and systems.
7. **Greater Productivity:** RPA enables staff to concentrate on strategic and creative tasks that call for human judgment and expertise by automating time-consuming jobs. Overall productivity and job satisfaction increase as a result.
8. **Regulatory Compliance:** RPA ensures compliance by regularly adhering to predetermined norms and procedures. Organizations may keep track of changes, keep audit trails, and show that they are following rules.
9. **Improved Client Experience:** RPA may speed up operations like order processing, support ticket resolution, and client on boarding, resulting in quicker response times and more customer satisfaction.
10. **Industry Adoption:** RPA has found success in a number of sectors, including manufacturing, healthcare, finance, and customer service. It can be applied to a variety of procedures thanks to its versatility.
11. **Future Potential:** RPA is anticipated to becoming increasingly more advanced as AI and cognitive skills advance. Bots will be able to manage complicated decision-making, unstructured data, and more difficult tasks thanks to cognitive automation.
12. **Skill Development:** With the popularity of RPA, there is now a requirement for individuals with knowledge of automation, process improvement, and bot management. The demand for experts with knowledge in RPA implementation and management is growing.

In conclusion, organizations now approach process automation differently as a result of the rise of robotic process automation. RPA offers a significant tool for enhancing operational efficiency and freeing up human workers to concentrate on jobs that call for creativity and critical thinking thanks to its cost effectiveness, scalability, and potential for mistake reduction. As RPA technology develops further, their effect on numerous industries is likely to expand further.

11.4 IMPACT OF ROBOTIC PROCESS AUTOMATION ON VARIOUS INDUSTRIES

By streamlining procedures, increasing effectiveness, decreasing errors, and raising overall production, robotic process automation (RPA) has had a substantial impact on a number of industries. Here are some examples of how RPA has impacted particular industries:

1. **Finance & Banking:** Data entry, account reconciliation, fraud detection, and customer on boarding are just a few of the tasks that financial organizations frequently use RPA for. It aids in transaction processing speed, accuracy improvement, and regulatory compliance assurance.
2. **Healthcare:** In the healthcare industry, RPA is used for tasks including invoicing, scheduling appointments, and claims processing. Electronic health records (EHR) systems can interact with RPA systems to improve data accuracy and lessen administrative overhead.
3. **Manufacturing:** Inventory management, order processing, and supply chain management in manufacturing all use RPA. It facilitates the improvement of supply chain visibility, lead time reduction, and production workflow optimization.
4. **Call Centers and Customer Service:** Routine customer interactions like answering questions, handling requests, and updating customer information are automated using RPA. As a result, customers receive quicker service and have a better overall experience.
5. **Retail and E-commerce:** RPA helps with price monitoring, order fulfillment, and inventory management. Retailers can use it to streamline pricing plans and guarantee prompt product availability.
6. **Human Resources:** RPA automates HR procedures like payroll processing, benefits administration, and employee on boarding. This lessens manual errors and frees up HR specialists to concentrate on important projects.
7. **Insurance:** RPA is utilized in insurance for underwriting, claims management, and policy administration. It speeds up claims resolution, shortens processing times, and improves data handling quality.
8. **Logistics and Transportation:** By automating processes like order processing, route optimization, and shipment tracking, RPA helps manage logistics operations. Both operational costs and supply chain visibility are improved.
9. **Legal Services:** RPA can automate compliance, document management, and contract review tasks in law companies. It improves the accuracy of legal documentation and gives solicitors more time to concentrate on more important tasks.
10. **Energy and Utilities:** RPA is used by energy and utility firms to handle jobs including processing invoices, reading meters, and managing outages. It enhances billing accuracy and aids in scheduling maintenance.
11. **Government and Public Services:** Administrative activities, permit processing, and regulatory compliance are handled by RPA in government organizations. It improves citizen services by speeding up the processing.
12. **Real Estate:** RPA facilitates the automation of tenant interactions, lease agreement processing, and property management operations. It increases the effectiveness of tenant interactions and real estate operations.
13. **Education:** RPA can automate administrative operations, student data management, and enrollment procedures in educational institutions. By decreasing administrative delays, it improves the learning environment for students.

As a result of automating repetitive and time-consuming jobs, streamlining processes, and allowing businesses to concentrate on strategic projects, robotic process automation has had a substantial impact on a variety of industries. RPA technology's versatility enables it to be customized to meet the needs of particular industries, enhancing efficiency and consumer experiences everywhere.

11.5 RELATIONSHIP BETWEEN AUTOMATION AND THE WORKFORCE

Automation and the worker have a complicated and changing connection. Automation has the potential to drastically alter sectors and the nature of employment, driven by technologies like robotics, artificial intelligence (AI), and machine learning. While automation can improve productivity, cut costs, and spur innovation, it also raises significant questions about job displacement, skill requirements, and the future of labor in general. Here is a list of the main elements:

1. **Job Displacement and Transformation:** Certain repetitive and routine work may be replaced by automation. Manual labor, data entry, and repetitive processing tasks are particularly vulnerable to automation. But automation also opens up new career paths for those interested in creating, running, and maintaining automated systems.
2. **Upskilling and Reskilling:** New talents are becoming more in demand as technology advances. To stay relevant in the evolving employment market, workers must continuously upgrade or reskill. Programming, data analysis, and digital literacy are examples of technical talents that are becoming more and more valuable.
3. **Human-machine Cooperation:** The idea of "augmented intelligence" emphasizes the possibility of productive human-machine cooperation. When repetitive jobs are handled by automation, people can concentrate on tasks that call for creativity, critical thinking, and emotional intelligence.
4. **Economic Impact:** Automation can boost economic growth and productivity. Concerns exist, though, regarding income inequality and employment polarisation, when low-skilled and high-skilled positions gain ground at the expense of middle-skilled positions.
5. **Job Quality and Well-Being:** Although automation can increase productivity, it's crucial to take into account the quality of new positions. The welfare of employees depends on new positions offering fair pay, benefits, and chances for career advancement.
6. **Adaptability and Lifelong Learning:** It becomes crucial to have the capacity to learn new skills and adjust to new technology. Individuals must adopt a lifelong learning philosophy if they want to maintain their competitiveness in the labor market.
7. **Ethical and Social Issues:** The impact of automation includes ethical issues including prejudice in algorithms, privacy issues, and the possibility of job loss in some regions. Social safety nets and regulations to deal with these issues are essential.
8. **Job Design and Creativity:** Routine tasks can be handled by automation, allowing for the restructuring of job responsibilities to place more emphasis on creativity, problem-solving, and strategic thinking. Work may be more meaningful and gratifying for employees.
9. **Industry-Specific Impact:** The effects of automation vary in intensity across various industries. Others, like healthcare and education, require a balance between technology and human connection while others, like manufacturing and logistics, have undergone considerable transformations.

10. **Remote Work and Digital Collaboration:** The growth of remote work and digital collaboration tools has boosted automation, enabling organizations to adjust to changing conditions and continue operating in difficult times.

In conclusion, there are several facets to the link between automation and the workforce. While automation has many advantages, it also presents problems that call for proactive policies and methods to enable a smooth and inclusive transition. The secret is to educate the workforce, upskill it, and encourage a cooperative approach to human-AI interaction.

11.6 IMPACT OF AUTOMATION ON THE JOB MARKET

There is a lot of discussion and interest about how automation will affect the job market. Robotics, artificial intelligence (AI), and machine learning-based automation have the potential to radically alter the nature of the workforce. Here is a breakdown of its effects:

1. **Job Displacement:** Automation has the potential to eliminate some jobs, especially those that need repetitive, regular tasks. Data entry, administrative, and manufacturing jobs are particularly prone. These sectors may experience temporary job losses as a result.
2. **Job Creation:** Automation may lead to the loss of some employment, but it also generates new positions. Professionals with expertise in disciplines like software development, data analysis, and robotics engineering are needed for the creation, maintenance, and administration of automated systems.
3. **Skills Gap:** As more jobs become automated, problem-solving, data analysis, and digital skills are becoming more important. The skills gap, where the available labor may not have the skills required for the new occupations produced by automation, is of significant concern.
4. **Employment Transformation:** Automation changes employment roles rather than completely eliminating them. As basic tasks become automated, many occupations will change, allowing workers to concentrate on work that calls for creativity, critical thinking, and emotional intelligence.
5. **Job Polarization:** High-skilled jobs and low-skilled jobs are growing at the expense of middle-skilled jobs in a trend known as job polarization. While low-skilled jobs necessitate a human touch and interpersonal abilities, high-skilled jobs require complicated problem-solving and decision-making.
6. **Industry Impact:** The impact of automation varies in strength across different sectors. While there has been tremendous automation in the manufacturing, logistics, and retail sectors, human connection and creativity are still more important in the healthcare, educational, and creative sectors.
7. **Transition Period:** There may be a period of uncertainty and job interruption during the transition to automation. Workers in industries that face automation require assistance to change into other professions or pick up new skills.
8. **Geographical Impact:** The effects of automation differ depending on the area. Areas with a high reliance on automatable sectors may see more job relocation. In order to resolve these differences, it is crucial to take into account regional policies and activities.
9. **Income Inequality:** Because automation disproportionately affects low-skilled labor, income inequality is a worry. Upskilling opportunities and income redistribution policies are crucial.

10. **Human-AI Collaboration:** Automation doesn't always mean that human workers will be entirely replaced. Humans and AI are increasingly working together, with AI tackling repetitive jobs and humans concentrating on higher-level decision-making.
11. **Job Happiness and Well-Being:** By removing tedious jobs, automation can increase job happiness. To ensure employee satisfaction and job satisfaction, however, the calibre of new roles must be taken into account.
12. **Lifelong Learning:** The importance of lifelong learning and professional upskilling is highlighted by automation. To stay employable, people must adapt to changing technologies and pick up new skills.
13. **Regulatory and Ethical Considerations:** It is crucial to have regulations and ethical standards in place to make sure that automation serves society as a whole and doesn't have any unfavorable effects like bias, discrimination, or overzealous monitoring.

In conclusion, there are numerous ways that automation is affecting the job sector. While automation promotes innovation and efficiency advantages, it also poses issues with job displacement, skill shortages, and income inequality. These difficulties can be lessened and a more sustainable employment market can be produced through educating, upskilling, and preparing the workforce for automation.

11.7 POTENTIAL STRATEGIES FOR WORKFORCE ADAPTATION

For people, businesses, and society as a whole, workforce adaptation to the changing landscape brought forth by automation is essential. Here are a few possible tactics that could make this adaption easier:

1. **Upskilling and Lifelong Learning:** Encourage people to develop new skills throughout their careers by fostering a culture of lifelong learning. In order to provide easily available upskilling programs, online courses, and workshops that concentrate on digital literacy, coding, data analysis, and other pertinent skills, educational institutions, corporations, and the government can cooperate together.
2. **Flexible Education Pathways:** Encourage flexible education routes that accommodate different learning preferences and styles. People can swiftly pick up specialized skills without committing to lengthy school programs thanks to online learning platforms, micro-credentials, and short-term courses.
3. **Reskilling Programs:** Put in place programs for reskilling workers whose employment are threatened by automation. These courses should teach the skills that are in demand in fields that are developing and in industries that are less vulnerable to automation.
4. **Industry Collaboration:** Industry and educational institutions working together may make sure that curricula reflect the capabilities that employers are looking for. Industry input can assist build curricula and narrow the skills gap.
5. **Offer Apprenticeships and Internships:** They give students first-hand exposure in cutting-edge fields. People can apply newly learned skills and develop appropriate work experience with the aid of practical training.
6. **The Development of Soft Skills:** Soft talents like communication, emotional intelligence, creativity, and critical thinking are challenging to automate and will always be in high demand. Training in these areas can improve a person's resilience and adaptability.
7. **Provide Career Counseling Services:** These are to assist people in identifying their talents, interests, and skill gaps. Counselors can direct clients towards appropriate job avenues and recommend pertinent upskilling opportunities.

8. **Inclusive Training Programs:** Ensure that programs for retraining and upskilling are accessible to all facets of society, including underrepresented groups, older employees, and those with disabilities. Accessibility issues must be resolved in order to adjust the workforce to be inclusive.
9. **Transition Support:** Provide services to help people transition, such as help with job placement, resume creation, interview prep, and networking opportunities. People that do this may shift into different professions or sectors more easily.
10. **Embrace Remote and Online Work:** Promote chances for remote and online work, which can let people access jobs outside of their immediate area and allow them to work from anyplace.
11. **Government Policies and Incentive Programs:** Governments can offer incentives to companies who spend money on employee training and development. Policies that aid in job transition programs and provide financial aid while retraining can also be successful.
12. **Employee-First Approach:** Companies should put their employees' growth and welfare first. Giving workers the chance to grow and change within the company can increase commitment and engagement.
13. **Multidisciplinary Teams:** Make cross-functional teams that promote cooperation amongst people with various backgrounds and talents. This improves problem-solving while also enabling knowledge and skill transfer.

In summary, workforce adaptation to automation requires a multi-faceted approach involving education, training, collaboration, and policy support. The goal is to empower individuals to navigate the changing job landscape and equip them with the skills and mindset needed to thrive in a technology-driven world.

Numerical Problems with Solutions

Problem 1. Workforce Efficiency Improvement: A company implemented automation and reduced its workforce by 30%. If the original workforce was 500 employees, how many employees are left after automation?

Solution: 500 employees * (1 – 0.30) = 350 employees.

Problem 2. Cost Savings Calculation: Automation reduced production costs by 15%. If the company's monthly production cost was $150,000, how much is the cost savings?

Solution: $150,000 * 0.15 = $22,500.

Problem 3. Return on Investment (ROI) Calculation: A company invested $100,000 in an automation project. The project resulted in annual savings of $30,000. Calculate the ROI.

Solution: ROI = (Annual Savings/Investment) * 100 = ($30,000/$100,000) * 100 = 30%.

Problem 4. Workforce Training Costs: To upskill its workforce for automation, a company spent $50,000 on training programs. If the average training cost per employee is $500, how many employees were trained?

Solution: $50,000/$500 = 100 employees.

Problem 5. Employee Productivity Increase: Automation improved employee productivity by 25%. If an employee used to produce 100 units per day, how many units are produced after the productivity increase?

Solution: 100 units * 1.25 = 125 units.

Problem 6. Automation Implementation Time: An automation project took 6 months to complete. If the project started on January 1st, when did it finish?

Solution: January 1st + 6 months = July 1st.

Problem 7. Job Displacement Rate: A factory automated certain tasks, leading to a reduction of 50 jobs out of an initial 200 jobs. Calculate the job displacement rate.

Solution: (Jobs Lost/Initial Jobs) * 100 = (50/200) * 100 = 25%.

Problem 8. Production Time Reduction: Automation reduced the production time of a product from 8 hours to 2 hours. Calculate the percentage reduction in production time.

Solution: [(Initial Time – Final Time)/Initial Time] * 100 – [(8 – 2)/8] * 100 = 75%.

Problem 9. Cost of Automation Equipment: A company invested $500,000 in purchasing and installing automation equipment. If the equipment has a lifespan of 10 years, calculate the annual depreciation.

Solution: Depreciation = Cost/Lifespan = $500,000/10 = $50,000 per year.

Problem 10. Revenue Increase due to Automation: Automation led to a 20% increase in production output. If the company's annual revenue was $2 million before automation, what is the new annual revenue?

Solution: $2 million * 1.20 = $2.4 million.

Short Questions and Answers

Q.1. What is automation?

Ans. Automation refers to using technology, such as robots or software, to perform tasks without human intervention.

Q.2. How is automation impacting the workforce?

Ans. Automation is changing job roles by replacing routine tasks and creating demand for new skills.

Q.3. What are some emerging trends in automation and the workforce?

Ans. Collaborative robotics, intelligent process automation, remote work, and AI-powered recruitment are some emerging trends.

Q.4. What are collaborative robots (cobots)?

Ans. Collaborative robots, or cobots, work alongside humans to enhance efficiency and safety in tasks like manufacturing.

Q.5. How is mathematics applied in cobots?

Ans. Mathematics is used for kinematics, trajectory planning, and sensor data analysis to ensure precise and safe human-cobot interactions.

Q.6. What is intelligent process automation (IPA)?

Ans. Intelligent process automation combines AI and automation to optimize business processes, reducing manual effort.

Q.7. How does IPA use mathematics?

Ans. Mathematics underpins algorithms that analyze data, optimize workflows, and make decisions in intelligent process automation.

Q.8. What is hyper-automation?

Ans. Hyper-automation refers to an extensive use of automation technologies, combining RPA, AI, and other tools to optimize complex processes.

Q.9. How is mathematics applied in hyper-automation?

Ans. Mathematics is used in algorithm design, machine learning models, and optimization techniques within hyper-automation.

Q.10. What impact has remote work had on automation?

Ans. Remote work has accelerated the adoption of digital collaboration tools and increased reliance on automation for remote task management.

Q.11. How does mathematics play a role in data analysis for remote work?

Ans. Mathematical statistics and data analytics help interpret remote work data to optimize processes and enhance productivity.

Q.12. What is reskilling and upskilling in the context of automation?

Ans. Reskilling involves learning new skills for different roles, while upskilling means enhancing existing skills to match automation demands.

Q.13. How can mathematics assist in reskilling and upskilling efforts?

Ans. Mathematics is essential for technical skills development, data analysis, and understanding automation technologies.

Q.14. What is augmented analytics?

Ans. Augmented analytics uses AI to assist in data preparation, insights generation, and decision-making in data analytics.

Q.15. How does mathematics apply to augmented analytics?

Ans. Mathematical models underpin data analysis algorithms and predictive modeling techniques in augmented analytics.

Q.16. How do autonomous vehicles and drones impact industries?

Ans. Autonomous vehicles and drones streamline logistics, transport, and delivery processes, increasing efficiency and reducing human error.

Q.17. What role does mathematics play in autonomous vehicles and drones?

Ans. Mathematics is essential for sensor fusion, path planning, and collision avoidance algorithms in autonomous systems.

Q.18. How does AI-powered recruitment work?

Ans. AI-powered recruitment uses algorithms to analyze resumes, assess candidates, and match them with suitable job roles.

Q.19. How does mathematics contribute to AI-powered recruitment?

Ans. Mathematical models help match candidate profiles with job requirements, improving the accuracy of candidate selection.

Q.20. What is Industry 4.0 or Smart Manufacturing?

Ans. Industry 4.0 involves the integration of IoT, AI, and automation to create smart and connected manufacturing processes.

Q.21. How is mathematics applied in smart manufacturing?

Ans. Mathematics is used in data analytics, optimization, and simulation models to enhance production efficiency in Industry 4.0.

Q.22. What are ethical considerations in automation?

Ans. Ethical considerations involve addressing bias, privacy, and the impact of automation on jobs and society.

Q.23. How does mathematics contribute to addressing bias in automation algorithms?

Ans. Mathematical analysis helps identify and mitigate bias by ensuring algorithms treat all data fairly.

Q.24. What is the digital divide?

Ans. The digital divide refers to the gap between those who have access to technology and those who don't, impacting automation adoption.

Q.25. How can mathematics help address the digital divide?

Ans. Mathematics informs data analysis that identifies underserved communities, helping target efforts to bridge the digital divide.

Q.26. How does automation affect job roles in healthcare?

Ans. Automation in healthcare can improve diagnostics, patient data management, and administrative tasks.

Q.27. How can mathematics improve healthcare automation?

Ans. Mathematics enhances algorithms for medical imaging analysis, patient data security, and treatment recommendations.

Q.28. What are some challenges in workforce adaptation to automation?

Ans. Challenges include skills gaps, job displacement, and ensuring equitable access to training opportunities.

Q.29. How can governments support workforce adaptation?

Ans. Governments can invest in education, provide funding for upskilling programs, and develop policies for a smooth transition.

Q.30. How can businesses encourage workforce adaptation?

Ans. Businesses can offer training, create pathways for career growth, and prioritize employee well-being during automation adoption.

Q.31. What is the future outlook for automation and the workforce?

Ans. The future will likely involve a blend of human-AI collaboration, continuous learning, and reimagined job roles.

Q.32. How does mathematics play a role in predicting job market trends due to automation?

Ans. Mathematical modeling and data analysis inform predictions about which jobs are likely to be affected by automation trends.

Q.33. What is cognitive augmentation?

Ans. Cognitive augmentation enhances human cognitive abilities using technology, aiding in tasks like memory recall and decision-making.

Q.34. How does mathematics contribute to cognitive augmentation?

Ans. Mathematical models guide the development of algorithms that enhance human cognitive functions through technology.

Q.35. How does automation impact the gig economy?

Ans. Automation can create new opportunities and platforms within the gig economy while also automating certain gig tasks.

Q.36. How can mathematics contribute to optimizing gig economy platforms?

Ans. Mathematical optimization techniques can enhance matching algorithms and pricing strategies in gig economy platforms.

Q.37. What role does mathematics play in developing algorithms for automation?

Ans. Mathematics is foundational for designing algorithms that automate tasks, process data, and make decisions.

Q.38. How does automation affect job roles in the retail sector?

Ans. Automation in retail can lead to streamlined inventory management, checkout processes, and personalized customer experiences.

Q.39. How can mathematics improve retail automation's impact?

Ans. Mathematics underpins algorithms that analyze customer data, optimize pricing strategies, and manage supply chains efficiently.

Q.40. How does automation impact job roles in the education sector?

Ans. Automation in education can lead to personalized learning, administrative efficiency, and remote education solutions.

Q.41. How can mathematics contribute to optimizing education automation?

Ans. Mathematical models guide the development of adaptive learning algorithms and predictive analytics for student success.

Q.42. How does automation impact job roles in the entertainment industry?

Ans. Automation in entertainment can streamline content creation, distribution, and personalized experiences for audiences.

Q.43. How can mathematics improve entertainment automation's impact?

Ans. Mathematics informs algorithms for content recommendation, user engagement analysis, and dynamic content generation.

Q.44. How does automation impact job roles in the agricultural sector?

Ans. Automation in agriculture can enhance precision farming, crop monitoring, and data-driven decision-making for farmers.

Q.45. How can mathematics contribute to optimizing agricultural automation?

Ans. Mathematical models guide sensor data analysis, yield prediction, and resource optimization in agricultural automation.

Q.46. What is the role of ethics in designing automation technologies?

Ans. Ethics guide the responsible design of automation technologies, considering their impact on individuals, society, and the environment.

Q.47. How can mathematics be used to quantify the impact of automation on the job market?

Ans. Mathematical analysis can assess trends, job displacement rates, and the correlation between automation adoption and job changes.

Q.48. What are the implications of automation for workplace diversity and inclusion?

Ans. Automation must be designed inclusively to ensure it benefits all groups equally and does not reinforce biases.

Q.49. How can mathematics help optimize the allocation of human and automated tasks in the workforce?

Ans. Mathematical optimization models can determine the most efficient distribution of tasks between human workers and automation.

Q.50. What are the social considerations when implementing automation?

Ans. Social considerations involve addressing job displacement, fostering a culture of continuous learning, and ensuring equitable access to automation benefits.

Multiple Choice Questions and Answers

Q.1. What is automation in the context of the workforce?

(*a*) The use of technology to replace all human labor.

(*b*) The process of using machinery and software to perform tasks with minimal human intervention.

(*c*) The act of outsourcing work to other countries.

(*d*) The practice of creating jobs through government programs.

Q.2. Which of the following is NOT an example of automation?

(*a*) Self-checkout kiosks in a grocery store. (*b*) Robots assembling cars in a factory.

(*c*) Human customer service representatives assisting clients.

(*d*) Chatbots handling customer inquiries on a website.

Q.3. What is the potential impact of automation on jobs?

(*a*) All jobs will be completely eliminated.

(*b*) Some jobs will be eliminated, but new jobs will also be created.

(*c*) Automation will only affect manufacturing jobs.

(*d*) Automation will only impact low-skilled jobs.

Q.4. What is reskilling in the context of automation?

(*a*) Replacing human workers with robots.

(*b*) Training employees to adapt to new job roles due to technological changes.

(*c*) Hiring new employees with specialized technical skills.

(*d*) Shifting to a more manual and labor-intensive work process.

Q.5. Which skills are likely to become more valuable in an automated workforce?

(*a*) Routine manual tasks.

(*b*) Soft skills like creativity, communication, and problem-solving.

(*c*) Specialized technical skills exclusively related to automation.
(*d*) Following directions precisely.

Q.6. What is the term for a blend of human and automated work processes?

(*a*) Total automation. (*b*) Robotic domination.
(*c*) Human-robot collaboration. (*d*) Job elimination.

Key to Multiple Choice Questions

1. (*b*) **2.** (*c*) **3.** (*b*) **4.** (*b*) **5.** (*b*) **6.** (*c*)

Projects Works

1. **Workforce Skills Analysis Tool:** Develop a tool that analyzes the skills of the current workforce and identifies gaps that need to be filled for the future. This could involve data collection, analysis, and recommendations for training programs.
   Hints: Create a web application that allows employees to input their skills and job roles. use machine learning algorithms to compare their skills against industry trends and suggest relevant courses for upskilling.
2. **Automated Employee Onboarding:** Design an automated system that streamlines the employee onboarding process, from document submission to training schedules.
   Hints: Develop a chatbot that interacts with new hires, provides necessary information, collects documents through secure upload, and schedules orientation sessions based on their availability.
3. **Predictive Workforce Planning:** Build a predictive model that forecasts workforce needs based on business growth projections and historical data.
   Hints: Gather data on past hiring patterns, business performance, and future projections. use machine learning algorithms to predict future staffing needs and provide insights to hr departments.
4. **Remote Work Productivity Tracker:** Create a tool to track remote employees' productivity, including time spent on tasks, breaks, and overall performance.
   Hints: Develop software that monitors keyboard/mouse activity, application usage, and tracks time spent on different tasks. It provides visual dashboards for both employees and managers to review productivity patterns.
5. **Employee Satisfaction Chatbot:** Design a chatbot that engages with employees to gather feedback, identify concerns, and suggest solutions to enhance job satisfaction.
   Hints: Develop a conversational ai that periodically interacts with employees, asking questions about their work experience, and uses sentiment analysis to gauge satisfaction levels. provide actionable insights to management.
6. **Robotic Process Automation (RPA) For HR Tasks:** Implement RPA to automate repetitive hr tasks such as payroll processing, leave requests, and data entry.
   Hints: Identify repetitive tasks in the hr department, then design and deploy RPA bots to perform these tasks. It ensures data security and accuracy through proper testing and monitoring.
7. **AI-Powered Talent Acquisition:** Create a system that uses ai to screen resumes, conduct initial candidate interviews, and recommend the best-fit candidates for specific roles.

Hints: Develop a platform that uses natural language processing (nlp) to scan resumes for relevant skills and experience. Utilize video interview AI to analyze candidate responses and rank them based on predefined criteria.

8. **Workplace Safety Automation:** Build a system that uses iot devices and ai to monitor workplace conditions for safety compliance, sending alerts and reports to both employees and management.

   Hints: Deploy sensors for monitoring temperature, humidity, noise levels, and more. use ai to analyze data and detect potential safety violations, triggering notifications and automated corrective actions.

9. **Upskilling Recommendation Engine:** Create a personalized recommendation engine that suggests upskilling opportunities based on an employee's current skills and career goals.

   Hints: Develop an algorithm that considers an employee's skills, job role, and desired career path. Match this information with available courses and certifications, offering tailored suggestions for skill enhancement.

10. **Remote Collaboration Enhancer:** Design a tool that uses ai to improve remote collaboration, suggesting optimal meeting times, facilitating virtual brainstorming sessions, and managing shared documents.

    Hints: Develop a platform that integrates with calendars to identify suitable meeting times across different time zones. Use AI to moderate virtual brainstorming sessions, encouraging equal participation. Integrate cloud-based tools for seamless document collaboration.

12

Interconnected Domains of Industrial Automation and Robotics

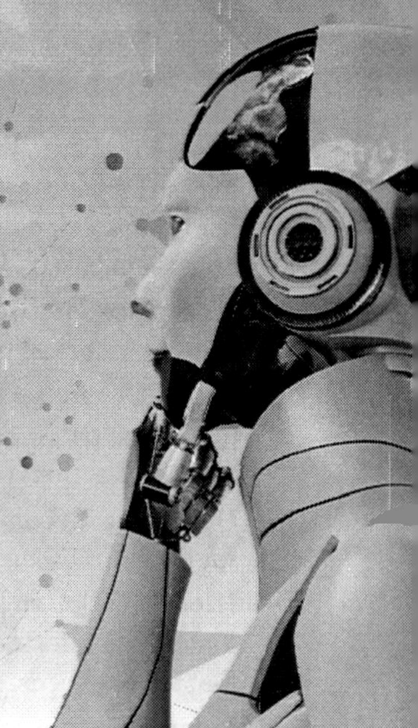

12.1 INTRODUCTION

Modern manufacturing, production, and several industries heavily rely on the interconnected domains of industrial automation and robotics. They entail the use of technology to direct and carry out functions that are typically performed by humans with the goal of enhancing efficiency, accuracy, safety, and productivity in various operations. Let's examine each of these ideas in more detail:

Automation used in industrial settings to automate numerous jobs, processes, and operations is referred to as industrial automation. The main objective is to minimize human involvement, minimize human mistake, and increase general effectiveness. To regulate and improve processes, automation can make use of a variety of technologies, including sensors, actuators, controls, and software. Robotics comprises the design, upkeep, and application of robots for a range of tasks. A robot is a machine that may be programmed to do activities automatically or somewhat autonomously. Robotics is employed in a variety of industries, including healthcare, logistics, entertainment, and more, in addition to industrial ones.

12.2 KEY ASPECTS OF INDUSTRIAL AUTOMATION AND ROBOTICS

Key aspects of industrial automation include:

1. **Control Systems:** These systems keep track of and manage a range of industrial process parameters, including temperature, pressure, speed, and others. They make sure that procedures follow the predetermined boundaries.
2. **Programmable Logic Controllers (PLCs):** PLCs are specialized computers that are used to manage processes and machinery. They have been programmed to carry out particular activities in response to data from sensors and other devices.
3. **Sensors and Actuators:** While actuators regulate physical processes, sensors collect data from the environment. Together, they make it possible for automated systems to adapt to environmental changes.
4. **Human-machine Interfaces (HMIs):** These give engineers and operators a way to communicate with automation systems. They enable input and display data for monitoring and control functions.

5. **Process Optimization:** By optimizing operations based on real-time data, automation can result in greater efficiency, less waste, and consistent quality.

Key aspects of robotics include:

1. **Manipulation:** Robots can manipulate items and carry out tasks thanks to their arms, grippers, and tools. They are very accurate in imitating human movements.
2. **Sensing:** To sense their surroundings and make judgments based on the information they gather, robots employ sensors like cameras, lidar, and touch sensors.
3. **Autonomy:** Autonomous robots have the ability to function without constant human supervision. To move around, communicate with objects, and carry out activities, they use sensors and programming.
4. **Collaborative Robots (Cobots):** These machines are made to cooperate with people in open-plan offices. They have safety mechanisms to guard against crashes and guarantee people's security.
5. **Applications:** Robotics is used in a variety of industries, including manufacturing (assembly, welding, painting), logistics (picking, packing, sorting), healthcare (surgery, rehab), and exploration (space, deep sea).

Robotics and industrial automation are related because robots are frequently used in automation to carry out tasks. By boosting productivity, enhancing safety, and facilitating the very precise manufacture of complex items, the combination of these technologies has revolutionized numerous sectors. Robotics and automation are continually being pushed to their limits by constant developments in sensors, AI, and machine learning.

12.3 AUTOMATION IN INDUSTRIAL SETTINGS

In industrial settings, automation is the use of machinery, technology, and control systems to run and manage diverse operations without a lot of direct human involvement. To streamline manufacturing, increase productivity, improve safety, and decrease human error, various technologies, sensors, software, and hardware must be integrated.

Let's examine automation in industrial settings in greater detail, including the technology and mathematics at play.

1. **Technologies:**
   (*a*) **Programmable Logic Controllers:** PLCs, or Programmable Logic Controllers, are specialized digital computers used to manage processes and machines. They take data from sensors as input, process it using logic and algorithms, and then output it to operate a variety of gadgets like motors, valves, and actuators.

   (*b*) **Actuators and Sensors:** Different actuators (temperature, pressure, proximity, etc.) and sensors (such as those mentioned above) gather real-time data from the industrial environment. Devices called actuators carry out operations in response to signals from sensors. They have the ability to open or stop valves, control the motion of machinery, etc.

   (*c*) **Human-Machine Interfaces:** HMIs (Human-Machine Interfaces) offer operators a graphical interface for monitoring and interacting with the automation system. They enable operators to enter commands and display real-time data and alarms.

(*d*) **Industrial Robots:** Robots are employed to carry out repetitive, hazardous, or precise activities. To correctly carry out duties, they are outfitted with sensors, actuators, and cutting-edge control algorithms.

(*e*) **SCADA (Supervisory Control and Data Acquisition) Systems:** SCADA systems collect information from numerous sensors and pieces of equipment, show how processes are carried out, and allow for remote monitoring and control.

(*f*) **Machine Vision Systems:** These systems utilise cameras and algorithms for image processing to check items for flaws, spot patterns, and direct robots in tasks that require visual recognition.

2. **Mathematics:**

(*a*) **Control Theory:** It is the study of how to create controllers that regulate industrial processes using mathematical methods. It covers ideas including stability analysis, feedback loops, and controller parameter tweaking.

(*b*) **Linear Algebra:** Linear algebra is used to model and examine the behavior of dynamic systems. The representation of system equations, transformations, and state variables is done using matrices and vectors.

(*c*) **Differential Equations:** They are used to simulate the behavior of dynamic systems over time. They aid in predicting how processes will alter in response to their present condition and outside inputs.

(*d*) **Probability and Statistics:** These are essential for the development of reliable automation systems. Statistics are used to analyze data trends and help decision-makers understand sensor reading uncertainties. Probability theory aids in this process.

(*e*) **Optimization:** To attain desired performance, the best set of control parameters are found using optimization techniques. This is especially crucial for processes that involve numerous restrictions and variables.

(*f*) **Signal Processing:** To filter and analyze sensor data, signal processing techniques are used. This lessens noise and aids in the extraction of useful information.

3. **Workflow:** These stages are frequently taken in the automation process:

(*a*) **Sensing:** Sensors gather information about temperature, pressure, and position from the industrial environment.

(*b*) **Processing:** In PLCs or other control systems, algorithms and control methods are used to process the acquired data.

(*c*) **Decision-Making:** Control systems decide how to change actuator parameters and actions based on the data that has been processed.

(*d*) **Actuation:** Actuators carry out the required operations, such as changing the position of valves, regulating the speed of motors, or relocating robotic arms.

(*e*) **Feedback Loop:** The control system can monitor the process and make modifications in real-time thanks to continuous feedback from the sensors.

(*f*) **Human Interaction:** By using HMIs to communicate with the automation system, operators can monitor the operation and take appropriate action as needed.

In order to optimize and manage complex processes, automation in industrial settings combines a variety of technology and mathematical principles. It increases manufacturing, production, and other industrial activities' efficiency, precision, and safety.

12.4 AUTOMATION IN ROBOTIC ARMS IN MANUFACTURING

Robotic arm automation is essential to production because it allows for accurate and effective handling of materials, assembly processes, and other jobs. The movements and actions of the robotic arm are controlled and optimized using a combination of technological and mathematical ideas. Let's explore the specifics of automation in manufacturing robotic arms:

1. **Technologies:**
   (*a*) **Robotic Arm Kinematics:** It is the study of the mathematical representation of the geometry, joints, and motions of an arm. It contains forward kinematics, which uses joint angles to calculate end-effector position, and inverse kinematics, which uses joint angles to arrive at a desired end-effector position.

   (*b*) **Sensors:** Different sensors, including encoders, accelerometers, and force/torque sensors, are used by robotic arms. Accelerometers help with motion control, encoders give data on joint positions, and force/torque sensors let you interact with the outside world.

   (*c*) **Actuators:** To move robotic joints, actuators such as hydraulic systems, pneumatic systems, and electric motors are used. Because they are precise and easy to regulate, electric motors are widely used.

   (*d*) **Computer Vision:** For visual feedback, object detection, and tracking, cameras and image processing techniques are used. When performing duties that call for picking and placing objects accurately.

   (*e*) **End-Effectors and Grippers:** Robotic arms frequently contain interchangeable end-effectors that may do certain operations like picking, putting, welding, or painting, such as grippers, vacuum cups, or specialized tools.

   (*f*) **Control Systems:** The robotic arm's movements are accurate and stable thanks to control algorithms like PID (Proportional-Integral-Derivative) controllers and more sophisticated techniques like adaptive control.

2. **Mathematics:**

(*a*) **Kinematics Equations:** Based on joint angles, forward kinematics calculates the location and orientation of the end-effector using trigonometric functions and matrices. In order to determine the joint angles for a specified end-effector location, inverse kinematics use mathematical approaches such as geometric methods or iterative solutions.

(*b*) **Trajectory Planning:** To create efficient and slick pathways for the robotic arm to go along, mathematics is used. The arm is made to move smoothly between waypoints using methods like cubic splines or Bézier curves.

(*c*) **Jacobian Matrix:** This mathematical formula connects the joint velocities with end-effector velocities. Tasks like obstacle avoidance and singularity analysis are made easier by being able to determine how changes in joint angles affect the motion of the end-effector.

(*d*) **Dynamics Equations:** They explain the forces and torques needed to produce particular movements or interactions. For tasks requiring the movement of large things or the maintenance of balance, this is crucial.

(*e*) **Control Theory:** To accomplish desired trajectories, control algorithms modify joint angles and torques using mathematical models. PID controllers and other feedback control loops aid in precise movement.

3. **Workflow:**
   (*a*) **Cameras and Sensors:** They give real-time information about the surroundings, objects, and the location of the robotic arm.
   (*b*) **Planning:** The control system creates a trajectory or path for the robotic arm to follow based on the desired task and the surroundings as seen by the user.
   (*c*) **Kinematics and Inverse Kinematics:** The control system determines the corresponding joint angles for each point along the trajectory using forward and inverse kinematics.
   (*d*) **Control:** To ensure that the robotic arm precisely follows the predetermined trajectory, the control system makes real-time adjustments to joint angles and torques using algorithms like PID controllers.
   (*e*) **Feedback:** Sensors give the control system information about the arm's real location, enabling it to make ongoing modifications to keep precision.
   (*f*) **Operation Execution:** The robotic arm carries out the assigned operation, which may include welding, putting together parts, or picking up objects.

Manufacturing is revolutionized by automation in robotic arms because it improves repeatability, speed, and precision. It increases overall production efficiency and lessens the need for humans to perform dangerous or repetitive operations. Robotic arms are capable of carrying out difficult tasks with a high degree of reliability thanks to the integration of advanced mathematics and technology.

12.5 AUTOMATION IN ROBOTIC ASSEMBLY LINES

A key component of contemporary manufacturing is automation in robotic assembly lines, which allows for efficient and accurate construction of complicated items. To optimize the assembly process, this entails integrating robotics, control systems, sensors, and mathematics.

Let's explore the details of automation in robotic assembly lines:

1. **Technologies:**
   (*a*) **Robotic Arms:** To assemble components, industrial robotic arms with various grippers and end-effectors are employed. They can perform repetitive jobs and difficult duties with great precision.
   (*b*) **Conveyor Systems:** A continuous assembly process is made possible by the use of conveyor belts or other material handling systems to move products and components between workstations.
   (*c*) **Vision Systems:** To check components, confirm proper assembly, and direct robots during activities that need visual feedback, cameras and machine vision algorithms are used.
   (*d*) **Sensors:** Robots can detect the presence of components, measure forces during assembly, and assure appropriate alignment with the use of proximity sensors, force/torque sensors, and other types of sensors.
   (*e*) **Human-Machine Interfaces (HMIs):** Through user-friendly interfaces, operators communicate with the assembly line, monitoring the process and making modifications as necessary.
   (*f*) **Control Systems:** To ensure flawless assembly, sophisticated control systems coordinate the motions of robotic arms, conveyor systems, and other equipment.

2. **Mathematics:**
   (*a*) **Kinematics and Inverse Kinematics:** Algorithms and mathematical models determine the joint angles necessary to precisely place robotic arms during assembly. The end-effector's position must be determined using inverse kinematics in order to obtain the required final assembly configuration.
   (*b*) **Path Planning:** To avoid impediments and maximise efficiency, robotic arms should follow pathways that are planned mathematically. Commonly utilised algorithms are A* (A-star) and RRT (Rapidly-exploring Random Trees).
   (*c*) **Trajectory Generation:** Creating trajectories that are smooth and devoid of collisions requires the use of mathematical techniques such as cubic splines, Bezier curves, or polynomial interpolation.
   (*d*) **Geometric Algorithms:** Collision detection and avoidance algorithms use geometric calculations to avoid collisions between robotic arms and components during assembly.
   (*e*) **Pose Estimation:** To enable precision assembly, the pose (position and orientation) of components can be precisely estimated using sensor data and mathematical algorithms.
   (*f*) **Control Theory:** During assembly, the robotic arms operate smoothly, precisely, and with the necessary amount of force thanks to control techniques like PID controllers and adaptive control.
3. **Workflow:**
   (*a*) **Handling of Components:** Workstations or the conveyor system are loaded with components as they arrive. Their position and presence are detected by sensors.
   (*b*) **Perception:** To ascertain the orientations and placements of components, vision systems and sensors inspect the parts.
   (*c*) **Path and Trajectory Planning:** The control system determines the best pathways and trajectories for robotic arms based on the planned assembly sequence and component placements.
   (*d*) **Kinematics and Inverse Kinematics:** To accomplish the desired assembly configuration, the control system computes joint angles and end-effector locations.
   (*e*) **Assembly:** In accordance with the assembly plan, robotic arms pick, place, tighten screws, weld, or glue components together.
   (*f*) **Feedback and Quality Control:** Sensors and vision systems check that the assembly is done correctly and look for any flaws or faults. Feedback loops guarantee that necessary adjustments are performed.
   (*g*) **Integration:** Completed sub-assemblies or products are transferred to ensuing workstations or testing facilities along the assembly line.

Robotic assembly lines that automate reduce cycle times, maximize consistency, and minimize errors to optimize output. Mathematical and cutting-edge technological integration provides exact coordination of several robotic arms, parts, and processes, leading to efficient and dependable assembly operations.

12.6 AUTOMATION IN COLLABORATIVE ROBOTS (COBOTS)

Cobots, also known as collaborative robots, are a subtype of industrial robots created to cooperate with people in a shared workspace. They have cutting-edge tools and security measures to guarantee a secure and productive communication.

Cobot automation utilizes a mix of technological and mathematical ideas. Let's examine the specifics:

1. **Technologies:**
   (*a*) **Cobots Sensors:** Cobots have a variety of sensors, including force/torque sensors, capacitive sensors, and vision systems. They can use these sensors to detect people or obstacles and change their course to avoid colliding with them.
   (*b*) **Force Control:** Cobots can feel external forces and react to them thanks to force/torque sensors. This is crucial for tasks involving sensitive handling or working with things of potentially different forms and sizes.
   (*c*) **Vision Systems:** Cobots can identify items, spot patterns, and locate targets for jobs like choosing and placing objects thanks to cameras and machine vision algorithms.
   (*d*) **Human-Machine Interfaces (HMIs):** Simple interfaces make it possible for operators to program and manage cobots. HMIs offer a mechanism to intuitively teach cobots jobs and keep an eye on their behavior.
   (*e*) **Control Modes:** Cobots provide a variety of control options, including impedance control, speed and force limiting, and safety-rated monitoring stops. These options make sure that when cobots encounter unforeseen barriers, they would slow down or halt.
   (*f*) **End-Effectors:** To carry out a variety of activities, cobots can be fitted with a variety of end-effectors, including grippers, vacuum cups, and specialized tools.
2. **Mathematics:**
   (*a*) In order to ensure a seamless and secure connection with the environment, mathematics is used to process data from force/torque sensors. The required torques and forces for actions like picking up things and exerting pressure are calculated by algorithms.
   (*b*) **Collision Avoidance:** Using sensor data, mathematical algorithms determine safe trajectories and modify cobot movements to prevent collisions with people or objects.
   (*c*) **Path Planning:** Math is employed to determine the best routes for cobots to take, taking into account barriers and human presence. Planning a collision-free path involves using algorithms like potential fields or rapidly-exploring random trees (RRT).
   (*d*) **Kinematics and Inverse Kinematics:** Like conventional robotic arms, cobots calculate joint angles and end-effector positions using kinematic and inverse kinematic calculations. This enables them to complete jobs precisely.
   (*e*) **Safety Zones:** Calculations are used to define safety zones surrounding cobots, which cause them to halt or slow down when people approach. These zones are established in accordance with the cobot's speed, reach, and force output.
3. **Workflow:**
   (*a*) **Training and Programming:** Operators can physically direct the cobot through desired motions or program cobots utilizing teach pendant interfaces. Even hand-guided training is possible with some cobots, in which the operator guides the cobot through the necessary manoeuvres.
   (*b*) **Task Definition:** It is done by operators either through programming or by using their intuition. Waypoints, target locations, and desired forces can all be specified here.
   (*c*) **Safety Features:** Establishing collaborative workspaces, safety zones, and force/torque limitations are some of the safety features that are specified as part of safety feature configuration.

(*d*) **Task Execution:** While environment monitoring is done by Cobots using sensors, the cobots modify its movements in response to any unforeseen pressures or obstructions.

(*e*) **Continuous Monitoring:** To ensure that interactions between cobots and people are safe, sensors and safety algorithms continuously monitor the workspace.

Collaborative robots increase productivity and safety by bringing automation closer to human workers. Cobots can carry out a variety of duties while interacting safely with people because to the integration of cutting-edge technologies and mathematical principles.

12.7 SAFETY IN ROBOTICS AND AUTOMATION

To ensure employee safety, safeguard equipment, and prevent accidents, safety in robotics and automation is of the utmost significance. To reduce risks and ensure a safe workplace, effective safety measures are essential as technology develops and robots are increasingly integrated into a variety of industries.

Here are some key aspects of safety in robotics and automation:

1. **Risk Assessment:** A thorough risk analysis should be carried out prior to the implementation of robotic and automation systems. This entails detecting prospective risks and evaluating their gravity, likelihood, and outcomes. The appropriate safety precautions and processes are determined with the aid of this assessment.
2. **Physical Safety:** Safety barriers and enclosures: By preventing direct access to robots and automated machinery, physical barriers like fences, walls, or cages can lower the danger of unintentional contact. Interlocks and safety sensors make sure that robotic systems turn off if access doors or barriers are unlocked. When a person enters a dangerous area, safety sensors like light curtains or pressure mats detect their presence and stop the machinery.

   Cobots, or collaborative robots, are created to work safely alongside people. In order to slow down or stop when they come into contact with objects or unexpected forces, they frequently feature force/torque sensors and collision detection systems.

   In case of an emergency, personnel can quickly halt robotic and automation systems thanks to readily available emergency stop buttons.
3. **Software and Control Safety:** Safe programming: Programming for robotic systems should strictly abide by safety regulations. This entails setting speed restrictions, establishing safe zones, and avoiding abrupt movements.

   Software Validation and Verification: To make sure automated systems act as planned and follow safety protocols, they should go through thorough testing and validation.

   Control systems should have fail-safe methods that make sure a safe state is reached in the event that a component fails or unanticipated occurrences take place.
4. **Education and Training:** For employees who engage with robotic and automated systems, appropriate education and training are crucial. They ought to be familiar with how to use, programme, and debug these systems in addition to how to react in an emergency.
5. **Maintenance and Inspections:** It's important to do routine maintenance and inspections on robotic and automation systems to look for wear, damage, or other issues that could endanger user safety. This entails looking for protruding bolts, frayed wires, and worn-out parts.
6. **Industry Standards and Regulations:** Adherence to industry standards and rules, such as ISO 10218 and ISO 13849 for industrial robots, aids in ensuring that safety precautions follow accepted best practices.

7. **Risk Mitigation:** Equipment failure risks can be decreased by implementing redundancy, adding backup systems, and employing safety-rated components.
8. **Constant Improvement:** Safety should be a process that never stops. Safety practises can be continuously improved by performing incident analyses, evaluating and upgrading safety protocols on a regular basis, and getting employee feedback.

Finally, when incorporating robotics and automation into different businesses, safety is a crucial factor. Robotic and automated systems can operate securely, protect users and prevent accidents by using a variety of physical safeguards, software controls, training, and adherence to standards.

12.8 SAFETY CONSIDERATIONS AND WORKING IN VARIOUS ENVIRONMENTS

Based on the particular environment and sector, different rules and considerations for safety apply when working with robots. Robotic systems are subject to different dangers and requirements in various industries.

Here are some safety tips and pertinent regulations for using robots in diverse settings:

1. **Industrial Manufacturing:** To prohibit access to dangerous locations, take into account:

   Physical Enclosures:
   - Appropriate training for employees who will be engaging with robots.
   - Frequently checking and maintaining robotic systems;
   - Emergency stop buttons and safety interlocks on access doors.

   Suitable Standards:
   - Industrial robot safety standards ISO 10218-1 and ISO 10218-2.
   - **ISO 13849:** Safety-related components of control systems for equipment.
   - **ANSI/RIA R15.06:** Robotic Industries Association (RIA) standard for robot safety.
2. **Healthcare and Medical:**

   Taking into account:
   - Ensuring safe interactions between robots and medical professionals; preventing robot interference with medical technology.
   - Infection prevention and control measures.
   - Adherence to medical standards and legislation.

   Relevant Standards:
   - **ISO 13482:** Personal care robot safety criteria.
   - **IEC 60601:** Safety requirements for medical electrical equipment.
   - Medical device FDA rules.
3. **Agriculture and Farming:**

   Considerations:
   - Safety of farm workers working alongside robots.
   - Navigation and obstacle avoidance to prevent collisions with people and objects.
   - Protection against weather and environmental conditions.

   Relevant Standards:
   - **ISO 12100:** Safety of machinery – General principles for design.
   - **ISO 25119:** Functional safety for agricultural machinery.

4. **Construction and Infrastructure:**

   Considerations:

   - Safeguarding construction workers from robotic equipment.
   - Collision avoidance systems to prevent accidents.
   - Maintenance and inspection procedures for robotic construction machinery.

   Relevant Standards:

   - **ISO 10218:** Safety requirements for industrial robots.
   - **EN 1525:** Safety of machinery – Robots for industrial environments.

5. **Logistics and Warehousing:**

   Considerations:

   - Safe coexistence of robots and human workers in shared spaces.
   - Collision avoidance and path planning in dynamic environments.
   - Training for workers to interact with automated systems.

   Relevant Standards:

   - **ISO 3691:** Safety requirements for industrial trucks.
   - **ANSI/ITSDF B56.5:** Safety standard for driverless industrial trucks and AGVs.

6. **Research and Education:**

   Considerations:

   - Safe interaction between researchers and experimental robots.
   - Training for students and researchers on safe robot operation.
   - Regular risk assessments for experimental setups.

   Relevant Standards:

   - **ANSI/RIA R15.06:** Robot safety standard by the Robotic Industries Association (RIA).
   - **ISO/TS 15066:** Guidelines for safe collaborative industrial robot operation.

It's crucial to follow general safety principles like adequate risk assessment, training, emergency response planning, and continual monitoring of robotic systems in addition to these industry-specific considerations and requirements. Each environment may need a special strategy to deal with its own problems and guarantee the security of people, things, and the environment.

Numerical Problems with Answers

Problem 1. Conveyor Speed Calculation: A conveyor belt is moving at a speed of 0.5 meters per second. If the distance between two objects on the belt is 2 meters, how much time does it take for the objects to move from one point to another?

Solution: Time = Distance/Speed = 2 meters/0.5 m/s = 4 seconds.

Problem 2. Robot Arm Reach Calculation: An articulated robot arm has a maximum reach of 1 meter. If the arm is fully extended horizontally, what is the height of the arm's wrist point from the base?

Solution: The arm's reach forms a right triangle. The height can be calculated using the Pythagorean theorem: $Height^2 = Reach^2 - Horizontal\ distance^2 = 1^2 - 1^2 = 0$. So, the height is 0 meters.

Problem 3. PLC Timer Calculation: A PLC program requires a timer to activate an output for 10 seconds. Calculate the preset and accumulated values for the timer in milliseconds.

Solution: Preset value = 10 seconds × 1000 ms/s = 10,000 ms. Accumulated value starts from 0 ms.

Problem 4. Robot Speed and Time Calculation: A robot travels a distance of 5 meters at a speed of 2 m/s. Calculate the time taken to cover the distance.

Solution: Time = Distance/Speed = 5 meters/2 m/s = 2.5 seconds.

Problem 5. Conveyor Load Calculation: A conveyor is designed to carry boxes weighing 20 kg each. If the conveyor is moving at a speed of 0.8 m/s and can handle a maximum load of 1000 kg, how many boxes can it carry simultaneously?

Solution: Maximum load = Number of boxes × Box weight. Number of boxes = Maximum load/Box weight = 1000 kg/20 kg = 50 boxes.

Problem 6. Control Systems: Calculate the gain of a proportional controller with an error of 30 and a proportional constant of 0.2.

Solution: Gain = Proportional Constant × Error = 0.2 × 30 = 6.

Problem 7. If a PID controller has a derivative time constant of 0.5 seconds and an error rate of change of 10, find the derivative term value.

Solution: Derivative Term = Derivative Time Constant × Rate of Change = 0.5 × 10 = 5.

Problem 8. Kinematics: A robotic arm with a link length of 0.4 meters is moving at an angular velocity of 2 rad/s. Calculate the linear velocity of the end effector.

Solution: Linear Velocity = Angular Velocity × Link Length = 2 × 0.4 = 0.8 m/s.

Problem 9. Given a 3-DOF robot arm with joint angles $\theta_1 = 30°$, $\theta_2 = 45°$, and $\theta_3 = 60°$, find the position of the end effector using forward kinematics.

Solution: X = 0.125 m, Y = 0.6495 m, Z = 0.1129 m.

Problem 10. Sensors and Actuators: A force sensor reads 80 N. Convert this force to dynes.

Solution: Force in Dynes = 80 N × 10^5 = 8 × 10^6 dynes.

Problem 11. Calculate the torque exerted by a motor with a force of 150 N at a distance of 0.3 meters from the axis of rotation.

Solution: Torque = Force × Distance = 150 N × 0.3 m = 45 Nm.

Problem 12. Robotics Applications: A robot arm moves from point A (2, 3) to point B (7, 10). Find the Euclidean distance between the two points.

Solution: Distance = $\sqrt{((7-2)^2 + (10-3)^2)} = \sqrt{50} \approx 7.07$.

Problem 13. Determine the workspace radius of a robot with a maximum reach of 1.5 meters and a joint angle range of 120°.

Solution: Workspace Radius = Max Reach × sin(θ/2) = 1.5 × sin(60°) ≈ 1.299 meters.

Problem 14. Feedback Control: A closed-loop control system has a proportional gain of 1.5. Calculate the steady-state error for a unit step input.

Solution: Steady-State Error = 1/Gain = 1/1.5 ≈ 0.667.

Problem 15. For a control system with a damping ratio of 0.8 and a natural frequency of 4 rad/s, calculate the settling time.

Solution: Settling Time = –ln(0.02)/(Damping Ratio × Natural Frequency) ≈ 1.942 seconds.

Short Questions and Answers

Q.1. What is industrial automation?

Ans. Industrial automation refers to the use of technology and control systems to operate machinery and processes with minimal human intervention.

Q.2. What are the benefits of industrial automation?

Ans. Benefits include increased efficiency, reduced human error, improved safety, higher production rates, and cost savings.

Q.3. Define robotics.

Ans. Robotics is the field that involves the design, construction, and operation of robots to perform tasks in various industries.

Q.4. What is a PLC?

Ans. A Programmable Logic Controller (PLC) is a digital computer used to control machinery and processes in industrial settings.

Q.5. What are sensors used for in automation?

Ans. Sensors gather data from the environment and provide input to automation systems for monitoring and control.

Q.6. Explain the term "end-effector."

Ans. An end-effector is a device or tool attached to the end of a robotic arm to perform specific tasks.

Q.7. What is SCADA?

Ans. SCADA (Supervisory Control and Data Acquisition) is a system used to remotely monitor, control, and manage industrial processes.

Q.8. What is the difference between industrial robots and cobots?

Ans. Industrial robots are designed for specific tasks, while cobots (collaborative robots) are designed to work alongside humans safely.

Q.9. What is a force/torque sensor used for in robotics?

Ans. A force/torque sensor measures forces and torques applied to a robot's end-effector, enabling precise interactions with the environment.

Q.10. Define kinematics in robotics.

Ans. Kinematics deals with the study of motion without considering the forces causing the motion.

Q.11. What is inverse kinematics?

Ans. Inverse kinematics calculates the joint angles required to achieve a desired position and orientation of a robot's end-effector.

Q.12. What is the role of computer vision in robotics?

Ans. Computer vision enables robots to process visual information, recognize objects, and navigate their environment.

Q.13. What is a HMI in automation?

Ans. HMI (Human-Machine Interface) is a graphical interface that allows operators to interact with and monitor automation systems.

Q.14. Explain the concept of "safety-rated monitored stop" in robotics.

Ans. Safety-rated monitored stop is a mode in which a robot's motion is safely stopped when humans approach its workspace.

Q.15. What is PID control?

Ans. PID (Proportional-Integral-Derivative) control is a feedback control mechanism used to regulate a system's behavior.

Q.16. What is a gripper in robotics?

Ans. A gripper is an end-effector used to grasp and hold objects with a robotic arm.

Q.17. Define trajectory planning.

Ans. Trajectory planning involves determining a smooth and collision-free path for a robot's end-effector to follow.

Q.18. What is a cobot's main feature?

Ans. A cobot's main feature is its ability to work safely alongside humans without requiring rigid physical barriers.

Q.19. Explain the concept of a "safety zone" in robotics.

Ans. A safety zone is an area around a robot where its speed or operation is reduced to ensure the safety of nearby humans.

Q.20. What is the purpose of redundancy in robotics?

Ans. Redundancy provides backup systems or components to ensure continued operation or safe shutdown in case of failures.

Q.21. What are the advantages of using robots in hazardous environments?

Ans. Robots can perform tasks in environments that are dangerous for humans, reducing risks to human health and safety.

Q.22. What is the significance of ISO 10218 in robotics?

Ans. ISO 10218 is an international standard that outlines safety requirements for industrial robots and their integration.

Q.23. What is a teach pendant used for?

Ans. A teach pendant is a handheld device that allows operators to program and control robotic movements manually.

Q.24. What is the difference between open-loop and closed-loop control?

Ans. Open-loop control uses preset instructions, while closed-loop control involves using sensors and feedback to adjust actions.

Q.25. What is the importance of safety interlocks in automation?

Ans. Safety interlocks ensure that access doors are closed before automated machinery can operate.

Q.26. What is an AGV?

Ans. An Automated Guided Vehicle (AGV) is a mobile robot used to transport ma... a controlled environment.

Q.27. What is the role of a PLC in process control?

Ans. A PLC processes input data from sensors and controls output devices to regulate industria... processes.

Q.28. Explain the concept of "deadlock" in automation.

Ans. Deadlock occurs when multiple processes or robots are waiting for each other to release resources, resulting in a standstill.

Q.29. What are industrial vision systems used for?

Ans. Industrial vision systems are used for quality control, product inspection, and guiding robotic actions based on visual information.

Q.30. Define collaborative robotics.

Ans. Collaborative robotics involves robots working alongside humans in shared spaces, promoting cooperation and safety.

Q.31. What is the purpose of force-limited robots?

Ans. Force-limited robots are designed to apply controlled forces when interacting with delicate objects or humans.

Q.32. What is the "teach and repeat" method in robotics?

Ans. The "teach and repeat" method involves manually guiding a robot through a task, recording the motions, and then having the robot repeat the actions.

Q.33. Explain the concept of "singularity" in robotics.

Ans. Singularity is a state where a robotic arm's joints align in a way that makes it difficult to determine a unique solution for its position.

Q.34. What are the safety considerations when using industrial lasers in automation?

Ans. Safety considerations include proper laser shielding, eye protection, and adherence to laser safety standards.

Q.35. What are the challenges of integrating robots in small and medium-sized enterprises (SMEs)?

Ans. Challenges include high initial costs, lack of technical expertise, and limited workspace.

Q.36. What is adaptive control in robotics?

Ans. Adaptive control involves adjusting a robot's control parameters based on changing conditions to optimize performance.

Q.37. What is the role of motion planning algorithms in robotics?

Ans. Motion planning algorithms calculate optimal paths for robots to move from one point to another, avoiding obstacles.

Q.38. What are the considerations for safe human-robot collaboration in a manufacturing setting?

Ans. Considerations include proper training, defining safety zones, using sensors for collision detection, and clear communication.

Q.39. What is meant by the "Internet of Things" (IoT) in industrial automation?

Ans. IoT refers to the interconnectedness of devices and sensors in an industrial setting to gather and exchange data for improved automation and efficiency.

Q.40. How do industrial robots contribute to precision manufacturing?

Ans. Industrial robots can perform repetitive tasks with high precision, reducing errors and ensuring consistent product quality.

Q.41. What is the role of predictive maintenance in automation?

Ans. Predictive maintenance uses data analysis to predict when equipment will need maintenance, reducing downtime and improving efficiency.

Q.42. Explain the concept of "lights-out manufacturing."

Ans. Lights-out manufacturing involves fully automated production that operates without human intervention, even during non-working hours.

Q.43. What safety measures should be in place when using autonomous mobile robots in warehouses?

Ans. Safety measures include obstacle detection, emergency stop buttons, and proper training for workers.

Q.44. How does automation enhance energy efficiency in industrial processes?

Ans. Automation can optimize energy usage by regulating equipment operation and minimizing waste.

Q.45. What is the role of simulation software in robotics?

Ans. Simulation software allows testing and optimizing robotic movements and actions in a virtual environment before implementation.

Q.46. What is the "just-in-time" production strategy and how does automation support it?

Ans. Just-in-time production aims to minimize inventory by producing goods only when needed. Automation ensures efficient and timely production to meet demand.

Q.47. Explain the concept of "robotic process automation" (RPA).

Ans. Robotic process automation involves using software robots to perform routine and rule-based tasks, improving efficiency in administrative processes.

Q.48. How do robots assist in hazardous material handling?

Ans. Robots can handle hazardous materials in environments with toxic gases, radiation, or other dangers, reducing risks to human workers.

Q.49. What is the role of machine learning in industrial automation?

Ans. Machine learning algorithms can analyze data from sensors and processes to make predictions, optimize operations, and detect anomalies.

Q.50. What is the impact of automation on job roles in various industries?

Ans. Automation can lead to the transformation of job roles, shifting the focus from routine tasks to higher-value activities like maintenance, programming, and decision-making.

Multiple Choice Questions and Answers

Q.1. What is the primary goal of industrial automation?

(*a*) Reduce human intervention
(*b*) Increase energy consumption
(*c*) Slow down production processes
(*d*) Enhance manual labor

Q.2. Which of the following is a key benefit of using robotics in manufacturing?

(*a*) Higher labor costs
(*b*) Reduced efficiency
(*c*) Increased production consistency
(*d*) Manual operation

Q.3. What does SCADA stand for?

(*a*) Systematic Control and Data Analysis
(*b*) Supervisory Control and Data Acquisition
(*c*) Scientific Computing and Data Automation
(*d*) Security and Control of Data Access

Q.4. Which type of industrial robot is designed to perform tasks in a limited workspace without changing positions?

(*a*) Cartesian robot
(*b*) SCARA robot
(*c*) Articulated robot
(*d*) Delta robot

Q.5. What is the purpose of a PLC (Programmable Logic Controller) in automation?

(*a*) Human resource management
(*b*) Complex mathematical calculations
(*c*) Process control and automation
(*d*) Entertainment

Q.6. Which type of sensor is commonly used in robotics to provide information about the robot's joint positions?

(*a*) Proximity sensor
(*b*) Ultrasonic sensor
(*c*) Potentiometer sensor
(*d*) Temperature sensor

Q.7. What is the term for a collaborative robot that works alongside humans without the need for safety barriers?

(*a*) Autonomous robot
(*b*) Barrier-free robot
(*c*) Humanoid robot
(*d*) Cobot (Collaborative robot)

Q.8. What is the advantage of using pneumatic actuators in industrial automation?

(*a*) Slower response time
(*b*) Lower energy efficiency
(*c*) High force and speed capabilities
(*d*) Limited range of motion

Q.9. Which programming language is commonly used for controlling industrial robots?

(*a*) Java
(*b*) C++
(*c*) Python
(*d*) Robot-specific languages (e.g., RAPID for ABB robots)

Q.10. What is the purpose of the "end effector" on a robot?

(*a*) To provide power to the robot
(*b*) To control the robot's motion
(*c*) To connect the robot to the power source
(*d*) To interact with the workpiece or environment

Q.11. What is the main advantage of using a PLC (Programmable Logic Controller) over traditional relay-based control systems?

(*a*) Lower cost (*b*) Faster response time
(*c*) Greater flexibility and ease of reprogramming
(*d*) Reduced power consumption

Q.12. Which type of sensor is commonly used to detect the presence or absence of an object without physical contact?

(*a*) Proximity sensor (*b*) Tactile sensor
(*c*) Force sensor (*d*) Gyroscopic sensor

Q.13. Which of the following is an example of a closed-loop control system?

(*a*) On/off switch (*b*) Relay logic
(*c*) PID controller (*d*) Manual joystick control

Q.14. In robotics, what does the term "degrees of freedom" refer to?

(*a*) The ability of a robot to perform complex tasks
(*b*) The number of sensors used in a robotic system
(*c*) The number of axes in which a robot can move independently
(*d*) The level of intelligence of a robot

Q.15. What is the purpose of a gripper in a robotic system?

(*a*) To provide power to the robot (*b*) To control the robot's motion
(*c*) To connect the robot to the power source (*d*) To grasp and manipulate objects

Q.16. Which type of industrial robot is known for its snake-like or flexible structure and is used for tasks in confined spaces?

(*a*) Cartesian robot (*b*) SCARA robot (*c*) Articulated robot (*d*) Snake robot

Q.17. What is the purpose of an HMI (Human-Machine Interface) in industrial automation?

(*a*) To physically control the robot
(*b*) To monitor and control industrial processes
(*c*) To perform complex mathematical calculations
(*d*) To simulate virtual reality environments

Q.18. Which component of an industrial robot is responsible for converting electrical signals into mechanical motion?

(*a*) Controller (*b*) Actuator (*c*) Sensor (*d*) End effector

Q.19. Which programming approach is used for teaching industrial robots specific tasks by physically moving them through desired motions?

(*a*) Offline programming (*b*) Simulation programming
(*c*) Manual programming (*d*) Teach pendant programming

Q.20. What is the term for the process of automatically guiding a robot to a desired position using feedback from sensors?

(*a*) Teleoperation (*b*) Supervised autonomy
(*c*) Visual servoing (*d*) Remote control

Q.21. What does the acronym "CNC" stand for in the context of manufacturing?

(*a*) Computerized Numerical Control (*b*) Control Network Circuit
(*c*) Centralized Network Communication (*d*) Computer Networking Consortium

Q.22. Which type of automation system involves the use of mechanical devices to perform tasks repetitively and without the need for electricity or electronics?

(*a*) Manual automation (*b*) Hard automation
(*c*) Soft automation (*d*) Flexible automation

Q.23. In the context of robotics, what is "end-of-arm tooling"?

(*a*) The last part of a robotic arm (*b*) The control panel for a robot
(*c*) The software interface of a robot
(*d*) The tools or equipment attached to a robot's arm

Q.24. What is the purpose of a sensor in an industrial automation system?

(*a*) To provide power to the system (*b*) To control the system's software
(*c*) To measure and collect data from the environment
(*d*) To process and analyze data

Q.25. Which communication protocol is commonly used for connecting industrial devices and machines in an automation network?

(*a*) Bluetooth (*b*) Wi-Fi (*c*) Ethernet
(*d*) NFC (Near Field Communication)

Q.26. What is the role of a "conveyor system" in industrial automation?

(*a*) Generating electricity for the factory
(*b*) Transporting goods or materials within a facility
(*c*) Controlling robotic arm movements
(*d*) Managing employee schedules

Q.27. What is the purpose of a "safety interlock" in an industrial automation setup?

(*a*) To monitor energy consumption (*b*) To improve production efficiency
(*c*) To ensure the safety of personnel and equipment
(*d*) To control robot programming

Q.28. Which type of industrial robot resembles a human arm with multiple joints and is commonly used for tasks that require high precision?

(*a*) Cartesian robot (*b*) SCARA robot (*c*) Articulated robot (*d*) Delta robot

Q.29. What does the term "PLM" stand for in the context of manufacturing and product development?

(*a*) Product Lifecycle Management (*b*) Programmable Logic Module
(*c*) Production Line Mapping (*d*) Precision Linear Motion

Q.30. Which of the following is a key challenge in implementing industrial automation and robotics?

(*a*) Reduced production speed (*b*) Limited flexibility
(*c*) Increased human involvement (*d*) Initial high cost and complexity

Key to Multiple Choice Questions

1. (*a*)	2. (*c*)	3. (*b*)	4. (*b*)	5. (*c*)	6. (*c*)	7. (*d*)	8. (*c*)	9. (*d*)
10. (*d*)	11. (*c*)	12. (*a*)	13. (*c*)	14. (*c*)	15. (*d*)	16. (*d*)	17. (*b*)	18. (*b*)
19. (*d*)	20. (*c*)	21. (*a*)	22. (*b*)	23. (*d*)	24. (*c*)	25. (*c*)	26. (*b*)	27. (*c*)
28. (*c*)	29. (*a*)	30. (*d*)						

Projects Works

1. **Automated Conveyor Belt System**: Design and build a conveyor belt system that can transport objects from one point to another automatically, incorporating sensors and control mechanisms for sorting and stopping.
2. **Robotic Arm Pick and Place**: Create a robotic arm that can pick up objects from one location and place them in another, demonstrating precise control and positioning.
3. **Smart Home Automation System**: Develop a system to control various home appliances using a smartphone app, showcasing the integration of IoT (Internet of Things) and automation.
4. **Automated Greenhouse Management**: Design an automated greenhouse system that regulates temperature, humidity, and watering based on sensor data, optimizing plant growth.
5. **Traffic Light Control System**: Build a traffic light control system that dynamically adjusts signal timings based on traffic flow to reduce congestion.
6. **Robotic Vacuum Cleaner**: Construct a robot capable of autonomously navigating a room and vacuuming the floor, utilizing sensors for obstacle avoidance.
7. **Remote-Controlled Drone**: Develop a drone that can be controlled remotely to perform tasks such as aerial photography, surveillance, or delivery.
8. **Autonomous Agricultural Robot**: Create a robot that can autonomously navigate through fields, identify and remove weeds, and optimize crop management.
9. **Automated Food Delivery System**: Build a system that uses robots to deliver food within a restaurant or a designated area, enhancing customer experience.
10. **Industrial Sorting System**: Design a machine that can sort different objects based on their properties, using sensors and actuators for efficient sorting.
11. **Robotic Welding System**: Construct a robot capable of performing automated welding tasks, ensuring precision and consistency in weld quality.
12. **Firefighting Robot**: Develop a robot equipped with sensors and firefighting mechanisms that can autonomously navigate through a simulated fire scenario.
13. **Automated Inventory Management System**: Create a system that uses RFID tags and sensors to track inventory levels, manage stock, and trigger reorder alerts.
14. **Robotic Surveillance System**: Design a robot equipped with cameras and sensors that can patrol an area, detect intruders, and send real-time alerts.
15. **Pick and Place Robotic Arm for Electronics Assembly**: Build a robotic arm capable of picking electronic components and placing them accurately on a PCB, demonstrating precision assembly.
16. **CNC Machine Automation**: Upgrade a CNC (Computer Numerical Control) machine with automated loading, tool changing, and monitoring capabilities.

17. **Autonomous Underwater Vehicle (AUV)**: Create a robot capable of navigating underwater environments, collecting data, and performing tasks such as underwater exploration or inspection.
18. **Humanoid Robot for Assistance**: Develop a humanoid robot that can assist users by performing tasks such as fetching objects, opening doors, and providing information.
19. **Robotic Car Parking System**: Design a system where robots autonomously park and retrieve cars in a parking lot, optimizing space utilization.
20. **Industrial Robotic Painting System**: Build a robotic system that can accurately apply paint to surfaces, demonstrating precise control and consistency in coating.

13 Robotics and Automation Technologies used in Modern Warfare

13.1 INTRODUCTION

Modern warfare increasingly employs robotic and automated technologies for a range of operations, including reconnaissance, surveillance, logistics, communication, and combat. These innovations can boost military might, increase effectiveness, and lessen hazards to live soldiers. It is significant to emphasize that the employment of robots and automation in warfare presents moral and legal issues, as well as worries about the possibility that autonomous systems may decide whether someone would live or die without human intervention. There are continuous discussions concerning the application of international rules and agreements like the Geneva agreements since they have not yet kept up with these advancements. Concerns also exist regarding the likelihood of arms races and the destabilizing impact of these technologies in international conflicts. The following sections go over some of the ways that robots and automation are used in modern warfare.

13.2 UNMANNED AERIAL VEHICLES (UAVs)

Precision airstrikes, intelligence collection, and reconnaissance are all carried out with drones like the MQ-9 Reaper. Without putting human pilots at risk, they can deliver real-time data and monitoring in hazardous or isolated locations. Drones, also referred to as unmanned aerial vehicles (UAVs), are widely used in contemporary warfare for a variety of purposes. These uses can be divided into the categories of surveillance, reconnaissance, obtaining intelligence, and, in some situations, combat. Numerous mathematical ideas, including as payload-specific computations, control systems, and navigation, are used in UAV operations.

Techniques and Instruments

1. **Navigation and Control Systems:**

   The Global Positioning System (GPS): It is essential to the precise positioning and navigation of UAVs.

   Inertial Measurement Unit (IMU): IMUs use gyroscopes and accelerometers to assess changes in orientation and velocity. For the exact control and stabilization of UAVs, they are crucial.

Control Algorithms: To modify the UAV's fly route, maintain altitude, and stabilize it in the presence of disturbances, control algorithms based on mathematics are utilized, such as PID (Proportional-Integral-Derivative) controllers.

2. **Payloads:**

   Cameras: For surveillance, reconnaissance, and target identification, high-resolution cameras and image systems are employed.

   Sensors: UAVs can incorporate a range of sensors, including radar, LiDAR, and thermal imaging, to carry out specialised missions.

   Communication Equipment: UAVs can be fitted with relays for communication or electronic warfare jammers.

3. **Applications in the Battlefield:**

   Surveillance and Reconnaissance: Using real-time imagery and intelligence provided by UAVs, military forces can analyse enemy positions, keep track of troop movements, and survey the battlefield.

   Intelligence Gathering: UAVs can be used in electronic warfare to intercept and gather signals intelligence (SIGINT) and electronic intelligence (ELINT).

   Some Armed Unmanned Aerial Vehicles (UAVs): These are fitted with guided weapons for accurate airstrikes against high-value targets.

   Electronic Warfare: UAVs can interfere with the electronic, radar, and communication systems of the enemy.

   Logistics and Resupply: UAVs may deliver equipment and supplies to troops in distant or hazardous locations.

Mathematics Behind UAV Operations

1. **Flight Dynamics**: Principles of aerodynamics and control theory are involved in UAV flight. To model and manage UAV flight, equations of motion such as Newton's laws and Euler angles are used.
2. **Path Planning**: The best flying paths are planned using algorithms like A* and Dijkstra's algorithm while avoiding obstacles and following mission requirements.
3. **Image Processing**: The visual information captured by UAV cameras is interpreted using image analysis and computer vision algorithms. This can involve image stitching, object detection, and pattern recognition.
4. **Kinematics**: The position, velocity, and acceleration of the UAV are described by kinematic equations. For control and navigation, these equations are crucial.
5. **Kalman Filtering**: Sensor fusion, which combines data from numerous sensors to precisely estimate the UAV's position and orientation, uses Kalman filters.
6. **Remote Sensing**: When it comes to sensors, remote sensing approaches use mathematical models to decipher the information obtained from devices like radar or LiDAR.
7. **Targeting and Guidance**: When UAVs are deployed in battle, mathematics is used to calculate trajectories, intercept targets, and discharge guided weapons precisely.

As UAV technology develops, more sophisticated mathematical models and algorithms are incorporated to boost their effectiveness and capabilities in combat. Modern warfare has benefited greatly from the use of unmanned aerial vehicles (UAVs), which offer a considerable edge in situational awareness and mission execution.

13.3 UNMANNED GROUND VEHICLES (UGVs)

These are utilized in urban warfare for operations including bomb disposal, mine clearance, and reconnaissance. UGVs can enter perilous zones that could be too risky for human soldiers to enter. Unmanned Ground Vehicles (UGVs) are autonomous or remotely operated vehicles that are utilized in contemporary warfare for a variety of functions like reconnaissance, surveillance, logistics, and dangerous jobs like bomb disposal. For successful military operations, the employment of UGVs entails a combination of methods, tools, applications, and mathematics.

Techniques and Instruments

1. **Navigation and Control Systems**:

   GPS and Inertial Navigation: In order to determine their position, velocity, and orientation, UGVs commonly employ GPS.

   Path Planning: Algorithms are used to plan effective paths that avoid obstacles, such as A* or Dijkstra's algorithm.

   Obstacle Detection and Avoidance: To identify impediments in their route and make navigational decisions, UGVs use a variety of sensors, including LIDAR, ultrasonic sensors, and cameras.

   Remote Control Interfaces: Military personnel can operate a variety of UGVs remotely.
2. **Manipulator Arms**: Some UGVs include manipulator arms that are controlled precisely during operations like bomb disposal using mathematical models.
3. **Sensors**: For particular purposes, UGVs may incorporate sensors like cameras, thermal imagers, chemical detectors, and ground-penetrating radar.
4. **Communication Equipment**: Real-time data sharing between UGVs and control centers is made possible via communication technologies.
5. **Payloads**: A variety of payloads, like as sensors, communication relays, and explosive ordnance disposal instruments, can be transported by UGVs.

Applications in the Battlefield

1. **Reconnaissance and Surveillance**: Real-time data about enemy positions, the environment, and other aspects of the battlefield are gathered by UGVs.
2. **Explosive Ordnance Disposal (EOD)**: Unexploded ordnance or explosive devices can be identified, handled, and securely disposed of with the help of UGVs.
3. **Search and Rescue**: UGVs can reach disaster-affected areas, find survivors, and transmit important information.
4. **Logistics Support**: UGVs deliver materials, tools, and ammunition, easing the load on human staff.
5. **Barrier and Obstacle Removal**: UGVs with bulldozer blades or other tools can pave the way for military forces by laying down routes.

Mathematics behind UGV Operations

1. **Kinematics and Dynamics**: UGVs use mathematical models, such as equations for location, velocity, and acceleration, to describe their motion. Understanding how UGVs move and respond to outside pressures is aided by kinematic and dynamic models.
2. **Path Planning Algorithms**: Algorithms for path planning use mathematical methods to determine the most direct path for UGVs to take while avoiding hazards. These algorithms take mission objectives and variables like terrain type and obstacle size into account.
3. **Sensor Fusion**: Sensor fusion techniques integrate data from many sensors used by UGVs to produce a more accurate depiction of the environment.
4. **Control Theory**: To stabilize UGVs and carry out particular tasks, such as manipulator arm control, control systems based on mathematical models, such as PID (Proportional-Integral-Derivative) controllers, are utilized.
5. **Localization and Mapping**: Simultaneous Localization and Mapping (SLAM) algorithms are used to create maps of the environment and determine the position of the UGV inside those maps.

UGVs accomplish duties that would otherwise be too risky or time-consuming while improving situational awareness, lowering hazards to human soldiers, and contributing to contemporary combat. UGV operations' performance in these applications is largely dependent on their mathematical foundations.

13.4 AUTONOMOUS WEAPONS SYSTEMS

Without any direct human involvement, these systems can locate and interact with targets. Despite being debatable, some countries have produced and employ them. The field of military technology known as autonomous weapons systems (AWS) is contentious and developing quickly. AWS are intended to recognize and interact with targets without the direct involvement of a person, and they create serious moral and legal questions. Here are specifics on the methods, tools, uses, and mathematics of AWS in contemporary warfare:

Techniques and Instruments

1. **Target Identification and Classification**: Machine Learning: To analyze sensor data and find possible targets, AWS frequently uses machine learning methods like deep neural networks. This calls for the use of methods like sensor fusion, audio analysis, and image recognition.
2. **Sensors**: To gather information about the surroundings and prospective targets, AWS may be fitted with a variety of sensors, including cameras, LIDAR, radar, and acoustic sensors.
3. **Autonomous Navigation**: AWS employ sophisticated navigation algorithms that incorporate GPS, inertial navigation, and obstacle avoidance to move within a predefined area or in the direction of a destination.
4. **Decision-Making Algorithms**: When deciding whether to engage a target after analyzing sensor data, AWS uses decision-making algorithms. These algorithms might include intricate rules of engagement and factors that aim to reduce collateral damage.
5. **Communication Systems**: As a part of a networked system, AWS are frequently used, enabling them to interact with one another, ground control stations, and sometimes human operators.

Applications in the Battlefield

1. **Surveillance and Reconnaissance**: AWS can be used to detect enemy positions and movements during autonomous surveillance and reconnaissance missions.
2. **Air Defense**: AWS with anti-aircraft systems can autonomously identify and take down intruding drones or planes.
3. **Ground Defense**: The use of AWS for autonomous ground defence against invading troops, vehicles, and other threats.
4. **Border Security**: By monitoring for and responding to unauthorized intrusions, AWS may patrol and secure borders.
5. **Anti-Ship Missiles**: Autonomous anti-ship missile systems are capable of independently locating and identifying hostile vessels.
6. **Counter-UAV Systems**: Unauthorized drones or UAVs can be found and stopped using AWS.

Mathematics behind AWS Operations

1. **Machine Learning and Neural Networks**: Deep learning models classify objects, interpret sensor data, and make judgments based on patterns and training data using neural networks.
2. **Computer Vision**: Computer vision algorithms examine image data from cameras to find and follow targets. It is possible to use methods like feature extraction, object recognition, and optical flow.
3. **Path Planning and Control**: In order to navigate and engage targets while avoiding obstacles, AWS uses mathematical models for path planning and control systems.
4. **Sensor Fusion**: AWS may use sensor fusion algorithms to combine data from various sensors (such as cameras, LIDAR, and radar) to produce a situational awareness image that is more complete.
5. **Probability and Decision Theory**: AWS bases its decisions on probabilistic logic, taking into account both the possibility of collateral harm and the likelihood that a target is hostile.

It is significant to emphasize that AWS create significant ethical and legal issues, notably in relation to responsibility, the risk for indiscriminate injury, and compliance with international rules, such as the Geneva Conventions. The use and regulation of AWS in warfare is a topic of discussion among several nations and international organizations.

13.5 LOGISTICS AND SUPPLY CHAIN AUTOMATION

Military supply chains and logistics are managed through automation, which increases their effectiveness and efficiency. The delivery of materials and equipment can be made more efficient through automated warehouses and logistics systems. Modern warfare depends heavily on logistics and supply chain automation, which makes it possible to transfer troops, supplies, and equipment quickly and effectively. A variety of methods, tools, software, applications, and mathematics are used in this automation to assist military activities.

Techniques and Instruments

1. **Supply Chain Management Systems**:Military organizations organize, monitor, and regulate the flow of resources using specialized supply chain management software. These systems include algorithms for inventory control, order optimization, and demand forecasting.

2. **Transportation Management Systems (TMS)**: By taking into account variables like route planning, load optimization, and scheduling, TMS technologies aid in the optimization of the transportation of goods and personnel.
3. **Fleet Management**: Automated management and tracking systems are employed for fleets of military vehicles. This includes tracking using GPS, planning maintenance, and keeping track of fuel usage.
4. **Barcode and RFID Technology**: Asset tracking and inventory management are facilitated by the use of bar-coding and RFID (Radio-Frequency Identification) technologies.
5. **Aerial Delivery Systems**: To distribute supplies and equipment to distant or inaccessible sites, aerial delivery systems employ precision airdrops and parachutes.
6. **Warehouse Automation**: To improve efficiency, military warehouses use automation technologies such conveyor systems, automated storage and retrieval systems (AS/RS), and robotic picking.
7. **Block chain and Secure Communication**: The integrity and security of logistics data are ensured using secure communication and block chain technology, especially in mission-critical operations.

Applications in the Battlefield

1. **Logistics Planning and Management**: The delivery of supplies, equipment, and resources to military units in the field is planned and managed using automation techniques.
2. **Supply Chain Visibility**: Real-time visibility into the movement of commodities is made possible by automated technologies, improving situational awareness and facilitating quicker decision-making.
3. **Maintenance and Repair**: Automated maintenance management systems facilitate the planning and execution of upkeep and repair tasks for military machinery and vehicles.
4. **Medical Logistics**: Automation facilitates the delivery of medical personnel, equipment, and supplies to combat medics and field hospitals.
5. **Personnel Deployment**: Automated systems help with military personnel rotation and deployment, guaranteeing effective scheduling and adherence to deployment plans.

Mathematics behind Logistics and Supply Chain Automation

1. **Optimization Algorithms**: Vehicle routing, inventory distribution, and load planning are just a few of the supply chain tasks that can be optimized using linear programming and integer programming.
2. **Statistical Analysis**: To forecast demand, monitor performance, and optimize resource allocation, statistical techniques such as regression analysis and time series analysis are employed.
3. **Graph Theory**: Routing and network design issues, such as finding the shortest route for supply delivery, are modeled and solved using graph theory.
4. **Queueing Theory**: Queueing theory is useful for analyzing and maximizing wait times, particularly when waiting periods are required for military units to receive supplies or services.
5. **Probability and Simulation**: In order to evaluate the effects of various logistics methods and pinpoint potential hazards or bottlenecks, probabilistic models and simulation are used.

6. **Data Analytics and Machine Learning**: To mine and analyze massive databases for insights into logistics operations and trends, data analytics and machine learning are used.

Automation of the logistics and supply chain is essential for preserving the preparedness and efficiency of military units. The mathematical and technological underpinnings of automation ensure that resources are utilized efficiently, lowering operational costs and enhancing the military's capacity to adjust to changing battlefield situations.

13.6 ROBOTIC EXOSKELETONS

Exoskeletons can enhance the physical capabilities of soldiers, allowing them to carry heavy loads and reducing fatigue. Robotic exoskeletons, also known as powered exposits, are wearable devices designed to enhance the physical capabilities of soldiers in modern warfare. These exoskeletons can assist with carrying heavy loads, reduce fatigue, and improve overall performance. Here are the details of the techniques, instruments, applications in the battlefield, and the mathematics behind robotic exoskeletons:

Techniques and Instruments

1. **Exoskeleton Structure**: Mechanical frames with joints that imitate the wearer's natural range of motion make up robotic exoskeletons. They comprise power sources, actuators, and sensors.
2. **Sensors**: Force sensors and inertial measurement units (IMUs) are used to track the wearer's motions and alter the exoskeleton's support as necessary.
3. **Actuators**: The necessary force to support the wearer's movements is provided by electric motors or hydraulic actuators. Pneumatic systems for precise control may be among them.
4. **Power Sources**: Robotic exoskeletons are frequently powered by rechargeable batteries or portable power packs to ensure longer use in the field.
5. **Control Systems**: Cutting-edge control algorithms guarantee that the exoskeleton moves in sync with the wearer's movements. Machine learning is used by several exoskeletons for adaptive control.
6. **Load Carriage Systems**: A load carrier system may be included in exoskeletons to support the weight of equipment like backpacks, weapons, and ammunition.

Applications in the Battlefield

1. **Load Carriage and Equipment Support**: Robotic exoskeletons make it less physically taxing for soldiers to carry heavy loads and equipment, making it simpler for them to convey necessary supplies.
2. **Fatigue Reduction**: Exoskeletons enable soldiers to perform for longer periods of time without becoming exhausted by assisting wearers' motions mechanically.
3. **Enhanced Mobility**: Exoskeletons can improve a soldier's mobility and agility, enabling them to navigate difficult terrain, scale barriers, and move more skillfully.
4. **Rehabilitation**: In order to help injured soldiers regain their strength and mobility during rehabilitation, exoskeletons are used.
5. **Rescue Operations**: Exoskeletons can help search and rescue personnel in disaster-stricken areas by helping them lift rubble and conduct operations.

Mathematics behind Robotic Exoskeletons

1. **Biomechanics**: It is essential to comprehend the biomechanics of the human body, including joint angles and muscle forces, in order to create exoskeletons that imitate the movements of the body.
2. **Control Theory**: To help the wearer's motions, control algorithms are employed to synchronize the actuators on the exoskeleton. PID controllers, or proportional-integral-derivative controllers, are frequently used.
3. **Kinematics and Dynamics**: The motion of the exoskeleton's joints and how they relate to the wearer's joints are described by kinematic equations. Dynamic models take movement's forces and torques into account.
4. **Force and Torque Calculations**: Based on sensor data, mathematical models determine the forces and torques required to help the wearer's motions, providing the right degrees of assistance.
5. **Power Consumption**: Exoskeletons need power to function. For creating trustworthy systems, power consumption and battery life calculations are crucial.
6. **Machine Learning and Adaptation**: Advanced exoskeletons might employ machine learning to modify their behavior in accordance with the wearer's movements, the environment, and the demands of the task.

Robotic exoskeletons are one type of human enhancement technology that can greatly improve a soldier's physical capabilities and lower the danger of harm during combat and other military missions. The safe and efficient use of these gadgets in modern combat depends on understanding the underlying mathematical and engineering principles.

13.7 REMOTE-CONTROLLED VEHICLES

These have a variety of uses, including patrolling and reconnaissance. They are manageable by soldiers at a secure distance. Unmanned vehicles that can be operated remotely are known as remote-controlled vehicles, and they are useful weapons in contemporary warfare. These vehicles are employed in a variety of military operations, such as combat, surveillance, and reconnaissance. Here are specifics on the methods, gear, battlefield uses, and math entailed by remote-controlled vehicles utilized in contemporary warfare:

Techniques and Instruments

1. **Communication Systems**: Secure and reliable communication technologies are essential for remote-controlled vehicles to receive instructions and send operators real-time data. Usually, encrypted data transmission and radio frequency (RF) communication are involved.
2. **Sensor Systems**: In order to offer situational awareness and gather important data, these vehicles are outfitted with a variety of sensors, including cameras, radar, microphones, thermal imaging, LIDAR, and webcams.
3. **Navigation and Control**: GPS is used by remote-controlled vehicles for accurate positioning and navigation. Control algorithms make sure that the vehicle obeys driver commands precisely.
4. **Remote Operation Interfaces**: To remotely control the movement, cameras, and sensors of the vehicle, operators utilize specialized consoles or control stations.

5. **Power Sources**: Energy for propulsion and onboard systems is provided by power systems, which frequently use rechargeable batteries. Vehicle design must take energy management into consideration.

Applications in the Battlefield

1. **Reconnaissance and Surveillance**: To analyze adversary positions and operations, real-time video feeds and data from remote-controlled vehicles are employed in intelligence collection.
2. **Explosive Ordnance Disposal (EOD)**: These vehicles are used to find explosive devices and properly dispose of them, lowering the hazards to human EOD crews.
3. **Search and Rescue**: Remote-controlled vehicles can enter dangerous or inaccessible places during military operations and disaster assistance to look for survivors.
4. **Target Identification**: These vehicles are able to find and recognize possible targets, such as enemy combatants or covert dangers.
5. **Electronic Warfare**: Remote-controlled vehicles may be employed in electronic warfare to obstruct adversary communications and conduct electronic surveillance.

Mathematics Behind Remote-Controlled Vehicles

1. **Kinematics and Dynamics**: Accurate control and course planning are made possible by mathematical models that accurately characterize the motion of the vehicle, including position, velocity, and acceleration.
2. **Control Algorithms**: Algorithms that transform operator commands into vehicle movements are developed using control theory, assuring stability and precise navigation.
3. **Image Processing**: Camera data is interpreted by algorithms for image processing and computer vision, enabling functions like object detection, tracking, and recognition.
4. **Localization and Mapping**: Simultaneous Localization and Mapping (SLAM) algorithms are used to create maps of the surroundings and calculate the position of the vehicle inside those maps.
5. **Path Planning and Collision Avoidance**: To create safe navigation while choosing the best routes and avoiding obstacles, mathematical models and algorithms are used.
6. **Signal Processing**: To enhance the quality of sensor data, signal processing techniques including noise reduction and feature extraction are used.

Due to its capacity to operate in hazardous or difficult-to-reach regions while reducing the risk to human soldiers, remote-controlled vehicles are useful tools in modern warfare. These vehicles' efficiency and safety in military operations depend on the mathematics and technology that support them.

13.8 AUTONOMOUS GROUND VEHICLES

Some military vehicles have autonomous driving features, which could increase safety by removing the need for human drivers. Unmanned vehicles with autonomous navigation and control systems, or autonomous ground vehicles (AGVs), are utilized in modern warfare for a variety of tasks such as reconnaissance, supply, and fighting. Because they are built to operate without direct human input, these vehicles pose fewer dangers to actual soldiers. The specifics of the methods, tools, battlefield applications, and mathematics underlying autonomous ground vehicles in contemporary warfare are as follows:

Techniques and Instruments

1. **Navigation and Control Systems:**

   GPS and Inertial Navigation: AGVs use inertial navigation systems for accurate control and orientation and GPS for placement.

   Path Planning: To plan effective paths that avoid obstacles, algorithms like A* or Dijkstra's algorithm are used.

   Obstacle Detection and Avoidance: To identify and avoid obstacles in their route, AGVs are fitted with sensors like LIDAR, ultrasonic sensors, and cameras.

   Communication Equipment: Because they are frequently a part of a networked system, AGVs can interact with other vehicles, control towers, and operators.
2. **Sensors:** AGVs use cameras, LIDAR, radar, and other environmental sensors to collect data about their surroundings in real time.
3. **Payloads:** AGVs are capable of transporting a range of payloads, including sensors, communication relays, and occasionally guns or ammunition.

Applications in the Battlefield

1. **Reconnaissance and Surveillance:** AGVs are employed to acquire intelligence by delivering real-time information about the locations of adversaries, the surrounding area, and potential dangers.
2. **Logistics and Supply:** AGVs bring supplies, gear, and ammunition to military formations, lightening the load on human soldiers in terms of logistics.
3. **Disposal of Explosive Ordnance (EOD):** AGVs are used to locate, handle, and properly dispose of explosive devices and unexploded ordnance.
4. **Security and Border Patrol:** By spotting and responding to unauthorised breaches, AGVs may patrol and secure borders.

Some AGVs are employed in electronic warfare to obstruct adversary communication systems and gather electronic intelligence.

Mathematics behind Autonomous Ground Vehicles

- **Kinematics and Dynamics:** AGVs use mathematical models, such as equations for location, velocity, and acceleration, to explain their movements. These concepts are employed for control and navigation.
- **Path Planning Methods 2:** Algorithms for path planning use mathematical methods to choose the best route while avoiding hazards, taking into account the topography and mission objectives.
- **Sensor Fusion:** Sensor fusion methods combine data from several sensors to produce a more accurate picture of the environment and to aid in decision-making.
- **Control Theory:** To efficiently stabilize AGVs and carry out particular duties, control systems based on mathematical models, such as PID (Proportional-Integral-Derivative) controllers, are used.
- **Localization and Mapping**: Algorithms like Simultaneous Localization and Mapping (SLAM) are used to create maps of the environment and estimate the AGV's position within those maps.

- **Probability and Decision Theory**: AGVs make decisions based on probabilistic reasoning, considering factors like the likelihood of encountering an obstacle or identifying a target.
- **Machine Learning and Adaptive Control**: Advanced AGVs may use machine learning techniques to adapt their behavior based on experience and changing conditions.

Modern military operations are significantly improved by autonomous ground vehicles, which also reduce dangers to human people and increase efficiency across a range of activities. For AGVs to be used in warfare safely and effectively, mathematics and technology are crucial.

13.9 AUTOMATED CYBERSECURITY

Cyber threats and attacks are detected and handled by automated systems. For the sake of safeguarding military networks and data, this is essential. Modern warfare uses sophisticated methods, tools, and applications to safeguard military networks and systems from cyber attacks, including automated cybersecurity. To identify cyberattacks, take action, and reduce their impact, these systems employ a variety of mathematical and computational techniques. The specifics of the methods, tools, battlefield applications, and mathematical underpinnings of automated cybersecurity in contemporary combat are as follows:

Techniques and Instruments

- **Intrusion Detection Systems (IDS):** These systems continuously scan network traffic for odd patterns or recognized attack signatures. To find possible risks, they employ machine learning and pattern recognition algorithms.
- **Firewalls and Network Security:** Incoming and outgoing network traffic is filtered by firewalls. To stop unauthorized access and data leaks, access control lists (ACLs) and rule-based systems are put in place.
- **Endpoint Protection:** Automated endpoint protection systems can identify and stop malware and other threats on specific devices using methods including behavior analysis, signature-based detection, and machine learning.
- **Security Information and Event Management (SIEM):** To provide real-time monitoring and threat detection, SIEM technologies gather and analyze data from diverse sources. These systems frequently identify security incidents using intricate correlation techniques.
- **Vulnerability Scanners:** These tools evaluate network and system vulnerabilities automatically. They priorities remediation using quantitative tools to assess potential threats.
- **Automated Incident Response:** These tools respond to security issues by using predetermined playbooks and algorithms to isolate affected systems and alert security professionals, for example.

Applications in the Battlefield

- **Network Security:** Automated cybersecurity is used to safeguard military communication networks and data, ensuring the privacy, accuracy, and accessibility of sensitive data.
- **Protection of Critical Infrastructure:** Cyberattacks on vital infrastructure, like power grids and water supply systems, are a worry in contemporary warfare. These essential services are protected by automated mechanisms.

- **Information Warfare and Electronic Countermeasures:** Automated cybersecurity is used in electronic warfare to find and block hostile cyberthreats including jamming and deception.
- **Secure Communications:** To secure military communications and avoid eavesdropping, automated encryption and decryption technologies are deployed.

Mathematics behind Automated Cybersecurity

- **Statistical Analysis:** Statistical techniques are used to spot patterns that can point to security breaches and to spot anomalies in network data.
- **Machine Learning:** Based on past data and behavior patterns, machine learning algorithms, such as supervised and unsupervised learning, are used to identify and categorize risks.
- **Cryptography:** For secure communication and data security, cryptographic techniques, including mathematical algorithms like RSA and AES, are used.
- **Graph Theory:** Models based on graph theory are used to analyze network topologies and find patterns that could be signs of intrusions or weaknesses.
- **Pattern Recognition:** In network traffic and system records, recognized attack signatures are found using pattern recognition techniques.
- **Queuing Theory:** Queuing theory is applied to assess network performance and response times, which are crucial for timely detection and response to cyber threats.

Modern warfare requires automated cybersecurity to protect military networks and infrastructure from online assaults. In an ever-changing cybersecurity environment, the mathematics and computational techniques used are crucial for quick threat identification, reaction, and mitigation.

13.10 DRONE SWARMS

Small drone squadrons can be employed for operations like target detection, observation, and even coordinated strikes. They can overpower hostile defenses and offer useful information. Due to its capacity to overwhelm defenses, gather intelligence, and conduct precision attacks, drone swarms, or a coordinated group of drones working together, have grown more and more relevant in modern warfare. These drone swarms make use of a variety of methods, tools, and mathematical ideas. The specifics of drone swarms in contemporary combat are as follows:

Techniques and Instruments

- **Communication and Coordination:** In order to coordinate their actions, drone swarms rely on communication networks. To keep swarm members in sync, communication protocols and encryption techniques are used.
- **Swarm Algorithms:** Swarm algorithms control how each drone in the swarm behaves. These algorithms, which frequently draw their inspiration from swarm behaviour in nature, decide on formation, flight trajectories, and activities based on specified rules.
- **Sensors:** To collect data and communicate with other swarm members, drones in a swarm are outfitted with a variety of sensors, such as cameras, LIDAR, GPS, and other environmental sensors.
- **Autonomous Operate:** Each drone in the swarm needs to be able to operate itself autonomously in order to adapt to shifting circumstances and keep the desired formation.

Drone swarms can transport a variety of payloads, such as cameras, sensors, communication relays, and, in some cases, weapons.

Applications in the Battlefield

- **Drone Swarms:** These are able to cover enormous regions and can give real-time data and imagery for reconnaissance and information gathering.
- **Swarms:** These are capable of tracking and identifying moving targets or possible threats, providing constant monitoring capabilities.
- **Electronic Warfare:** By flying in formations that cause interference or by intercepting signals, drone swarms can interfere with enemy communication systems and obtain electronic intelligence (ELINT).
- **Precision Strikes:** Some drone swarms have armaments on board and can launch precise attacks on adversary targets with little collateral damage.
- **Area Denial and Air Defense:** Swarms can be employed to prevent enemy aircraft from landing in particular locations or to thwart approaching dangers by outnumbering hostile air defences.

Mathematics behind Drone Swarms

- **Swarm Behavior Modeling:** Differential equations and discrete algorithms are used in the mathematics of swarm behavior modeling to represent the interactions and motions of individual drones.
- **Path Planning and Control:** Mathematical methods for path planning and control make sure that drones keep their formations, stay clear of obstacles, and adjust to shifting environmental conditions.
- **Sensor Fusion:** To create a comprehensive situational awareness picture for the entire swarm, algorithms for sensor fusion combine input from several sensors.
- **Trajectory Optimization:** Algorithms for trajectory optimization are used to identify the drones' most effective flight courses and formations.
- **Communication Protocols:** To enable trustworthy data flow between members of the swarm and the central control system, communication protocols are created utilizing network theory.
- **Probability and Decision Theory:** These are used by drone swarms to make coordinated decisions based on the swarm's collective data and predetermined rules of engagement.

The use of drone swarms in modern warfare has many benefits, including improved situational awareness, pinpoint strikes, and the capacity to control or disrupt airspace. They depend on mathematics and algorithms for their deployment and operation to be successful.

13.11 AUTONOMOUS NAVAL SYSTEMS

For activities like mine hunting and underwater surveillance, unmanned boats and submarines are employed.

Modern naval warfare increasingly uses autonomous naval systems, such as unmanned surface boats (USVs) and autonomous underwater vehicles (AUVs), for a variety of purposes. These systems improve naval operations by utilizing a variety of methods, tools, and mathematical concepts. The specifics of autonomous naval systems in contemporary combat are as follows:

Techniques and Instruments

- **Navigation and Control Systems:** For accurate location and navigation, autonomous naval systems use sonar-based localization systems, inertial navigation, and GPS.

  These devices can carry out missions automatically, dodging obstacles and maintaining on route, thanks to control algorithms.
- **Sensors:** Sensors are essential for data collection and perception. While AUVs use sonar, cameras, and a variety of environmental sensors, USVs include sonar, radar, AIS (Automatic Identification System) receivers, and LIDAR as options.
- **Communication Systems:** To transfer data, receive commands, and relay information back to control centers or other naval assets, these systems need communication equipment. They may communicate through radio, satellite, or underwater acoustic waves.
- **Payloads:** Depending on their mission profiles, autonomous naval systems are outfitted with payloads. This can include torpedoes, mine detectors, and environmental monitoring sensors.

Applications in the Battlefield

- **Mine Detection and Clearance:** To reduce dangers to human personnel, autonomous naval systems are used for mine detection and clearance.
- **Anti-Submarine Warfare:** Using sonar and other sensors, AUVs are employed in anti-submarine warfare to track and find submarines.
- **Surveillance and Reconnaissance:** These systems maintain constant watch on adversary ships, gather intelligence, and maintain maritime domain awareness.
- **Defense of Naval Assets:** USVs can be used to defend naval assets by supplying perimeter defense and catching intruders.
- **Environmental Monitoring:** Autonomous systems are employed to gather environmental information, evaluate the conditions underwater, and aid in scientific investigation.

Mathematics behind Autonomous Naval Systems

- **Kinematics and Dynamics:** Mathematical models are used to describe the motion and dynamics of autonomous naval systems, including position, velocity, and acceleration.
- **Path Planning and Control:** To plan and carry out missions while avoiding obstacles and maintaining intended trajectories, path planning techniques and control theory are utilized.
- **Sensor Fusion:** Sensor fusion techniques are used to combine input from many sensors, ensuring precise target tracking and ambient perception.
- **Underwater Acoustics:** Advanced mathematics are utilized for underwater acoustic communications, including wave propagation in underwater environments, channel modeling, and signal processing.
- **Navigation and Localization:** In order to determine the position and orientation of autonomous systems in underwater environments, methods like Kalman filters and particle filters are used.
- **Probability and Decision Theory:** By using probabilistic reasoning and taking into consideration the ambiguities in sensor data and mission objectives, these systems arrive at judgments.
- **Communication Protocols:** Data transmission between autonomous systems and command centers is based on mathematical principles.

Modern naval warfare depends on autonomous naval technologies because they give naval forces the ability to operate more effectively while posing fewer risks to their human members. These systems' successful operation in difficult marine conditions depends on the mathematics and algorithms they employ.

13.12 COMMUNICATION AND CONTROL

In military operations, automation is utilised to streamline communication and control systems, allowing for cooperation between various troops and systems. The foundation of contemporary warfare is comprised of communication and control technologies, which make it easier to coordinate forces, share information, and carry out military actions. These systems use a range of methods, tools, and mathematical concepts to guarantee efficient communication and command in the combat zone. Below are the specifics:

Techniques and Instruments

1. **Communication Systems:**
   - A wide range of technologies, including radios, satellite communication, data linkages, and secure networks, are used in communication systems in modern combat.
   - To secure communication and safeguard sensitive data, encryption techniques including public-key infrastructure (PKI) and advanced encryption standards (AES) are used.
2. **Command and Control (C2) Systems:**
   - These systems give military units a hierarchical structure that enables commanders to give commands and track the position of forces in real time.
   - To assist in the decision-making and planning processes, decision support tools such as Geographic Information Systems (GIS) and computer-aided planning are utilized.
3. **Data Fusion and Intelligence Systems:**
   - To produce a complete picture of the battlefield, data fusion techniques integrate data from multiple sources, including sensors, reconnaissance, and surveillance.
   - Data analysis and intelligence gathering are accomplished through the use of artificial intelligence and machine learning technologies.
4. **Navigation and Positioning**: Target localization, navigation, and situational awareness are aided by the exact position data that GPS and inertial navigation systems give military assets.
5. **Jamming and Electronic Warfare (EW)**: To interfere with adversary communication and control systems, electronic countermeasures such as signal manipulation and jammer systems are employed.

Applications in the Battlefield

1. **Command and Control:** With the use of communication and control systems, commanders are able to efficiently allocate resources, coordinate actions, and transmit orders in real time.
2. **Information Collection and Reconnaissance:** These systems facilitate reconnaissance and surveillance, obtaining up-to-date data on adversary movements, topography, and possible dangers.

3. **Target Acquisition and Engagement:** By exchanging target intelligence, directing weaponry, and permitting coordinated attacks, communication systems help with target acquisition and engagement.
4. **Supply Chain Management and Logistics:** Military logistics systems depend on coordination and communication to oversee the allocation of materials, tools, and staff.
5. **Air Traffic Control and Air Defense:** These systems are essential for managing military aviation, monitoring both military and civilian air traffic, and supplying air defense.

Mathematics behind Communication and Control

1. **Network Theory:** Taking into account elements like network topology, routing, and resilience, network theory is used to describe and analyze communication networks.
2. **Signal Processing:** To improve transmission signal quality, lower noises, and find anomalies in data, signal processing techniques are used.
3. **Information and Probability Theory:** Information theory measures information content and transmission efficiency in communication networks, whereas probability theory is utilized to represent uncertainty.
4. **Control Theory:** The design of control systems that sustain unit cohesion while stabilizing and regulating military assets depends heavily on control theory.
5. **Coding Theory:** Reliable data transmission over noisy communication channels is ensured by the employment of error-correcting codes.
6. **Cryptography:** To prevent sensitive data from being intercepted or accessed by unauthorized parties, encryption algorithms are based on mathematical principles.
7. Coordinate systems and geographic information systems (GIS) are two examples of geographic mathematics, which is essential for location-based applications and navigation.

Systems of communication and control are essential to the accomplishment of contemporary military operations. These systems' mathematical and technological underpinnings give armed personnel the ability to efficiently plan, coordinate, and carry out missions in a secure manner.

Numerical Problems with Solutions

Problem 1. You are designing an autonomous ground vehicle for reconnaissance in a military operation. The vehicle's mission requires it to cover a circular area with a radius of 1,000 meters. If the vehicle's maximum speed is 20 meters per second, how long will it take to complete the mission?

Solution: Time (t) = Distance (d)/Speed (v), t = 2 * π * 1000 meters/20 meters per second = 314.16 seconds

Problem 2. A military drone is flying at an altitude of 500 meters. It needs to capture images of a target on the ground with a resolution of 10 centimeters per pixel. Calculate the minimum focal length of the drone's camera lens required to achieve this resolution.

Solution: Resolution = (Altitude * 100)/(Focal Length * Sensor Size) 10 cm/pixel = (500 meters * 100)/(Focal Length * Sensor Size), Focal Length = (500 meters * 100)/(10 cm/pixel * Sensor Size)

Problem 3. An autonomous underwater vehicle (AUV) is performing a mission at a depth of 200 meters. The AUV's pressure hull is designed to withstand a maximum pressure of 500 bar. Calculate the total pressure the hull is experiencing at this depth.

Solution: Pressure (P) = Density (ρ) * Gravitational Acceleration (g) * Depth (d)

$$P = (1{,}030 \text{ kg/m}^3) * (9.81 \text{ m/s}^2) * 200 \text{ meters}$$

$$= 2{,}046{,}600 \text{ pascals (Pa) or } 2.046 \text{ bar}$$

Problem 4. A military robot's battery has a capacity of 50,000 watt-hours (Wh). If the robot operates at an average power consumption of 500 watts, how long can it operate on a single charge?

Solution: Operating Time (t) = Battery Capacity (C)/Power Consumption (P)

$$t = 50{,}000 \text{ Wh}/500 \text{ watts} = 100 \text{ hours.}$$

Problem 5. You are designing a communication network for a swarm of 20 autonomous drones. Each drone can transmit data at a rate of 2 megabits per second (Mbps). Calculate the total network bandwidth required to support these drones.

Solution: Total Bandwidth = Number of Drones * Individual Drone's Data Rate

Total Bandwidth = 20 drones * 2 Mbps = 40 Mbps

Problem 6. A military robot needs to navigate from point A to point B on a battlefield grid. The grid is 10 × 10 units, and the robot can move up, down, left, or right. Determine the shortest path from A to B while avoiding obstacles.

Solution: This problem can be solved using path planning algorithms such as Dijkstra's algorithm, A*, or breadth-first search (BFS).

Problem 7. A surveillance drone is equipped with a camera that can identify enemy tanks. If the drone's camera has a 20x zoom, what is the maximum distance from which it can identify a tank that has a width of 3 meters?

Solution: Use the concept of angular resolution. Given the width of the tank and the zoom level, you can calculate the maximum distance at which it can be identified.

Problem 8. An autonomous ground vehicle uses LIDAR and radar for obstacle detection. The LIDAR can detect obstacles up to 50 meters, and the radar can detect obstacles up to 100 meters. Calculate the effective detection range when both sensors are used together.

Solution: Combine the information from both sensors to determine the effective detection range. The effective range is not simply the sum of individual ranges due to the overlapping coverage.

Problem 9. A military drone needs to maintain a certain altitude while flying in turbulent conditions. The drone's altitude control system is overshooting its setpoint. Calculate the control system's PID parameters to reduce overshoot.

Solution: Use the Ziegler-Nichols method or another PID tuning method to calculate the appropriate proportional, integral, and derivative gains.

Problem 10. A mobile robot's GPS system has an accuracy of 5 meters. The robot's actual position is (x, y) = (100, 200) meters. Determine how much error correction is needed to adjust the robot's position.

Solution: Calculate the error in position by comparing the GPS-reported position to the actual position. Then determine the control input needed to correct this error.

Short Questions and Answers

Q.1. What is the primary purpose of military robotics and automation?

Ans. The primary purpose is to enhance military capabilities, reduce human risk, and improve efficiency in various operations.

Q.2. What are unmanned aerial vehicles (UAVs) commonly known as?

Ans. Drones.

Q.3. What's the advantage of using autonomous underwater vehicles (AUVs) in naval operations?

Ans. AUVs can perform underwater tasks autonomously, reducing the need for human divers and minimizing risks.

Q.4. How do path planning algorithms assist military robots?

Ans. Path planning algorithms help robots navigate complex environments, avoid obstacles, and reach their destinations.

Q.5. What is sensor fusion in the context of robotics?

Ans. Sensor fusion combines data from multiple sensors to create a more accurate representation of the environment.

Q.6. Why is AI essential in military robotics?

Ans. AI enables robots to make autonomous decisions, adapt to changing conditions, and perform complex tasks.

Q.7. What's the purpose of reconnaissance drones in warfare?

Ans. Reconnaissance drones gather intelligence and provide real-time information about enemy positions and activities.

Q.8. What are the advantages of using exoskeletons in the military?

Ans. Exoskeletons enhance soldiers' physical capabilities, reduce fatigue, and assist in carrying heavy loads.

Q.9. What is the primary application of unmanned ground vehicles (UGVs) in modern warfare?

Ans. UGVs are used for tasks such as bomb disposal and reconnaissance in dangerous areas.

Q.10. How do drone swarms function in coordinated military operations?

Ans. Drone swarms work together to achieve mission objectives, often by overwhelming defenses or gathering intelligence.

Q.11. What is the benefit of using autonomous navigation systems in military vehicles?

Ans. Autonomous navigation reduces the need for human drivers, enhancing safety and operational efficiency.

Q.12. What's the main objective of cyber-physical systems in the military?

Ans. Cyber-physical systems integrate computing, networking, and physical processes to improve overall military capabilities.

Q.13. How do logistics automation systems support military operations?

Ans. These systems help plan and manage the distribution of supplies, equipment, and personnel to military units.

Q.14. What technology is used to secure military communications and data in the digital battlefield?

Ans. Advanced encryption and secure communication protocols.

Q.15. What role does electronic warfare play in modern military operations?

Ans. Electronic warfare includes jamming enemy communication systems and gathering electronic intelligence.

Q.16. What mathematical concept is fundamental in controlling autonomous systems?

Ans. Control theory, which includes algorithms like PID controllers.

Q.17. How are remote-controlled vehicles operated in the field?

Ans. Operators use specialized consoles or control stations to remotely control vehicle movements and sensors.

Q.18. What is the primary objective of aerial delivery systems in the military?

Ans. To safely deliver supplies and equipment to remote or inaccessible locations via airdrops.

Q.19. Why is localization and mapping essential for autonomous systems?

Ans. It helps the system understand its position within its environment and build maps of the surroundings.

Q.20. What role do automated maintenance management systems play in military operations?

Ans. They help plan and execute maintenance and repair activities for military vehicles and equipment.

Q.21. What is the primary purpose of the Automatic Identification System (AIS) in maritime operations?

Ans. AIS is used for tracking and identifying vessels, enhancing maritime situational awareness.

Q.22. How do drones communicate with operators and other drones in real-time?

Ans. Drones use communication networks, including radio links and data links.

Q.23. What is the concept of sensor fusion in autonomous systems?

Ans. Sensor fusion combines data from multiple sensors to provide a comprehensive view of the environment.

Q.24. Why is inertial navigation important for autonomous vehicles?

Ans. Inertial navigation systems provide precise orientation and position data without relying solely on external signals like GPS.

Q.25. What type of control system is commonly used in autonomous vehicles and robots?

Ans. PID (Proportional-Integral-Derivative) control systems.

Q.26. What mathematical model describes the motion of robotic arms in industrial automation?

Ans. Forward and inverse kinematics models.

Q.27. How are UAVs used for surveillance and reconnaissance in modern warfare?

Ans. UAVs provide aerial surveillance and intelligence gathering capabilities, allowing for real-time data collection.

Q.28. What is the primary role of autonomous naval systems in modern naval warfare?

Ans. To perform tasks such as mine detection, anti-submarine warfare, and surveillance without human intervention.

Q.29. What is the primary application of drone swarms in military operations?

Ans. Drone swarms are used for various tasks, including surveillance, target tracking, and electronic warfare.

Q.30. How is unmanned aerial refueling (UAR) technology used in the military?

Ans. UAR allows for the mid-air refueling of drones, extending their operational range.

Multiple Choice Questions and Answers

Q.1. What is the primary purpose of using robotics and automation in modern warfare?

(*a*) Enhancing leisure activities
(*b*) Reducing human involvement in conflict
(*c*) Promoting international cooperation
(*d*) Expanding economic growth

Q.2. Which of the following is a common application of Unmanned Aerial Vehicles (UAVs) in modern warfare?

(*a*) Agricultural monitoring
(*b*) Marine biology research
(*c*) Surveillance and reconnaissance
(*d*) Wildlife conservation

Q.3. What do Autonomous Underwater Vehicles (AUVs) primarily operate in?

(*a*) Air
(*b*) Space
(*c*) Water
(*d*) Underground

Q.4. What technology is frequently used for obstacle detection and mapping in robotics?

(*a*) Radar
(*b*) LIDAR
(*c*) Sonar
(*d*) Infrared

Q.5. How do robotic exoskeletons assist military personnel?

(*a*) By providing camouflage
(*b*) By enhancing physical strength and endurance
(*c*) By transmitting encrypted messages
(*d*) By intercepting enemy communications

Q.6. What role does Artificial Intelligence (AI) play in military robotics?

(*a*) AI controls military robots autonomously
(*b*) AI designs weapons
(*c*) AI produces energy for robotic systems
(*d*) AI coordinates human soldiers

Q.7. In a military drone's control system, what does GPS stand for?

(*a*) Global Positioning System
(*b*) General Protection Service
(*c*) Geographic Programming Software
(*d*) Global Pathfinding System

Q.8. What is sensor fusion in the context of military robotics?

(*a*) Combining different types of robots
(*b*) Merging sensor data from multiple sources
(*c*) Confusing enemy sensors
(*d*) Designing sensors for military robots

Q.9. Which technology is used to disrupt or manipulate enemy communication systems in electronic warfare?

(*a*) GPS
(*b*) Encryption
(*c*) Electronic countermeasures
(*d*) LIDAR

Q.10. What is the primary function of Autonomous Ground Vehicles (AGVs) in modern warfare?

(*a*) Providing entertainment
(*b*) Transportation of supplies
(*c*) Artillery bombardment
(*d*) Humanoid communication

Q.11. What mathematical principles are used in encryption techniques for secure military communication?

(*a*) Algebraic geometry
(*b*) Cryptography
(*c*) Number theory
(*d*) Quantum mechanics

Q.12. Why are Unmanned Ground Vehicles (UGVs) often used for explosive ordnance disposal (EOD) tasks?

(*a*) To provide camouflage
(*b*) To maximize human interaction
(*c*) To reduce risks to human EOD teams
(*d*) To simulate enemy movements

Q.13. What is the primary purpose of using AI in military robotics?

(*a*) Designing robotic exoskeletons
(*b*) Enhancing leisure activities
(*c*) Enabling autonomous decision-making and adaptive behavior
(*d*) Assisting in diplomatic negotiations

Q.14. How do communication protocols for underwater vehicles differ from those for aerial drones?

(*a*) They use semaphore flags
(*b*) They primarily use satellite communication
(*c*) They utilize underwater acoustic communication
(*d*) They rely on telepathy

Q.15. What are electronic countermeasures used for in military operations?

(*a*) Promoting electronic warfare
(*b*) Enhancing communication systems
(*c*) Disrupting or manipulating enemy electronic and communication systems
(*d*) Preventing enemy attacks

Q.16. What is the main application of drone swarms in military operations?

(*a*) Providing entertainment for troops
(*b*) Transporting supplies to remote areas
(*c*) Overwhelming the enemy with coordinated actions
(*d*) Conducting espionage missions

Q.17. How are autonomous naval systems used in anti-submarine warfare?

(*a*) To deliver supplies to submarines
(*b*) To create artificial reefs
(*c*) To track and detect enemy submarines
(*d*) To perform maintenance on submarines

Q.18. What type of sensors are commonly used for target tracking in military applications?

(*a*) Video cameras
(*b*) Infrared sensors
(*c*) Radar and LIDAR sensors
(*d*) Sonar sensors

Q.19. What is the primary role of communication and control systems in modern warfare?
(*a*) Enhancing military camouflage
(*b*) Providing entertainment for troops
(*c*) Coordination, data exchange, and command execution
(*d*) Reducing the need for human soldiers

Q.20. What is the primary advantage of using AI in autonomous navigation for military robots?
(*a*) Increased cost of operation
(*b*) Improved camouflage
(*c*) Adaptation to complex and dynamic environments
(*d*) Reduced need for maintenance

Key to Multiple Choice Questions

1. (*b*)	**2.** (*c*)	**3.** (*c*)	**4.** (*b*)	**5.** (*b*)	**6.** (*a*)	**7.** (*a*)	**8.** (*b*)	**9.** (*c*)
10. (*b*)	**11.** (*b*)	**12.** (*c*)	**13.** (*c*)	**14.** (*c*)	**15.** (*c*)	**16.** (*c*)	**17.** (*c*)	**18.** (*c*)
19. (*c*)	**20.** (*c*)							

Project Works

1. **Autonomous Reconnaissance Drone**: Design a reconnaissance drone capable of autonomous flight, data collection, and real-time threat detection. Implement computer vision and machine learning for object recognition.
2. **Robotic Exoskeleton for Soldiers**: Develop a wearable exoskeleton to enhance the physical capabilities of soldiers, such as strength, endurance, and mobility.
3. **AI-Based Target Tracking System**: Create a target tracking system that uses AI and computer vision to identify and track enemy vehicles or troops on the battlefield.
4. **Automated Anti-Drone System**: Build a system that autonomously detects and neutralizes enemy drones using radar, LIDAR, and anti-drone technologies.
5. **Unmanned Ground Vehicle (UGV) for EOD**: Design a UGV for explosive ordnance disposal tasks, incorporating advanced sensors, manipulators, and remote control capabilities.
6. **Swarm Robotics for Surveillance**: Develop a swarm of small drones or ground robots that can work together for surveillance and reconnaissance, utilizing swarm algorithms for coordination.
7. **AI-Based Decision Support System**: Create an AI-powered system that assists military commanders in making tactical decisions based on real-time data and predictive analytics.
8. **Secure Communication Network for Military**: Design a highly secure communication network using encryption, blockchain, and quantum-resistant cryptography to protect sensitive military information.
9. **Automated Mine Detection and Clearance**: Build a system that uses robotics, AI, and sensor fusion to detect and clear landmines safely.
10. **Underwater Autonomous Vehicles for Naval Warfare**: Develop AUVs for anti-submarine warfare, equipped with advanced sonar and AI capabilities for tracking and engaging enemy submarines.
11. **AI-Enhanced Military Logistics**: Create an AI-driven logistics system to optimize supply chain management, distribution, and resource allocation in military operations.

12. **Electronic Warfare Simulation**: Develop a simulation environment to train military personnel in electronic warfare techniques, including jamming and signal interception.
13. **Path Planning and Obstacle Avoidance for UAVs**: Implement advanced path planning algorithms for UAVs to navigate complex terrains and avoid obstacles using LIDAR and computer vision.
14. **Robotic Mine Clearing in Urban Environments**: Design robots capable of navigating urban settings to locate and safely neutralize explosive devices, enhancing urban warfare safety.
15. **AI-Enabled Military Strategy Game**: Create a computer game that simulates modern warfare scenarios, allowing players to develop and execute AI-driven military strategies.
16. **Autonomous Supply Drones**: Develop autonomous supply drones that can deliver critical resources to troops in the field, reducing the need for manual resupply missions.
17. **Aerial Drone Defense System**: Create an anti-drone system using AI and advanced sensors to detect, track, and neutralize enemy drones in real time.
18. **Secure AI-Driven Communication for Soldiers**: Develop a wearable communication device that uses AI for speech recognition and secure encrypted communication for soldiers in the field.
19. **Swarm Robotics for Area Denial**: Design a swarm of autonomous robots that can deny access to specific areas and disrupt enemy forces through coordinated actions.
20. **Autonomous Electronic Countermeasures**: Create a system that can autonomously detect and counter enemy electronic warfare tactics, including jamming and signal manipulation.

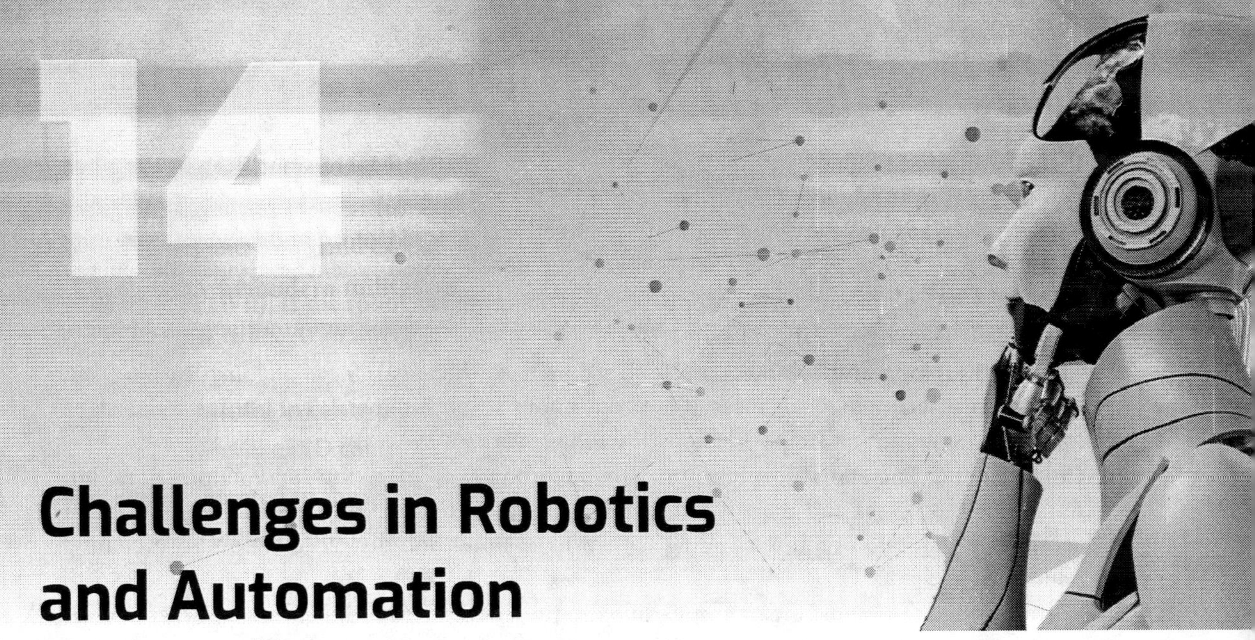

14 Challenges in Robotics and Automation

14.1 INTRODUCTION

Automation and robotics have made incredible strides in recent years, revolutionizing industries and changing many aspects of our lives. These innovations have the potential to revolutionize a variety of industries, including manufacturing, healthcare, logistics, and agriculture. To fully realize the potential of robotics and automation, academics, engineers, and politicians must overcome a number of important hurdles in addition to these game-changing advantages. The main difficulties that these fields encounter are highlighted in this introduction.

1. **Integration and Complexity of Technology:** It is frequently necessary to combine several technologies, including artificial intelligence, computer vision, mechanical engineering, and control systems, when creating complex robots and automation systems. It is quite difficult to coordinate various technologies and guarantee their flawless interplay.
2. **Safety and Reliability:** It is increasingly important to ensure that robots operate safely around people and in a variety of contexts as they become more powerful and autonomous. Building failsafe procedures to stop accidents from happening and designing systems that can recognize and react to unforeseen scenarios are difficult tasks.
3. **Ethical and Social Implications:** Automation may result in changes to employment patterns and the loss of jobs. It can be difficult to strike a balance between effectiveness and the protection of livelihoods. Additionally, concerns regarding privacy, responsibility, and decision-making are raised by the ethical use of robots, particularly in industries like healthcare and law enforcement.
4. **Adaptation to Dynamic Contexts:** A lot of real-world applications take place in dynamic and unpredictable contexts. It can be difficult to program robots to adapt to changes, variances, and unstructured situations.
5. **Human-Robot Interaction:** Effective collaboration between humans and robots involves a thorough grasp of human behavior and communication. For seamless integration, it's crucial to guarantee natural interactions and intuitive interfaces.
6. **Cost and Scalability:** Although automation and robots might result in long-term cost benefits, the costs of setup and integration can be pricey. It is difficult to make sure that these technologies are usable and scalable, particularly for small and medium-sized businesses.

7. **Regulation and Standardization:** Regulatory frameworks are frequently outpaced by the quick development of technology. It can take a long time and effort to create the proper rules and regulations to ensure safety, interoperability, and ethical use.
8. **Sustainability:** Robotics and automation's effects on the environment, especially the energy used and waste produced during production and operation, are a concern. The issue of creating ecologically friendly systems calls for creative solutions.
9. **Lifelong Learning and Adaptation:** Robots may need the capacity for ongoing learning and adaptation as technology advances. It is a continuous struggle to create algorithms and designs for robotic systems that support lifelong learning.
10. **Cultural and Societal Acceptability:** The incorporation of robots and automation into numerous businesses and daily life calls for a change in cultural attitudes as well as societal acceptance. Adoption requires dealing with worries about job displacement and overcoming reluctance to change.

Automation and robotics hold great promise but also present numerous difficulties. In order to meet these difficulties, interdisciplinary cooperation, ongoing research and development, and a proactive strategy for influencing the moral, social, and technological environment of the future are all necessary. We can only fully utilize robotics and automation for the advancement of civilization by making these efforts.

14.2 TECHNICAL CHALLENGES FACED IN ROBOTICS

Undoubtedly, a vast range of technical issues must be resolved in order to produce sophisticated and effective robotic systems. Some of the main technological difficulties are listed below:

1. **Sensing and Perception:** Using sensors like cameras, LiDAR, radar, and more, robots must accurately sense and comprehend their surroundings. Effective robot operation requires the development of strong perception algorithms that can cope with fluctuations in lighting, weather, and object appearances.
2. **Localization and Mapping:** Accurate mapping of the environment is necessary for navigation and interaction, as is keeping an accurate estimate of the robot's position within the map. To solve this problem, methods like simultaneous localization and mapping (SLAM) are used.
3. **Path Planning and Navigation:** Robots need to be able to design the best routes across complicated and dynamic surroundings, avoiding hazards, and taking into account variables like speed, safety, and energy efficiency.
4. **Control and Manipulation:** For operations like grasping, manipulation, and assembly, precise control of robotic limbs and end effectors is required. It is a huge difficulty to achieve dexterity and precise motor control while taking into consideration sensor uncertainties.
5. **Autonomy and Decision-Making:** For autonomous operation, it is essential to create algorithms that enable robots to make intelligent judgments in real-time based on data from sensors and their awareness of the surroundings.
6. **Integration of Machine Learning and Artificial Intelligence Techniques:** Using these techniques, robotics can perform tasks like object recognition, gesture comprehension, and natural language processing. It takes a lot of work to get these algorithms to work consistently across many situations.
7. **Human-Robot Interaction:** Creating robots with advanced vision and natural language processing skills is necessary in order for them to comprehend and react to human gestures, facial emotions, and commands.

8. Robots frequently run on batteries or sparse power sources, which results in their energy efficiency. Extending their operational time and minimizing their environmental impact require optimizing their energy consumption and developing effective motion and control systems.
9. **Multi-Robot Coordination:** Coordinating the actions of several robots to accomplish a shared objective presents difficulties in terms of communication, teamwork, and resolving conflicts. Coordination in changing contexts is difficult to ensure effectively.
10. **Software and Hardware Integration:** Creating robotic systems requires the integration of sophisticated hardware elements with software algorithms. It can be difficult to deal with hardware limits and ensure flawless software and hardware interaction.
11. **Variability Adaptation:** Robots frequently deal with variations in the real environment, such as shifting illumination, topography, and object appearances. For reliable operation, algorithms must be created that can adjust to these fluctuations.
12. **Lengthy-Term Autonomy and Learning:** Creating algorithms for lifelong learning is necessary to provide robots the ability to continuously learn and adapt to new activities and situations over lengthy periods of time.
13. **Safety and Redundancy:** For human-robot interaction, it's crucial to create safe robots that can manage unforeseen circumstances and correct mistakes. Redundancy and fail-safe techniques must be included.
14. **Incorporating New Materials and Mechanisms:** Technological developments in the fields of materials science and robotics mechanics can produce more effective and adaptable robotic systems. It is difficult to incorporate these new components and mechanisms into designs while keeping reliability.

Collaboration between professionals in diverse domains, such as robotics, artificial intelligence, computer vision, control systems, and more, is necessary to address these technical issues. By overcoming these obstacles, robotic systems will become more powerful, flexible, and dependable, which will revolutionize industries and enhance our daily lives.

14.3 PERCEPTION OF TECHNICAL CHALLENGES

As it entails the capacity of robots to comprehend and interpret their surroundings through sensory data, perception is a fundamental difficulty in robotics. For activities including navigation, object manipulation, human-robot interaction, and more, effective perception is essential. Here are some of the main technical difficulties in vision that robotics must overcome:

1. **Sensor Fusion:** To learn about their surroundings, robots frequently use a variety of sensors, including cameras, LiDAR, radar, and ultrasonic sensors. Advanced algorithms are needed to integrate data from various sensors and combine them to provide a coherent depiction of the world.
2. **Noise and Uncertainty:** Due to factors including sensor errors, ambient fluctuations, and occlusions, sensor data is inherently noisy and unpredictable. It is crucial to create algorithms that can manage and reduce this noise while delivering precise information.
3. **Real-Time Processing:** To enable prompt decision-making and action, many robotic activities necessitate real-time perception. It can be difficult to develop algorithms that can handle enormous amounts of sensory input quickly enough to facilitate real-time operation.

4. **Semantic Understanding:** To interpret raw sensor data and extract meaningful information, such as identifying objects, people, and their spatial relationships, advanced computer vision and machine learning techniques must be developed.
5. **Dynamic Environments:** Robots frequently work in settings with moving parts, shifting lighting, and lively situations. It is a difficult effort to adjust to these changes while preserving accurate perception.
6. **3D Perception:** Activities like navigation and manipulation require an understanding of the three-dimensional structure of the environment. A technical challenge is creating algorithms that can precisely assess depth, object forms, and distances.
7. **Object Tracking:** For tasks like following a moving target or interacting with objects, it is crucial to continuously track the motion and identification of objects throughout time. It is difficult to maintain tracking accuracy when there are obstructions and abrupt changes in velocity.
8. **Scene Understanding:** Robots must be able to comprehend the context and semantics of a scene in addition to object recognition. Recognizing connections among the scene's objects and characters between objects, scene layouts, and even predicting future actions or events.
9. **Perception in Unfavorable Situations:** Situations with poor lighting, inclement weather, or cluttered backgrounds make perception difficult. It's crucial to create algorithms that can manage these situations and deliver accurate data.
10. **Data Efficiency:** A lot of labeled data is needed to train perception models. It is difficult to create methods that can effectively train models with less input while maintaining high accuracy.
11. **Transfer Learning and Generalization:** Research in transfer learning and generalization techniques is necessary to teach robots to see and comprehend novel situations that may differ greatly from their training data.
12. **Privacy and Ethical Issues:** Robotic perception systems may entail the collection and processing of visual or sensor data in ways that cause privacy issues. creating systems that respect privacy and follow moral principles
13. **Human-Robot Interaction:** Accurate perception of human behaviors and intentions is necessary for robots to be able to recognize and react to human gestures, expressions, and commands.
14. **Cognitive Load:** The computing required by perception algorithms can be substantial. It can be difficult to strike a balance between the computing load, the necessity for real-time processing, and energy efficiency.

Continuous advancement in fields like computer vision, sensor technology, machine learning, and robotics are necessary to meet these problems. The development of more capable and intelligent robotic systems that can function well in a variety of contexts and circumstances will be aided by advancements in perception.

14.4 MOBILITY FACED IN ROBOTICS AND AUTOMATION

Robotics and automation systems' ability to move about and interact with their surroundings is referred to as their mobility. For functions including navigation, exploration, surveillance, and delivery, mobility is essential.

Some of the major technical obstacles to mobility in robotics and automation are listed below:

1. **Getting Around in Unstructured Environments:** Advanced path planning and obstacle avoidance algorithms are needed to navigate across areas with barriers, variable terrains, and dynamic factors like moving people or vehicles.

2. **Localization and Mapping:** For a robot to operate safely and effectively, it must be able to localize itself in its environment and create accurate maps of that area. Real-time dependable and accurate localization is difficult to achieve, particularly in conditions without GPS or indoors.
3. **Dynamic Obstacle Avoidance:** Handling dynamic impediments, such pedestrians or other vehicles, that can arrive abruptly is difficult. To prevent crashes, robots must foresee these impediments and respond to them immediately.
4. **Great-Distance Mobility:** Some applications, like self-driving cars or drone deliveries, call for robots to move over great distances. Long-term safety, energy efficiency, and dependable navigation are difficult to guarantee.
5. **Multi-Terrain Adaptation:** Robots may have to move through a variety of terrains, such as uneven ground, stairs, bodies of water, or urban settings. It is difficult to create the machinery and control schemes needed for robots to adjust to various terrains.
6. **Moving through Tight and Constricted Places:** Some activities, such as disaster recovery scenarios or exploring subsurface structures, need robots to move through small, constricted places. It is difficult to create robots that can move and navigate in these areas.
7. **Difficulties with Airborne Mobility:** Drones and airborne vehicles encounter difficulties with stability, wind resistance, 3D space collision avoidance, and energy efficiency. It is essential to guarantee controlled and safe flight under varied circumstances.
8. **Manipulation While Mobile:** Some robots must carry out operations while moving, such as picking up objects as they walk. The simultaneous coordination of manipulation and movement calls for careful management and planning.
9. **Energy Efficiency:** To increase operational duration and lessen the need for regular refueling or charging, mobility systems must be energy-efficient. It is crucial to create propulsion systems and control systems that use less energy.
10. **Multi-Robot Coordinating:** In environments with numerous robots operating simultaneously, coordinating to prevent collisions and optimize tasks can be difficult, especially in surroundings that are complicated or busy.
11. **Localization in GPS-Denied Environments:** Using alternative localization methods like optical odometry or SLAM is necessary when operating in locations where GPS signals are weak or unavailable, such as indoor spaces.
12. Environments might alter over time as a result of things including weather, construction, or natural occurrences. Robots must possess the flexibility to adjust their maps and plans in response to these changes.
13. **Robustness to Sensor Variability:** The accuracy and dependability of different sensors can vary. Robust navigation depends on designing mobility systems that can deal with sensor errors and uncertainty.
14. **Human Interaction and Safety:** It's important to provide safe human interaction in communal areas. It is difficult to create algorithms and behaviors that enable robots to comprehend and react to human intentions and gestures.

It takes interdisciplinary cooperation between specialists in robotics, control systems, perception, mechanics, and artificial intelligence to address these technical issues in mobility. By overcoming these obstacles, robotic and automation systems will be able to operate and navigate in a variety of dynamic contexts.

14.5 MANIPULATION FACED IN ROBOTICS

A difficult and crucial component of robotics is manipulation, or the capacity of robots to interact with and manipulate items in their environment. Robots can accomplish activities like assembly, object manipulation, and environment interaction thanks to effective manipulation.

The following are some of the main technological difficulties in manipulation:

1. **Dexterity and Fine Motor Control:** Manipulating objects of various shapes, sizes, and weights requires fine control of robot arms and end effectors (such as grippers or hands). It is crucial to provide mechanisms and control systems that support dexterity and fine motor control.
2. **Grasping and Object Recognition:** Robots must be capable of identifying items in varied contexts and determining the best way to grab them. It is difficult to create algorithms that let robots find good grip spots and change to fit item shapes.
3. **Adaptive Grasping:** In the actual world, objects might have a variety of shapes and characteristics. Robots must modify their grasping techniques to accommodate various items, particularly those with erratic forms or ambiguous characteristics.
4. **Force and Tactile Sensing:** To ensure safe and controlled interactions with objects and the environment, effective manipulation frequently entails sensing forces and tactile information. It is challenging to create sensors and algorithms for tactile and force feedback.
5. **Contact Modeling and Control:** Robots must comprehend and simulate the physical interactions at contact locations while engaging with objects. It is difficult to create precise contact models and control methods to manage interactions like pushing, sliding, and rolling.
6. Accurate perception is essential for proper handling of objects and their characteristics (size, shape, material). A technical challenge is integrating sensor data with computer vision to guide manipulation tactics.
7. **Bimanual and Multifinger Manipulation:** It can be challenging to coordinate the motions of many robot arms or fingers to carry out simultaneous actions, such as opening a jar while holding the lid.
8. **Object Pose Estimation:** For manipulation, it is essential to know the exact location and orientation of objects in the environment. It is crucial to create algorithms that precisely estimate object poses from sensor data.
9. **Handling Uncertainty:** Variable friction or item characteristics are just two examples of uncertainty that might exist in objects and situations. It can be difficult to create manipulation tactics that can manage uncertainty and change with the environment.
10. **In-Hand Manipulation:** Repositioning and adjusting an object when it is already in the gripper or hand of the robot requires intricate movements. creating algorithms for current tasks
11. **Tool utilize and Tool Changes:** Robust planning, coordination, and control are needed to enable robots to utilize tools for specialized tasks and transition between different tools or end effectors.
12. **Interaction with Soft and Malleable Items:** Compared to manipulating rigid items, soft, malleable, or fragile objects call for a new set of tactics. It is difficult to develop methods for handling these kinds of items.
13. **Safety and Collision Avoidance:** To ensure safe manipulation, collisions with objects, people, and the environment must be avoided. It is essential to develop collision detection and avoidance methods.

14. **Learning and Adaptation:** Creating learning algorithms that can generalize from experience is necessary to provide robots the ability to learn new manipulation techniques or adjust to changes in their environment.

It takes knowledge of disciplines including control theory, mechanics, computer vision, machine learning, and human-robot interaction to address these technological issues in manipulation. More adaptable and capable robotic systems that can carry out a variety of activities in a variety of conditions will be made possible by advancements in manipulation capabilities.

14.6 ONGOING RESEARCH EFFORTS IN ROBOTICS AND AUTOMATION

Robotics and automation research is moving forward, addressing problems, and creating new opportunities for applications across industries.

The following significant areas of current research are listed:

1. **Robotic Perception and AI:** Using cutting-edge computer vision algorithms, deep learning, and sensor fusion, researchers are attempting to improve robots' perception abilities. This entails enhancing scene comprehension, item recognition, and instantaneous decision-making based on sensory data.
2. Collaboration between humans and robots is the subject of extensive research. This collaboration should be safe and natural. To efficiently collaborate with robots and people, this requires creating algorithms for natural language processing, gesture recognition, and shared control.
3. **Autonomous Vehicles:** Research into autonomous vehicles includes anything from drones and autonomous ships to self-driving cars. To assure safety and regulatory compliance, ongoing efforts focus on improving perception systems, path planning algorithms, and human-aware navigation techniques.
4. **Soft Robotics:** Soft robots can adapt to complex shapes and interact with humans and delicate situations. The development of flexible actuators, compliant mechanisms, and control schemes for these adaptable robots is the main goal of research in soft robotics.
5. **Swarm Robotics:** This field investigates how numerous, relatively simple robots can work together to complete challenging tasks. Swarm algorithms, communication techniques, and emergent behaviors are being researched for uses in environmental monitoring and catastrophe response.
6. **Robotic Healthcare:** The healthcare industry is investing in research into telemedicine, assistive technology for the aged, and surgical robots. Robotic prosthetics, wearable robots, and robot-assisted procedures are all being improved by researchers.
7. **Agricultural Robotics:** Robots are being created for agricultural jobs like planting, harvesting, and crop monitoring. Automation is being studied to improve agricultural procedures, save labor costs, and boost crop yields.
8. **Industry 4.0 and Smart Manufacturing:** In this field of study, robotics and automation are integrated into the manufacturing process. Research is currently being done on technologies including flexible production systems, AI-driven quality control, and collaborative robots (cobots).
9. **Environmental Exploration:** Robotics is utilized to explore environments that are dangerous or difficult for people to navigate. The creation of underwater drones for marine exploration drones for environmental monitoring, and robots for disaster response are all now being researched.

10. **Ethical and Societal Implications:** As robots are used more frequently in daily life, researchers are looking into the moral, legal, and social ramifications of this. This covers concerns about losing jobs, privacy, bias in AI, and creating regulatory frameworks.
11. **Energy Efficiency and Sustainability:** Researchers are aiming to make automation systems and robots more sustainable and energy-efficient. This entails creating algorithms that are energy-efficient, utilizing renewable energy sources, and minimizing the negative effects of manufacturing processes on the environment.
12. Robotic learning research is focused on giving robots the ability to learn from their mistakes and adapt to new tasks and settings. Techniques for transfer learning, imitation learning, and reinforcement learning are being investigated.
13. **Bioinspired Robotics:** Taking cues from the natural world, scientists are creating robots with characteristics found in insects and animals. This comprises biohybrids, robots that can mimic complicated behaviors, and robots with nimble movement.
14. **Long-Term Autonomy:** Creating robots that can function independently for long stretches of time while adjusting to shifting conditions is a difficult task. Mechanisms for self-healing, self-renewal, and long-term decision-making are the subject of research.

These ongoing projects demonstrate the transdisciplinary and dynamic nature of robotics and automation. To spur innovation, fix problems, and bring about dramatic changes in numerous industries and facets of daily life, collaboration between academics, engineers, and policymakers is crucial.

14.7 ONGOING DEVELOPMENT IN ROBOTICS AND AUTOMATION

The subject of robotics and automation is still evolving, and these changes are having a significant impact on several businesses and society at large.

These major locations of continuing development are listed below:

1. **Artificial Intelligence Integration:** Deep learning, reinforcement learning, and natural language processing are three AI techniques that robotics and automation are increasingly using. Robotic decision-making, perception, and adaptability are improved by this combination.
2. **Cobots and Collaborative Automation:** Cobots are robots that can collaborate with people in a shared workspace. Cobots are currently being developed to be safer, easier to program, and capable of a larger variety of cooperative duties.
3. **Autonomous Vehicles:** Research and development into autonomous vehicles, such as self-driving automobiles, drones, and unmanned aerial vehicles (UAVs), is ongoing. The development of these technologies is being fueled by advancements in navigation, safety, and perception technology.
4. **Medical Robotics:** As they advance in sophistication, medical robots can help with minimally invasive surgeries, recovery, and diagnostic procedures. Enhancing teleoperational capabilities, haptic feedback, and precision are the main areas of ongoing development.
5. Automated warehouses and distribution centers are in more demand as a result of the growth of e-commerce. The utilization of mobile robots, automated guided vehicles (AGVs), and AI-driven inventory management are all current advancements.
6. **Advanced Sensing and Perception:** The development of sensor technology, such as enhanced RGB-D cameras, tactile sensors, and LiDAR, has benefited robotics. These advancements improve robots' capacity to observe and engage with their surroundings.

7. **Soft Robotics and Biohybrids:** Soft robot development, inspired by living things, is advancing. These robots are well suited for a variety of applications, including exploration and healthcare, because they can manoeuvre through complicated situations and interact with delicate things.
8. **Energy Efficiency and Sustainability:** Scientists and engineers are aiming to improve the energy efficiency and sustainability of robotics and automation systems. This entails creating energy-efficient actuators, lightweight structures, and robotic component recycling techniques.
9. **Telepresence and Teleoperation:** Robots may now be óperated remotely more easily, allowing professionals to control them in dangerous or remote locations. Real-time communication, haptic feedback, and immersive interfaces are the main areas of development.
10. **Agricultural Automation:** Crop planting, monitoring, and harvesting are all made easier by automation technologies. Continual improvement comprises integrating drones, AI-powered analytics, and precision agriculture techniques.
11. **Ethics and Regulations:** As robotic technology develops, attempts are being made to create ethical standards and legal frameworks to guarantee the secure, accountable, and responsible use of robots in a variety of applications.
12. **Micro- and Nano-Robotics:** Studies are being done on tiny robots that can function at small scales, as inside the human body or in cramped spaces. These tiny and tiny robots have the potential to be used in environmental monitoring and medicinal procedures.
13. **Brain-Computer Interfaces:** New developments in this field are making it possible for robots and the human brain to communicate directly. Prosthetics, assistive robots, and human augmentation are three fields where this technology has applicability.
14. **Education and Skill Development:** As robotics becomes more common, initiatives are being made to create educational materials and programs to instruct the upcoming generation of roboticists and automation specialists.
15. **Block Chain and Automation:** For applications like supply chain management, safe data exchange among robots, and assuring transparency in automated operations, the integration of block chain technology with automation is being researched.

These ongoing breakthroughs highlight the dynamic nature of robotics and automation, with interdisciplinary collaborations and technological advances continuously expanding the realm of what is feasible. We can anticipate even more profound changes in how robotics and automation systems affect our lives as research and innovation proceed.

14.8 FUTURE OF ROBOTICS

Robotics' potential to transform industries, improve our quality of life, and solve some of society's most pressing problems is enormous. The following are some significant developments and prospects for the future of robotics:

1. **AI-Powered Autonomy:** As artificial intelligence develops, especially machine learning and deep learning, robots will become more autonomous and adaptable. Robots will undertake a larger range of jobs without human assistance, learn from their interactions, and make complicated judgments in real-time.
2. **Human-Robot Collaboration:** Cobots, or collaborative robots, will be more common in offices and will help people complete repetitive, precise, or physically demanding activities. These partnerships will be more fluid and effective because to improved safety features and user-friendly interfaces.

3. **Medical and Healthcare Robotics:** With applications in surgery, diagnostics, rehabilitation, and telemedicine, robotics will continue to revolutionize healthcare. Robotic prosthetics, remote operations, and minimally invasive procedures will advance in sophistication, enhancing patient outcomes and accessibility to healthcare.
4. **Autonomous Vehicles and Drones:** Transportation and logistics will change as a result of self-driving automobiles, drones, and aerial vehicles. These innovations will result in safer roads, less clogged traffic, effective delivery, and even brand-new modes of urban transportation like flying taxis.
5. **Personal Assistants and Service Robots:** Advanced service robots, which help with housework, aged care, and customer service, will become a crucial part of our daily lives. These robots will be able to comprehend spoken language, identify emotions, and cater to human needs.
6. Robots will be essential in the exploration of severe settings, such as deep-sea missions, space missions, and disaster-affected areas. They will contribute important information, carry out dangerous duties, and help with search and rescue operations.
7. **Sustainable and Eco-Friendly Robotics:** By maximizing resource consumption, lowering waste, and supporting environmentally friendly practices in sectors including waste management, manufacturing, and agriculture, robotics will promote sustainability initiatives.
8. **Neurotechnology and Brain-Computer Interfaces:** By combining robotics with brain-computer interfaces, direct brain-machine connection will be made possible. This will result in advancements in prosthetics, assistive technology, and improving human skills.
9. **Robotics in Education and Research:** As it continues to influence the next generation of scientists, engineers, and inventors, robotics will play a crucial role in both education and research. Programming, engineering principles, and problem-solving techniques will be taught via interactive educational robots.
10. **Ethics and Regulation:** Ethical issues and legal frameworks will change as robotics is more fully incorporated into society. The use and regulation of robots will be influenced by discussions about robot rights, employment displacement, privacy, and accountability.
11. **Bioinspired and Soft Robotics:** Research in biomimetic and soft robotics will produce robots with properties similar to those of living things, such as flexibility and capability. These robots will execute jobs that were previously difficult for conventional robots, navigate hard situations, and communicate politely with people.
12. Robotics will gain from global networking and data sharing because it will enable robots to interact across distances and learn from one another's experiences. New powers and skills will grow faster as a result.
13. **AI-Enhanced Creativity:** AI-driven robots will advance the creative industries of music, design, and art. They'll come up with fresh concepts, produce works of art, write music, and make original contributions to the creative process.

Overall, innovation, teamwork, and the capacity to tackle difficult problems will define the future of robotics. Robots will become essential members of our society as interdisciplinary research and technological advancements continue, revolutionizing business, healthcare, education, and how we connect with the outside world.

14.9 SPECULATION ON THE FUTURE OF ROBOTICS AND AUTOMATION

There are various hypothetical routes in which robotics and automation could develop in the upcoming years, despite the fact that the future is unclear and subject to a wide variety of possibilities:

1. **AI Super intelligences:** With the rapid development of AI, it is conceivable that extremely sophisticated AI systems—referred to as super intelligences—might emerge. These beings may be more intelligent than humans and have the capacity to solve challenging issues, find new information, and accelerate technological advancements.
2. **Totally Autonomous Economies:** In a hypothetical future, totally autonomous production systems would be integrated, with robotics and automation controlling every aspect of production, distribution, and service provision. New economic paradigms and a redefining of the notion of ownership may result from this.
3. **Universal Robotics:** In a hypothetical future, a variety of jobs across industries could be accomplished by universal robots. There would be less need for specialized robots for each individual application if these robots could be reprogrammed or given modules to adapt to different jobs.
4. **Human Augmentation:** Combining robots and automation with human biology might result in a hypothetical future in which human skills are augmented. Robotic prosthetics, brain-computer interfaces, and exoskeletons could enable people to carry out tasks beyond
5. **AI Ethics and Governance:** As AI and robotics advance in sophistication, it is critical to address ethical issues and create governance frameworks. To assure the ethical development and use of AI and robotics, there may be specialized AI ethics boards and worldwide rules.
6. **Robotic Colonization and Exploration:** In the realm of science fiction, robots might be crucial to efforts to colonize space and explore the universe. In space conditions, autonomous robots could be used to construct dwellings, gather resources, and carry out research.
7. **Conscious AI and Machine Emotions:** In the future, it's possible that AI systems may be able to feel emotions. Despite being very theoretical, this raises moral concerns regarding how these AI creatures should be treated and what rights they should have.
8. **Ethical and Emotional Companion Robots:** It is possible that robots will one day be able to recognize and react to human emotions. These robots could offer companionship, aid, and support to people, especially in circumstances where there is little opportunity for human interaction.
9. **Quantum Robotics:** Quantum mechanics and computation could fundamentally alter robotics and automation. Speculatively, complicated calculations needed for robot control, optimization, and machine learning could be greatly accelerated by quantum computers.
10. **Global Robotic Networks:** In a hypothetical future, a worldwide network of connected robots and automation systems may work together smoothly to carry out activities like resource management, infrastructure upkeep, and disaster response on a global scale.

It's critical to approach these hypothetical situations with a healthy mix of excitement and skepticism. Even though some of these possibilities might result in beneficial improvements, they also pose risks, pose moral quandaries, and have societal repercussions that would need to be carefully addressed and managed. Robotics and automation will change as a result of a confluence of technology developments, moral considerations, and societal decisions.

14.10 ENVISIONING POSSIBILITIES: HUMAN-LIKE ANDROIDS AND SWARM ROBOTICS

Future robotics and automation scenarios are made exciting by imagining possibilities like human-like androids and swarm robotics. These ideas demonstrate how robotics has the capacity to revolutionize many facets of our life while also posing challenging ethical questions:

1. **Human-Like Androids**:

   Companion and Carer Robots: Humanoid androids could be used as companions for the elderly or people who don't engage in much social interaction. They might offer emotional support, help with daily chores, and even have deep conversations.

   Education and Entertainment: Androids can be used in classrooms to engage students in conversational learning about history, languages, and other subjects. Additionally, they might act as interactive storytellers or performers.

   Customer Service and Interaction: Companies might employ androids to provide customer service, increasing interactions and offering a more individualized experience.

   Avatar and Telepresence: Using androids as avatars, people might experience many locales virtually, communicate with others, and take part in events as if they were physically present.

2. **Swarm Robotics**:

   Precision Agriculture: Swarm robots could be used for precision agriculture, effectively managing pests, fertilising crops, and monitoring them jointly.

   Search and Rescue: Swarms of robots might quickly search vast areas in disaster scenarios to find survivors, evaluate damage, and communicate information to responders.

   Environmental Monitoring: Swarms of drones or aquatic robots might follow changes in wildlife populations, collect information on pollution levels, and monitor ecosystems.

   Infrastructure and Construction: Swarm robots could work together to build bridges, buildings, and other structures, streamlining construction procedures and accelerating construction time.

   Collective Decision-Making: Swarm robots may use emergent behaviors to reach decisions as a group, such as choosing the most efficient route for navigation or resolving challenging optimization issues.

While these possibilities are intriguing, they also bring up significant moral and pragmatic questions:

1. **Ethical Considerations for Human-Like Androids**:

   Privacy and Consent: Androids with advanced sensors present privacy and consent issues, particularly in public settings.

   Emotional Bonding: Establishing emotional bonds with androids may have psychological and social repercussions, potentially obfuscating the boundaries between relationships between humans and machines.

   Identity and Human Rights: If androids display sophisticated AI and emotions, debates on their legal standing, rights, and obligations may start.

2. **Challenges with Swarm Robotics**:

   Coordination: Complex algorithms and communication systems are needed to coordinate a large number of robots so that they can operate efficiently without colliding or interfering.

Scalability: A difficult task is ensuring that swarm behaviours and efficiency scale up as the number of robots rises.

Swarm robots must be trained to avoid harming or overstressing any creatures, ecosystems, or people they come into contact with.

It will be crucial to carefully analyze the implications as these possibilities materialize. In order to create a future where human-like androids and swarm robotics benefit our society and well-being, innovation must be carefully balanced with ethical and social responsibilities.

14.11 UNIVERSAL ROBOTIC ASSISTANTS

In order to assist and work with humans across a variety of fields, universal robotic assistants must be thought of as versatile, adaptive robots. These assistants would be created with the ability to interact with their surroundings intelligently, gain knowledge from their experiences, and offer helpful assistance in a variety of circumstances. The following are some considerations and potential uses for ubiquitous robotic assistants:

Key Possibilities

1. **Multi-Task Capabilities:** With their array of tools, sensors, and end effectors, universal robotic assistants would be able to carry out a variety of jobs, from cooking and cleaning to maintaining and providing care.
2. **Flexibility:** By interacting with people and their surroundings, these assistants would be able to learn new abilities and behaviors and adapt to various contexts and tasks.
3. **Natural Language Interaction:** Universal robotic assistants will be able to comprehend and react to commands given in natural language, facilitating smooth communication and teamwork.
4. **Personalization:** These assistants might be made to fit each person's interests and requirements, delivering personalized services like appointment reminders, shopping assistance, or companionship.
5. **Cooperative Work with People:** These robots would cooperate with people to complete jobs that call for both human intuition and robotic precision.
6. **Remote Control:** Because universal robotic assistants could be controlled from a distance, people could use them even if they weren't physically there.

Considerations and Challenges

1. **Ethical and Privacy Concerns:** Robust robots that can help with a variety of tasks pose concerns regarding privacy, consent, and the proper use of data obtained through interactions.
2. **Job Displacement:** Even though these robots could help people, there is concern about how they can affect the employment situation in some sectors. It's crucial to strike a balance between human employment and automation.
3. **Learning and Adaptation:** It is a difficult undertaking to create algorithms that allow these robots to quickly pick up new skills and adapt to different settings and tasks.
4. **Safety and Reliability:** To avoid mishaps and undesirable behaviors, it is essential to make sure that these aides function securely and dependably in a variety of situations.

5. **Human-Robot Interaction:** Successful collaboration requires the creation of interfaces and behaviors that are simple for people to use and comprehend.
6. **Technical Difficulty:** It is extremely difficult to build robots with the necessary sensors, actuators, and AI capabilities to serve as universal assistants.
7. **Cultural and Social Differences:** These robots' behavior and skills would need to take into account both individual preferences and cultural and social standards.
8. **Accessibility and Affordability:** For these robots to be widely used, it will be crucial to make them accessible and inexpensive to a variety of people.

Universal robotic assistants have the power to completely change the way we use technology and go about our daily lives. Although the idea is intriguing, it will take a thorough approach to development, regulation, and ethical concerns to make sure that these robots improve society and our quality of life.

Numerical Problems with Solutions

Problem 1. A robotic arm needs to pick up an object and place it at a target location. If the arm's end effector can move with a precision of 0.1 mm, how close can it place the object to the target location?

Solution: The arm can place the object within 0.1 mm of the target location.

Problem 2. A swarm of 25 robots is tasked with covering an area of 500 square meters for environmental monitoring. If each robot covers an equal area, how much area does each robot cover?

Solution: Each robot covers an area of 500/25 = 20 square meters.

Problem 3. An autonomous vehicle travels at a speed of 60 km/h. If it needs to stop within 5 meters to avoid a collision, what is its maximum safe deceleration?

Solution: Deceleration = $(0 - v^2)/(2 * x)$, where v = initial velocity, x = stopping distance. Deceleration = $(0 - (60 * 1000/3600)^2)/(2 * 5) = -3.7 \text{ m/s}^2$.

Problem 4. A robotic arm exerts a force of 200 N to lift an object weighing 50 kg. What is the acceleration of the object?

Solution: F = m * a, where F = force, m = mass, a = acceleration. Acceleration = F/m = 200/50 = 4 m/s^2.

Problem 5. An automated factory produces 500 units of a product in 8 hours. What is the production rate in units per hour?

Solution: Production rate = Total units produced/Time taken = 500/8 = 62.5 units/hour.

Problem 6. A robot arm has a reach of 1.5 meters. If it needs to pick up an object located 1 meter away, what is the minimum angle the arm joint needs to rotate through?

Solution: The angle is given by the inverse cosine of (1/1.5) = 48.19 degrees.

Problem 7. An autonomous drone uses 50 watts of power to fly for 30 minutes. What is the total energy consumption in watt-hours?

Solution: Energy consumption = Power × Time = 50 × (30 * 60) = 90,000 watt-hours.

Problem 8. A robotic vehicle navigates through a maze in a zigzag pattern. If its distance covered is 150 meters and its displacement is 80 meters, what is the total path length?

Solution: The total path length is 150 meters (distance covered), as displacement measures only the straight-line distance.

Problem 9. An industrial robot has a speed of 2 meters per second and takes 5 seconds to complete a task. What is the total distance traveled during the task?
Solution: Total distance traveled = Speed × Time = 2 × 5 = 10 meters.

Problem 10. A robot arm with a load of 20 kg needs to be accelerated to a speed of 2 m/s². What force is required to achieve this acceleration?
Solution: F = m * a = 20 * 2 = 40 N.

Problem 11. An autonomous car travels 150 km with an energy consumption of 25 kWh. What is its energy efficiency in kWh/km?
Solution: Energy efficiency = Energy consumed/Distance traveled = 25/150 = 0.1667 kWh/km.

Problem 12. A robotic arm with a payload of 10 kg exerts a torque of 30 Nm. What is its angular acceleration?
Solution: Torque = Moment of inertia × Angular acceleration, so Angular acceleration = Torque/ Moment of inertia. (Moment of inertia for a point mass) = m * r^2 = 10 * $(0.5)^2$ = 2.5 kg*m^2. Angular acceleration = 30/2.5 = 12 rad/s^2.

Problem 13. An autonomous drone needs to ascend 100 meters in 20 seconds. What is its average vertical velocity?
Solution: Average vertical velocity = Vertical displacement/Time = 100/20 = 5 m/s.

Problem 14. A robot arm has a total of 5 joints. If each joint has a range of motion from –45 degrees to +45 degrees, what is the total degrees of freedom for the arm?
Solution: Total degrees of freedom = Number of joints × Degrees of freedom per joint = 5 × 90 = 450 degrees.

Problem 15. A robotic arm with a mass of 15 kg accelerates upwards at 2 m/s². What force does gravity exert on the arm?
Solution: Weight = mass × acceleration due to gravity = 15 × 9.81 = 147.15 N.

Problem 16. An automated factory has a throughput of 800 units per day. If it operates 24 hours a day, what is the production rate in units per hour?
Solution: Production rate = Throughput/Operating hours = 800/24 = 33.33 units/hour.

Problem 17. A robot arm needs to reach a target location 2 meters away. If its maximum speed is 1 m/s, what is the minimum time it takes to reach the target?
Solution: Time = Distance/Speed = 2/1 = 2 seconds.

Problem 18. An autonomous vehicle travels a distance of 250 km with an average speed of 100 km/h. How long does the journey take in hours?
Solution: Time = Distance/Speed = 250/100 = 2.5 hours.

Problem 19. A robotic arm exerts a torque of 40 Nm. If its moment of inertia is 5 kg*m², what is its angular acceleration?
Solution: Angular acceleration = Torque/Moment of inertia = 40/5 = 8 rad/s^2.

Problem 20. An automated assembly line produces 600 units in 4 hours. What is the production rate in units per minute?
Solution: Production rate = Total units produced/Time taken = 600/(4 * 60) = 2.5 units/minute.

Short Questions and Answers

Q.1. What is a major challenge in robotics?

Ans. Achieving dexterity and fine motor control for complex tasks.

Q.2. What is one challenge in automation?

Ans. Ensuring seamless integration of automated systems with existing processes.

Q.3. What is the challenge in robotic perception?

Ans. Handling noisy sensor data and uncertain environmental conditions.

Q.4. What's the difficulty in navigation for robots?

Ans. Navigating through dynamic and cluttered environments.

Q.5. What's a challenge in multi-robot coordination?

Ans. Coordinating actions without conflicts in a shared workspace.

Q.6. What's an ongoing challenge in AI and robotics?

Ans. Developing algorithms for lifelong learning and adaptation.

Q.7. What's a key obstacle in robotic manipulation?

Ans. Designing algorithms for adaptive and versatile grasping.

Q.8. What's a challenge in swarm robotics?

Ans. Coordinating large numbers of robots effectively.

Q.9. What's the challenge in energy efficiency for robots?

Ans. Optimizing robot motion and control for longer operational times.

Q.10. What's a challenge in autonomous vehicles?

Ans. Ensuring safe navigation in complex and unpredictable environments.

Q.11. What's a challenge in industrial automation?

Ans. Retrofitting existing factories for automation without disrupting operations.

Q.12. What's a challenge in soft robotics?

Ans. Designing actuators that can mimic natural movements.

Q.13. What's a difficulty in robot-human interaction?

Ans. Creating robots that can understand and respond to human emotions.

Q.14. What's a challenge in robotic ethics?

Ans. Determining responsibility in cases of robot errors or accidents.

Q.15. What's an obstacle in integrating AI with robotics?

Ans. Ensuring AI models generalize well across diverse scenarios.

Q.16. What's a challenge in underwater robotics?

Ans. Developing reliable communication in underwater environments.

Q.17. What's a difficulty in medical robotics?

Ans. Ensuring precise and safe control of robotic surgical instruments.

Q.18. What's a challenge in agricultural automation?

Ans. Designing robots that can adapt to various crop types and terrains.

Q.19. What's a challenge in space robotics?

Ans. Creating robots that can withstand extreme conditions and perform tasks in space.

Q.20. What's an ongoing challenge in robotics research?

Ans. Developing robust algorithms for real-time decision-making.

Q.21. What's a challenge in balancing automation and jobs?

Ans. Ensuring automation enhances job opportunities rather than replacing them.

Q.22. What's a challenge in developing humanoid robots?

Ans. Designing robots with balance and stability for bipedal locomotion.

Q.23. What's a difficulty in collaborative robots?

Ans. Ensuring safety features that allow humans and robots to work closely.

Q.24. What's a challenge in haptic feedback for robots?

Ans. Creating realistic touch sensations in human-robot interactions.

Q.25. What's a challenge in regulatory frameworks for robotics?

Ans. Defining legal responsibilities when robots make decisions autonomously.

Q.26. What's a challenge in aerial robotics?

Ans. Developing robust control algorithms for drones in windy conditions.

Q.27. What's a difficulty in automating creative tasks?

Ans. Replicating human creativity in fields like art and music.

Q.28. What's a challenge in ethics for autonomous vehicles?

Ans. Deciding how self-driving cars should make ethical decisions in potential accidents.

Q.29. What's a challenge in maintaining robot security?

Ans. Preventing hacking and unauthorized access to robots' control systems.

Q.30. What's a challenge in deploying robots in remote areas?

Ans. Ensuring reliable communication and power supply in isolated locations.

Q.31. What's a difficulty in underwater exploration with robots?

Ans. Dealing with high pressure and underwater currents.

Q.32. What's a challenge in integrating robots with IoT devices?

Ans. Ensuring seamless communication and interoperability.

Q.33. What's a challenge in adapting robots for hazardous environments?

Ans. Designing robots to withstand extreme temperatures, radiation, or toxins.

Q.34. What's a challenge in developing robotic exoskeletons?

Ans. Creating lightweight yet sturdy designs for wearable assistance devices.

Q.35. What's a challenge in biohybrid robots?

Ans. Ensuring compatibility and integration between biological and robotic components.

Q.36. What's a difficulty in achieving robot autonomy?

Ans. Designing algorithms that can handle unexpected and ambiguous situations.

Q.37. What's a challenge in underwater communication for robots?

Ans. Overcoming the limitations of wireless communication in underwater environments.

Q.38. What's a challenge in social acceptance of robots?

Ans. Addressing concerns about job displacement, privacy, and machine ethics.

Q.39. What's a challenge in automating intricate assembly tasks?

Ans. Developing robots with fine motor control for delicate operations.

Q.40. What's a difficulty in handling different terrains for robots?

Ans. Designing mechanisms that allow robots to navigate rough, uneven surfaces.

Q.41. What's a challenge in advancing prosthetic robotics?

Ans. Creating prosthetics that provide natural movements and sensations.

Q.42. What's a challenge in implementing robots in education?

Ans. Designing educational robots that engage and motivate learners effectively.

Q.43. What's a challenge in maintaining robot reliability?

Ans. Ensuring that robots can function reliably for extended periods without frequent breakdowns.

Q.44. What's a challenge in robotics for disaster response?

Ans. Developing robots that can navigate unstable and hazardous environments.

Q.45. What's a difficulty in integrating robots with social infrastructure?

Ans. Overcoming regulatory hurdles and adapting to existing urban environments.

Q.46. What's a challenge in preventing bias in AI-powered robots?

Ans. Developing AI models that are fair and unbiased in their decision-making.

Q.47. What's a challenge in robotics for entertainment?

Ans. Creating interactive and engaging robotic experiences that captivate audiences.

Q.48. What's a challenge in automating complex decision-making?

Ans. Developing AI algorithms that can make morally and ethically sound decisions.

Q.49. What's a challenge in maintaining robot accuracy in repetitive tasks?

Ans. Preventing wear and tear on robot components that can lead to decreased precision.

Q.50. What's a challenge in user-friendly robot programming?

Ans. Designing intuitive interfaces that allow non-experts to easily program and control robots.

Multiple Choice Questions and Answers

Q.1. What is a key challenge in achieving precise control of robot arms and end effectors?

(*a*) Power consumption (*b*) Path planning

(*c*) Dexterity and fine motor control (*d*) Communication latency

Q.2. Which challenge involves handling uncertain sensor data and noisy environmental conditions?

(*a*) Navigation
(*b*) Multi-robot coordination
(*c*) Perception
(*d*) Localization

Q.3. What challenge arises when coordinating the movements of multiple robots to avoid conflicts?

(*a*) Robot-human interaction
(*b*) Swarming
(*c*) Bipedal locomotion
(*d*) Remote operation

Q.4. What's a key challenge in developing algorithms for robots to adapt to new tasks and environments?

(*a*) Battery life optimization
(*b*) Vision system calibration
(*c*) Learning and adaptation
(*d*) Control system stability

Q.5. Which challenge involves ensuring the safety of humans working alongside robots?

(*a*) Energy efficiency
(*b*) Automation integration
(*c*) Human-robot interaction
(*d*) Swarm robotics

Q.6. What's a challenge in designing robots for extreme space environments?

(*a*) Navigation in cluttered spaces
(*b*) Energy-efficient locomotion
(*c*) Handling soft objects
(*d*) Withstanding extreme conditions

Q.7. What's a challenge in integrating robots with the Internet of Things (IoT) devices?

(*a*) Ensuring safety for human-robot interactions
(*b*) Designing energy-efficient actuators
(*c*) Enabling seamless communication and interoperability
(*d*) Developing robot ethics guidelines

Q.8. Which challenge relates to the efficient use of energy in robotic systems?

(*a*) Bipedal locomotion
(*b*) Navigation in dynamic environments
(*c*) Energy efficiency
(*d*) Emotional interaction with humans

Q.9. What's a challenge in developing algorithms for robots to make morally sound decisions?

(*a*) Social acceptance
(*b*) Learning and adaptation
(*c*) Ethical decision-making
(*d*) Path planning

Q.10. Which challenge involves creating robots that can mimic natural movements and interactions?

(*a*) Soft robotics
(*b*) Multi-robot coordination
(*c*) Swarm robotics
(*d*) Industrial automation

Q.11. What challenge is posed by the need to prevent hacking and unauthorized access to robots?

(*a*) Navigation in cluttered environments
(*b*) Energy optimization
(*c*) Robot security
(*d*) Bipedal locomotion

Q.12. What's a challenge in integrating automation into existing factories without disrupting operations?

(*a*) Learning and adaptation
(*b*) Navigation in dynamic environments
(*c*) Retrofitting challenges
(*d*) Swarm robotics

Q.13. What challenge involves ensuring that AI models generalize well across diverse scenarios?

(*a*) Robot-human interaction
(*b*) Ethical decision-making
(*c*) Learning and adaptation
(*d*) Cross-validation

Q.14. What's a challenge in developing algorithms for robots to handle unexpected and ambiguous situations?

(*a*) Path planning
(*b*) Robot-human interaction
(*c*) Navigational efficiency
(*d*) Autonomy in uncertain conditions

Q.15. Which challenge pertains to creating robots that can navigate unstable and hazardous environments?

(*a*) Navigation in dynamic environments
(*b*) Multi-robot coordination
(*c*) Disaster response
(*d*) Perception in cluttered spaces

Q.16. What challenge involves preventing bias in AI-powered robots' decision-making?

(*a*) Bipedal locomotion
(*b*) Human-robot interaction
(*c*) Ethical decision-making
(*d*) Energy optimization

Q.17. What's a challenge in developing AI models that are fair and unbiased in their decisions?

(*a*) Swarm robotics
(*b*) Learning and adaptation
(*c*) Robot security
(*d*) Preventing bias

Q.18. Which challenge pertains to the legal responsibilities of robots in autonomous decision-making scenarios?

(*a*) Robot-human interaction
(*b*) Ethical decision-making
(*c*) Legal frameworks
(*d*) Industrial automation

Q.19. What's a challenge in achieving effective human-robot collaboration without conflicts?

(*a*) Swarming
(*b*) Navigation in cluttered environments
(*c*) Energy optimization
(*d*) Human-robot interaction

Q.20. What challenge involves developing algorithms for robots to handle a wide range of terrains?

(*a*) Learning and adaptation
(*b*) Navigation in dynamic environments
(*c*) Multi-robot coordination
(*d*) Versatile locomotion

Q.21. What's a challenge in preventing wear and tear on robot components that can lead to decreased precision?

(*a*) Robot security
(*b*) Reliability maintenance
(*c*) Ethical decision-making
(*d*) Energy optimization

Q.22. Which challenge involves developing algorithms for robots to make sound decisions in ambiguous scenarios?

(*a*) Robot-human interaction
(*b*) Learning and adaptation
(*c*) Path planning
(*d*) Ethical decision-making

Q.23. What challenge relates to ensuring seamless integration of automated systems with existing processes?

(*a*) Navigation in cluttered environments
(*b*) Retrofitting challenges
(*c*) Multi-robot coordination
(*d*) Energy efficiency

Q.24. What's a challenge in balancing automation and human employment?

(*a*) Navigation in dynamic environments
(*b*) Human-robot interaction
(*c*) Job displacement concerns
(*d*) Swarm robotics

Q.25. Which challenge pertains to addressing concerns about privacy, consent, and data usage?

(*a*) Perception in cluttered spaces
(*b*) Ethical decision-making
(*c*) Human-robot interaction
(*d*) Robot security

Q.26. What challenge involves creating interactive and engaging robotic experiences?

(*a*) Navigation in dynamic environments
(*b*) Entertainment robotics
(*c*) Path planning
(*d*) Energy optimization

Q.27. What's a challenge in developing reliable algorithms for real-time decision-making in robotics?

(*a*) Swarm robotics
(*b*) Learning and adaptation
(*c*) Navigation in cluttered environments
(*d*) Control system stability

Q.28. What's a difficulty in creating prosthetics that provide natural movements and sensations?

(*a*) Navigation in dynamic environments
(*b*) Perception in cluttered spaces
(*c*) Bipedal locomotion
(*d*) Achieving fine motor control

Q.29. Which challenge involves designing educational robots that engage and motivate learners effectively?

(*a*) Human-robot interaction
(*b*) Learning and adaptation
(*c*) Education integration
(*d*) Ethical decision-making

Q.30. What's a challenge in developing robots that can withstand extreme temperatures and radiation?

(*a*) Robot-human interaction
(*b*) Navigational efficiency
(*c*) Handling hazardous materials
(*d*) Withstanding extreme conditions

Q.31. Which challenge pertains to designing robots that can adapt to various crop types and terrains?

(*a*) Agricultural automation
(*b*) Perception in cluttered spaces
(*c*) Learning and adaptation
(*d*) Multi-robot coordination

Q.32. What challenge involves creating robots that can provide companionship and support for the elderly?

(*a*) Learning and adaptation
(*b*) Human-robot interaction
(*c*) Ethical decision-making
(*d*) Entertainment robotics

Q.33. What's a challenge in ensuring that robot behaviors align with cultural and social norms?

(*a*) Learning and adaptation
(*b*) Cultural integration
(*c*) Navigation in cluttered environments
(*d*) Human-robot interaction

Q.34. What challenge involves preventing robots from causing harm or stress to animals and ecosystems?

(*a*) Learning and adaptation
(*b*) Ethical decision-making
(*c*) Perception in cluttered spaces
(*d*) Ecological compatibility

Q.35. Which challenge pertains to the ethical considerations of robots making autonomous decisions?

(*a*) Robot-human interaction
(*b*) Ethical decision-making
(*c*) Learning and adaptation
(*d*) Navigation in dynamic environments

Q.36. What challenge involves designing algorithms for robots to navigate cluttered and dynamic environments?

(*a*) Navigation in dynamic environments
(*b*) Learning and adaptation
(*c*) Energy optimization
(*d*) Human-robot interaction

Q.37. What's a challenge in creating robots that can handle and manipulate delicate objects?

(*a*) Navigational efficiency
(*b*) Perception in cluttered spaces
(*c*) Fine motor control in manipulation
(*d*) Bipedal locomotion

Q.38. Which challenge involves addressing concerns about job displacement due to automation?

(*a*) Bipedal locomotion
(*b*) Job displacement concerns
(*c*) Energy optimization
(*d*) Multi-robot coordination

Q.39. What's a challenge in developing robots that can operate in hazardous or toxic environments?

(*a*) Robot-human interaction
(*b*) Handling hazardous materials
(*c*) Ethical decision-making
(*d*) Learning and adaptation

Q.40. What challenge relates to designing interfaces that allow non-experts to easily program and control robots?

(*a*) Learning and adaptation
(*b*) User-friendly programming
(*c*) Navigation in cluttered environments
(*d*) Ethical decision-making

Q.41. What's a challenge in preventing accidents and unwanted behaviors in robots?

(*a*) Learning and adaptation
(*b*) Safety and reliability
(*c*) Navigation in dynamic environments
(*d*) Bipedal locomotion

Q.42. Which challenge involves addressing concerns about robots' rights and responsibilities?

(*a*) Robot-human interaction
(*b*) Ethical decision-making
(*c*) Legal frameworks
(*d*) Navigation in dynamic environments

Q.43. What's a challenge in designing algorithms for robots to understand and respond to human emotions?

(*a*) Navigational efficiency
(*b*) Human-robot interaction
(*c*) Ethical decision-making
(*d*) Learning and adaptation

Q.44. Which challenge pertains to creating robots that can withstand physical interactions and impacts?

(*a*) Learning and adaptation
(*b*) Energy optimization
(*c*) Durability and impact resistance
(*d*) Robot-human interaction

Q.45. What challenge involves creating algorithms for robots to efficiently plan paths through cluttered spaces?

(*a*) Navigation in cluttered environments
(*b*) Learning and adaptation
(*c*) Energy optimization
(*d*) Human-robot interaction

Q.46. What's a challenge in ensuring that robot behaviors are intuitive for humans to understand and interact with?

(*a*) Learning and adaptation
(*b*) Designing natural movements
(*c*) Ethical decision-making
(*d*) Human-robot interaction

Q.47. What challenge involves creating algorithms for robots to learn and adapt from their interactions?

(*a*) Navigation in cluttered environments
(*b*) Learning and adaptation
(*c*) Robot-human interaction
(*d*) Energy optimization

Q.48. Which challenge pertains to developing robots that can handle and manipulate objects of varying shapes and sizes?

(*a*) Bipedal locomotion
(*b*) Fine motor control in manipulation
(*c*) Perception in cluttered spaces
(*d*) Human-robot interaction

Q.49. What's a challenge in creating robots that can function reliably for extended periods without frequent breakdowns?

(*a*) Learning and adaptation
(*b*) Energy optimization
(*c*) Maintenance of reliability
(*d*) Ethical decision-making

Q.50. What challenge involves creating algorithms for robots to collaborate with humans effectively without conflicts?

(*a*) Swarming
(*b*) Navigation in cluttered environments
(*c*) Human-robot collaboration
(*d*) Multi-robot coordination

Key to Multiple Choice Questions

1. (*c*)	**2.** (*c*)	**3.** (*b*)	**4.** (*c*)	**5.** (*c*)	**6.** (*d*)	**7.** (*c*)	**8.** (*c*)	**9.** (*c*)
10. (*a*)	**11.** (*c*)	**12.** (*c*)	**13.** (*c*)	**14.** (*d*)	**15.** (*c*)	**16.** (*c*)	**17.** (*d*)	**18.** (*c*)
19. (*d*)	**20.** (*d*)	**21.** (*b*)	**22.** (*b*)	**23.** (*b*)	**24.** (*c*)	**25.** (*c*)	**26.** (*b*)	**27.** (*b*)
28. (*d*)	**29.** (*c*)	**30.** (*d*)	**31.** (*a*)	**32.** (*b*)	**33.** (*b*)	**34.** (*d*)	**35.** (*b*)	**36.** (*a*)
37. (*c*)	**38.** (*b*)	**39.** (*b*)	**40.** (*b*)	**41.** (*b*)	**42.** (*c*)	**43.** (*b*)	**44.** (*c*)	**45.** (*a*)
46. (*d*)	**47.** (*b*)	**48.** (*b*)	**49.** (*c*)	**50.** (*c*)				

Project Works

1. **Automated Greenhouse Management System:** Develop a robotic system that monitors and controls environmental conditions in a greenhouse, adjusting factors like temperature, humidity, and lighting to optimize plant growth.
2. **Robotic Disaster Response**: Create a robot capable of navigating through disaster-stricken areas to locate survivors, assess damage, and relay critical information to rescue teams.
3. **Swarm Robotics for Pollution Cleanup:** Design a swarm of robots that collaborate to identify and collect pollution from water bodies, demonstrating effective coordination and environmental cleanup.

4. **Autonomous Delivery Drones:** Develop drones equipped with AI navigation systems to autonomously deliver packages to specified locations, addressing challenges of airspace management and obstacle avoidance.
5. **AI-Powered Healthcare Robotics:** Build a robotic system that assists medical professionals in surgical procedures by providing precise instrument manipulation and real-time imaging.
6. **Humanoid Robot Balance Control:** Design a humanoid robot capable of maintaining balance and stability while walking on different terrains, showcasing advancements in bipedal locomotion.
7. **Robotic Companion for Elderly**: Create a robot companion that provides emotional support, assists with daily tasks, and helps monitor health conditions of elderly individuals.
8. **Automated Farming System**: Develop an automated farming system that uses robotics and sensors to optimize planting, irrigation, fertilization, and harvesting processes.
9. **Autonomous Underwater Exploration**: Build an underwater robot capable of exploring and mapping underwater environments, addressing challenges of communication and navigation.
10. **Ethical Decision-Making AI for Self-Driving Cars**: Develop an AI system for autonomous vehicles that makes ethical decisions in potential accident scenarios, considering factors like risk to passengers and pedestrians.
11. **Robotic Exoskeleton for Rehabilitation**: Create a wearable robotic exoskeleton that assists individuals with mobility impairments in rehabilitation exercises and daily activities.
12. **AI-Powered Home Automation**: Build a smart home system that uses AI to optimize energy consumption, security, and comfort, overcoming challenges of interoperability and user interface.
13. **Virtual Reality Telepresence Robot**: Design a robot with a VR interface that allows users to remotely explore different locations and interact with people, bridging the gap between physical and virtual worlds.
14. **Autonomous Solar Panel Cleaning Robot**: Develop a robot that autonomously cleans solar panels to maintain their efficiency, addressing challenges of navigation on elevated surfaces and water conservation.
15. **Robotic Sorting and Recycling System**: Create a robotic system that identifies and sorts recyclable materials from mixed waste streams, contributing to waste reduction and efficient recycling.
16. **AI-Enhanced Language Translation Robot**: Build a robot capable of understanding and translating spoken language in real time, overcoming challenges of natural language processing and accuracy.
17. **Automated Warehouse Management System**: Develop an automated system that uses robots to efficiently manage and organize inventory in warehouses, optimizing storage and retrieval processes.
18. **Collaborative Robot Artist**: Design a collaborative robot-artist that creates unique artworks through interactions with human artists, showcasing the fusion of human creativity and robot precision.
19. **Robotic Security Guard**: Create a security robot equipped with surveillance cameras, facial recognition, and patrol capabilities to enhance security in public spaces.
20. **Autonomous Agricultural Drone Fleet**: Develop a fleet of agricultural drones that work together to monitor and analyze crops, identify pests, and optimize crop management practices.